Time-Frequency/Time-Scale Analysis

Wavelet Analysis and Its Applications

The subject of wavelet analysis has recently drawn a great deal of attention from mathematical scientists in various disciplines. It is creating a common link between mathematicians, physicists, and electrical engineers. This book series consists of both monographs and edited volumes on the theory and applications of this rapidly developing subject. Its objective is to meet the needs of academic, industrial, and governmental researchers, as well as to provide instructional material for teaching at both the undergraduate and graduate levels.

To those of us who are interested in the understanding, analysis, and processing of signals, wavelets provide us with a very powerful mathematical toolbox to solve problems that the Fourier approach does not perform well. In fact, the time-frequency and time-scale localization capability of wavelets for signal analysis plays a significant role in defining wavelet analysis as a truly interdisciplinary area of research. This book covers time-frequency and time-scale analysis of signals. First written in the French language by a leading expert and authority of the subject, this English edition is the product of the skillful translation by yet another wavelets expert. It is a very valuable addition to the book series. The series editor would like to thank both the author, Patrick Flandrin, and the translator, Joachim Stöckler, for this excellent contribution to the wavelets literature.

This is a volume in
WAVELET ANALYSIS AND ITS APPLICATIONS

Charles K. Chui, Series Editor

A list of titles in this series appears at the end of this volume.

Time-Frequency/Time-Scale Analysis

Patrick Flandrin
Laboratoire de Physique
Ecole Normale Supérieure
Lyon, France

Translated from the French by
Joachim Stöckler
Institut für Angewandte
Universität Hohenheim
Stuttgart, Germany

ACADEMIC PRESS

San Diego London Boston
New York Sydney Tokyo Toronto

This book is printed on acid-free paper. ∞

Copyright © 1999 by Academic Press

Temps-fréquence
Patrick Flandrin
© Hermès, Paris, 1993, 1998
Éditions Hermès
8, quai du Marché-Neuf
75004 Paris

All rights reserved.
No part of this publication may be reproduced or transmitted in any form or by any means, electronic or mechanical, including photocopy, recording, or any information storage and retrieval system, without permission in writing from the publisher.

ACADEMIC PRESS
525 B Street, Suite 1900, San Diego, CA 92101-4495, USA
http://www.apnet.com

ACADEMIC PRESS
24–28 Oval Road, London NW1 7DX, UK
http://www.hbuk.co.uk/ap/

Library of Congress Cataloging-in-Publication Data
Flandrin, Patrick
    [Temps-fréquence. English]
    Time-frequency/time scale analysis / Patrick Flandrin.
    p.   cm. — (Wavelet analysis and its applications ; v. 10)
    Translation of: Temps-fréquence.
    Includes bibliographical references and index.
    ISBN 0-12-259870-9 (alk. paper)
    1. Signal processing—Mathematics. 2. Time-series analysis.
3. Wavelets (Mathematics) I. Title. II. Series.
TK5102.9.F5513   1999
621.382'2—dc21   98-38814
                                                                CIP

Printed in the United States of America
98 99 00 01 02 IP 9 8 7 6 5 4 3 2 1

Contents

Preface xi

Foreword 1

Chapter 1. The Time-Frequency Problem 9

1.1. **The Time-Frequency Duality and Its Bars** 10
    1.1.1. Fourier Analysis 10
        Limitations 10. *Citations* 11.
    1.1.2. Heisenberg-Gabor Uncertainty Principle 12
        The time-frequency inequality 14.
        Interpretations 16.
    1.1.3. Slepian-Pollak-Landau Theory 18
        Concentrations 18. *Sampling* 19.
        The eigenvalue equation 21.
        Approximation of bandlimited signals 23.
        Approximative dimension of a signal 24.
        Inequality of the concentrations 24.

1.2. **Leaving Fourier?** 26
    1.2.1. Local Quantities 26
        Instantaneous frequency 27. *Group delay* 31.
        Interpretative remarks 31. *An example* 32.
    1.2.2. Nonstationary Signals 36
        Definition 36. *Generalizations* 38. *Variations* 40.

1.3. **Towards Time-Frequency: Several Approaches** 42
    1.3.1. The Time-Frequency Plane and Its Three Readings 43
        Frequency (time) 43. *Time (frequency)* 43.
        Time-frequency 43.
    1.3.2. Decompositions, Distributions, Models 43
        Decompositions 44. *Distributions* 44. *Models* 44.
    1.3.3. Moving and Joint, Adaptive and Evolutionary Methods 45
        Moving and joint 45. *Adaptive and evolutionary* 45.

Chapter 1 Notes 46

Chapter 2. Classes of Solutions 49

2.1. An Introduction with Historical Landmarks 50

2.1.1. Short-Time Fourier and Instantaneous Spectrum 50

Sonagram and spectrogram 51. *Restrictions* 52.

2.1.2. Atomic Decompositions 54

Gabor 54. *Variations* 55. *Wavelets* 56.

2.1.3. Pseudo-Densities 57

Wigner-Ville 58. *Page* 60. *Rihaczek* 60. *Extensions* 62. *Unification* 62.

2.1.4. The Parallel to Quantum Mechanics 64

Different concerns 64. *Some intersections* 65.

2.2. Atomic Decompositions 66

2.2.1. Projections and Bases – General Principles 66

Discrete bases 66. *Continuous bases* 67. *Frames* 69.

2.2.2. Time-Frequency Examples 70

Short-time Fourier 70. *Obstruction established by Balian–Low Theorem* 73. *Gabor and variants* 75.

2.2.3. Time-Scale Examples 76

Continuous wavelets 76. *Discrete wavelets* 80. *Multiresolution analyses and orthonormal bases* 83. *Pyramidal algorithms* 89. *Some wavelet bases* 92.

2.2.4. A "Detection-Estimation" Viewpoint 97

Ambiguity functions 97. *Atoms and matched filtering* 101.

2.3. The Energy Distributions 103

General setting 103. *Covariance principles* 104.

2.3.1. Construction of the Bilinear Classes 104

Time-frequency 104. *Time-scale* 107.

2.3.2. The Troika of Parameterizations-Definitions-Properties ... 109

Definitions 110. *Constraints* 114. *Cohen's class* 116. *Affine class* 132.

2.3.3. Results of Exclusion and Conditional Uniqueness 142

Wigner's Theorem 143. *Some results of exclusion* 145. *Some results on conditional uniqueness* 147.

2.4. The Power Distributions ... 150
2.4.1. From Deterministic to Random Signals ... 150
Decompositions and fluctuations 150.
Distributions and expectation values 151.
Cramér and beyond 152.
2.4.2. The Orthogonal (or Almost Orthogonal) Solutions ... 153
Karhunen decompositions 153.
Priestley spectrum 154.
Tjøstheim, Mélard, Grenier approach 156.
2.4.3. The Frequency Solutions ... 160
Harmonizable signals 160.
Wigner-Ville spectrum 165.
2.4.4. Some Links Between the Different Spectra ... 170
Continuous time 170. *Discrete time* 171.
Chapter 2 Notes ... 174

Chapter 3. Issues of Interpretation ... 183
3.1. About the Bilinear Classes ... 185
3.1.1. The Different Parameterizations ... 185
Time-frequency 187. *Time-time* 188.
Frequency-frequency 191. *Frequency-time* 191.
3.1.2. Parameterizations, Operators and Correspondence Rules ... 194
Why operators? 194.
The operator of time-frequency shifts 195.
Correspondence rules 197. *Kernels* 199.
Weyl calculus 200. *Moments* 201.
Dilations and ambiguities 205.
3.1.3. Time-Frequency or Time-Scale? ... 208
Fourier scale 208. *Mellin scale* 210.
Analysis and decision statistics 213.

3.2. The Wigner-Ville Distribution and Its Geometry 213
    3.2.1. Wigner-Ville versus Spectrogram 213
        Structure of the distributions 213.
        Pseudo-Wigner-Ville 215.
        Supports 216. *Localization to chirps* 217.
        Spectrogram and reassignment 221.
        Discretization 225.
    3.2.2. The Mechanism of Interferences 226
        Construction principle 228.
        A different perspective from the ambiguity plane 231.
        Inner and outer interferences 232.
        Approximation by the method of stationary phase 234.
        Singularities and catastrophes 237.
        Interferences, localization, and symmetries 243.
        Generalization to the s-Wigner distributions 245.
        Generalization to the affine distributions 247.
    3.2.3. Reduction of the Interferences 252
        Analytic signal 252.
        Wigner-Ville and atomic decompositions 252.
        Smoothing 254. *Coupled smoothing* 255.
        Separable smoothing 256. *Joint smoothing* 261.
        Variable and/or adapted smoothing 269.
        "Image" approaches 274.
    3.2.4. Usefulness of the Interferences 274
        Unitarity 274. *Phase information* 275.
        What is a component? 277.
    3.2.5. Statistical Estimation of the Wigner-Ville Spectrum 279
        Assumptions 280. *Classes of estimators* 281.
        Bias 282. *Variance* 283. *Examples* 284.

3.3. About the Positivity 289
    3.3.1. Some Problems Caused by the Nonpositivity 289
    3.3.2. Positivity by the Signal 293
        An example 293. *Hudson's theorem* 294.
        Random signals and positive spectra 295.
    3.3.3. Positivity by the Distribution 296
        Positive distributions 296. *Positive smoothing* 297.
        A stochastic interpretation 300.

Chapter 3 Notes 301

Contents

Chapter 4. Time-Frequency as a Paradigm ... 309

4.1. Localization ... 311

4.1.1. Heisenberg-Gabor Revisited ... 311
Example 1 311.   *Example 2* 313.

4.1.2. Energy Concentration ... 315
Problem formulation 315.
The general eigenvalue equation 316.
Restriction to ellipsoidal domains 316.
Interpretations and conjecture 323.

4.1.3. Other Time-Frequency Inequalities ... 323
L_p-norms 324.   *Localization and stationary phase* 325.

4.2. Signal Analysis ... 327

4.2.1. Time-Frequency, Time-Scale, and Spectral Analysis ... 327
Paving and marginal distributions 328.
The example of "$1/f$-noise" 329.
Analysis of self-similar processes 330.

4.2.2. Nonstationary Characteristics ... 334
Distance from the stationary case 334.
Demodulation 336.   *Local singularities* 338.
Evolutionary singularities 340.

4.3. Decision Statistics ... 342

4.3.1. Matched Time-Frequency Filtering ... 343

4.3.2. Maximum Likelihood Estimators for Gaussian Processes ... 345
Classical solution 345.
Time-frequency formulation 346.

4.3.3. Some Examples ... 347
Rayleigh channel 347.
Detection of chirps and Doppler tolerance 347.
Locally optimal detection 349.
Time-frequency jitter 350.
A broader class of time-frequency receptors 351.

Chapter 4 Notes ... 354

Bibliography ... 359

Index ... 381

Preface

What is a signal? What information can be drawn from a signal? How can we extract this information? How is a signal denoised? At which point does a *priori* knowledge about a signal enter and how can we utilize it? These are essential questions to which no answers could be given until today. In fact, they have been subject to considerable change during the past fifty or more years.

Between 1940 and 1960 signals were analog, and signal processing was primarily a part of physics. However, with the onset of microprocessors the analog signal lost its rank to the digital signal, and information theory irrupted. Fast computational algorithms (such as fast Fourier transform) allowed most filtering operations to be performed almost instantly, and this enlarged tremendously the scope of possible manipulations. Furthermore, signal processing also gained an advantage from all the new achievements in statistics. This occurred to such a degree that one could sometimes look upon signal processing as a part of statistics.

A third revolution (1970–1980) in this field occurred with the advent of methods and techniques of mathematical physics and quantum mechanics. This cleared the way for mathematicians to participate in scientific activities related to signal processing.

Today signal processing offers a forum to numerous disciplines and requires manifold knowledge, which can be combined only rarely by a single scientist.

P. Flandrin takes up this challenge with so much competence and a truly multidisciplinary vision. During the study of his book devoted to time-frequency signal analysis I once more admired the brightness of his spirit, the depth of his insight, the wealth of his knowledge, and an erudition that is never ponderous.

He devotes most of this book to the *time-frequency analysis* of signals. This analysis involves unfolding the signal in the time-frequency plane in the way best suited to comprehension. This operation can be compared to writing the score of a symphony as one listens to the music. There exists no unique solution, of course. However, P. Flandrin exposes a class of methods

(Cohen's class) and provides an exhaustive discussion of the qualities and defects of each of them. One of the most celebrated members of this class is the *Wigner-Ville transform*. It lays the foundation for the field in which quantum mechanics, the theory of pseudo-differential operators, and signal statistics meet.

This work by P. Flandrin invites us to become (or continue to be) researchers in this field; after having read and reread this book, one is well-prepared to participate in the exploration of the unaccessible instantaneous frequency ... .

French Academy of Sciences   Yves Meyer

Foreword

> "There must be, in the represented things, the insistent murmuring of the resemblance; there must be, in the representation, the ever possible recourse of the imagination."
>
> Michel Foucault

A signal is the physical carrier of some information. It can originate from a multitude of different sources (acoustics, radioelectronics, optics, mechanics, etc.). Beyond this diversity, however, the main object of interest is the observation of a *time-varying* quantity, which is collected at one or more sensors. This constitutes the basis on which a "signal processor" can perform operations for extracting some useful information. The facility of gaining and processing this information certainly depends on its readability. This is our motivation for leaving the immediate representational space, in which plain data are given. We pass, instead, to a transformed space containing the same information, in order to obtain a clearer picture of specific characteristics of the signal by "looking at" it from a particular angle.

The choice of a representation is crucial for the ultimate task of processing data, which often comprises several consecutive steps for solving a statistical *decision* problem (detection, estimation, classification, recognition, etc.). The pertinence of a representation is rooted in its capability to provide well-suited descriptors for this task. Viewed from a perspective of signal analysis, the representation should "tell" the user something about the structure of the signal. As long as one knows very little about the constitution of the signal *a priori*, the representation should require as few external specifications about this structure as possible. This situation falls in the category of *nonparametric* methods, which often employ a much larger representational space than used for the initial data. A better structuring of the transformed information should counterbalance the onset of redundancy. This is our preferred point of view in this book, and we shall make use of it in many *nonstationary* situations.

The necessity of dealing with nonstationary signals is rooted in a paradox. On one side, a huge amount of work has already been devoted to the stationary case. Today one can say that this case is well-analyzed from a theoretical perspective and enjoys a multitude of performing tools that are often linked to frequential descriptions. On the other side, it is also true that the real signals in their vast majority are nonstationary; expressed even more drastically, an integral part of the information (if not the dominant one) can be found in the nonstationary property itself (e.g., start and termination of events, drifts, ruptures, modulations). These two counterparts have led to the development of tools that are designed specifically to match nonstationary situations. Among them emerge the *time-frequency methods*, which explicitly consider the time-dependence of the frequential content of the signal. This book concentrates primarily on such methods.

The following describes the organization of the book:

- The first chapter ("The Time-Frequency Problem") addresses the description of some general concepts associated with the notions of time and frequency. We dwell on the relations between these two variables and encompass the problems resulting from their combined use. The limits of the physical interpretation of the Fourier transform and the existence of certain mathematical obstructions will be discussed. This forces us to search for possible substitutes that better match nonstationary situations. A first attempt to furnish a classification of the conceivable solutions is contained at the end of this chapter.

- After emphasizing in the first chapter a number of problems that arise when we search for a time-frequency description of signals, the second chapter ("Classes of Solutions") is devoted to the discussion of some conceivable approaches, including an extensive inventory of their properties. In a certain sense the presentation can be regarded as a "multiresolution" advance. The main features of a general panorama are first explored from an historical perspective. Then we introduce more details of the principal approaches. These can be grouped into three large sets comprising the (linear) "atomic" decompositions, (bilinear) energy distributions, and power distribution functions. In each case we make every effort to begin with general principles for constructing general classes of solutions. Only afterwards do we specify the form or the properties resulting from some additional constraints.

- The objective of the third chapter ("Issues of Interpretation") is a more profound study of the potential, the characteristics and/or the limitations of several time-frequency tools that have already been constructed. We attempt to give some guidance concerning their application (which representation should be chosen?) and facilitate their interpretation (which information can be extracted?). Because not all approaches can be covered

Foreword

in detail, we focus on the *bilinear* representations (Cohen's class and affine class) in this chapter.

- The time-frequency representations do more than offer an arsenal of adaptive methods for nonstationary signals: they manifest a new paradigm. It is from this perspective that the last chapter ("Time-Frequency as a Paradigm") attempts to illustrate, by some typical examples, how an explicitly joint description can amount to a new vision of several classical problems in signal analysis and signal processing, and how it can lead to solutions that have "natural" interpretations.

As we do not aim to write an encyclopedic treatise on the very general theme of "time-frequency," we had to make our choices as to which subjects to include and which to leave aside. They result from two different motivations: one is the concern for competence (describing facts that one believes to know), and the other one is the desire to offer a large panorama of subjects that were unevenly covered in the few existing surveys (which means writing a book that one would have liked to exist already). These were the main reasons for selecting the material found in this book. Furthermore, we emphasize mostly general principles (omitting most of the algorithms and applications), and concentrate on the nonparametric bilinear approaches (to the detriment of linear methods and/or parametric modeling).

We have described here what the reader will be able to find in this book. It might also be helpful to point out those subjects not treated and which other sources can be consulted instead.

Only little or nothing is said about algorithms (in any case there are no "codes" included in a "ready to use" format). However, after the first edition of this book was published in French in 1993, the situation concerning algorithms that put the time-frequency methods to work changed quite a bit. We would like to refer to a public "toolbox" written in Matlab and freely accessible via the internet under the address

```
http://www.physique.ens-lyon.fr/ts/tftb.html
```

This collection of routines became available thanks to the efforts of O. Lemoine, F. Auger, and P. Gonçalvès, and was supported by CNRS through its successive GdRs TdSI and ISIS. The interested reader can thus access over a hundred functions that allow one to create or manipulate nonstationary signals. These include efficient algorithms for the computation of the time-frequency representations that are presented from a mostly theoretical point of view in this book. Moreover, they offer different postprocessing methods. The collection of these functions is documented in due form in a reference guide as well as in a tutorial that has been sorted by examples.

We would like to mention that this toolbox was used to generate most of the figures in Chapter 3 of this English edition.

We also do not develop any "concrete" applications. These issues alone would necessitate the writing of another book. While awaiting someone tackling this job, one might meanwhile consult the surveys by Basseville, Flandrin, and Martin (1992) and Hlawatsch and Boudreaux-Bartels (1992), or the more recent collections in Loughlin (1996), Akay (1997), Feichtinger and Strohmer (1998), and Mecklenbräuker and Hlawatsch (1998). We also do not deal with model-based parametric methods. Again, this is a vast field (and, what is more, has a different philosophy), which should be considered as well. For undertaking such an effort one should look at the article by Grenier (1987).

The subjects included here receive a different degree of attention. As already mentioned, *bilinear time-frequency* methods constitute the central part of the book. Nevertheless, we grant a significant portion to the *linear time-scale* methods (wavelet theory), without trying to be as exhaustive as the description in Daubechies (1992), for instance. The importance of wavelet theory (which is far from being confined to the time-frequency problem) justifies our decision to give a short account of its basic principles. On the other hand, we hope that the coexistence of the time-frequency and time-scale approaches in one book will help improve comparative understanding and facilitate complementary use. Conversely, some important issues in the theory of bilinear representations have been left aside, such as the signal *synthesis* problem. Here we refer the reader to the general presentations by Hlawatsch (1985; 1986) and Hlawatsch and Krattenthaler (1992; 1998), which were written by the investigators themselves. The same is done with the bivariate extensions (more specifically directed at image methods (see Zhu et al., 1992)) and the multilinear (or "higher–order") methods that have flourished recently (see, e.g., Amblard and Lacoume, 1992, or, Nikias and Petropulu, 1993).

Let us finally remark that many of the proofs are not given in a mathematically rigorous form. We preferred rather to stress a "sensible" understanding of the results and their physical interpretation.

Although it has become almost trivial (at least well understood and accepted) to talk about signals in a time-frequency setting today, it is worthwhile to remember that this was not always obvious. Not very long ago these descriptions ranged from being either a theoretical exception or a heuristical mishmash. It is, therefore, my great pleasure to salute the memory of B. Escudié, who was one of the believers during the early days. Remembrance of him stays alive in all who knew him. I am thankful to him for introducing me to a research area that was not marked out clearly at that time, and which in due course proved its importance.

Foreword

Some of the work presented in this book, and which work I performed later, has greatly profited from other collaborations that I wish to mention here. In particular, building the theory of the Wigner-Ville spectrum and its estimation was done with W. Martin, the "geometry of the interferences" was explored together with F. Hlawatsch, and the introduction and investigation of time-scale distributions obtained by affine smoothing was work done together with O. Rioul, and later with P. Gonçalvès. Thanks to all of them, as well as to P. Abry, F. Auger, Y. Biraud, J. Gréa, J. Sageloli, and J. P. Sessarego for other fruitful cooperations. I also include in my thanks my former colleagues from the Laboratoire de Traitement du Signal de l'*ICPI Lyon*, M. Chiollaz, J. P. Corgiatti, T. Doligez, N. Gache, and M. Zakharia.

There were other influences, which I do not underestimate; these are felt in a more diffuse way. In particular, I am referring to the workgroups "Adaptatif/Evolutif" and "Non stationnaire," which I conducted together with O. Macchi and M. Basseville within the GdR 134 CNRS *Traitement du Signal et Images*, and to the work on comparing the time-frequency methods performed there. I also think of several encounters with the "wavelet community" (with A. Grossmann and Y. Meyer in the first place), where we could discuss our more complementary rather than differing points of view.

P. Abry, R. G. Baraniuk, G. F. Boudreaux-Bartels, and O. Michel were kind enough to reread and criticize preliminary versions of the manuscript (especially the first two chapters). I profited from numerous remarks that they made, and I wish to express my thanks to them. I am also grateful to Y. Meyer, who granted me the honor of writing the preface.

Finally, I wish to express my gratitude to P. Duvaut and D. Garreau, who asked me to contribute to their book series *Traitement du Signal*, and whose friendly pressure led to the existence of this book. Thanks also to J. Dellon, who went without his beloved computer for several months, in order to let me accomplish this project.

While almost all material in this translation is taken from the original "Temps-fréquence" from 1993, several small corrections and additions have been incorporated. Concerning the corrections, I especially wish to thank the translator, J. Stöckler, whose numerous remarks during the translation process helped clarifying some details that were still missing in the first French edition. These corrections are also incorporated in the second French edition. There is only a small number of additions, which are intended mainly to give account of some more recent developments in the field since 1993. This makes it easier to situate the book in the present literature on this subject. Those additions can be found mainly in the concluding remarks at the end of each chapter, and they refer to an updated and extended bibliography. Most figures have been redrawn and slightly enhanced.

Lyon
July 1993 and May 1998

Patrick Flandrin

Further introductory remarks.

1. Each chapter starts with a short résumé to serve as both an outline and a guideline for the reader. At the end of each chapter a significant number of notes and remarks are to be found. They are concerned mainly with comments on the subject that can be omitted during a first reading, precise bibliographic references, and brief sketches of further issues that cannot be covered in this book. In order to keep the text uniform (it is intended to be self-contained apart from a few exceptions) and to avoid overloading it, we include references to the original work in these notes. They contain pointers to cumulative References that have been placed at the end of the book.

2. The time-frequency distributions are denoted mostly by the initials of those authors who defined them (example: B for Bertrand, CW for Choï-Williams, etc.). Other frequently used notations are listed in the Nomenclature.

3. In most illustrations of the time-frequency representations the x-axis denotes time, the y-axis denotes frequency. The values of the representation are either drawn as a contourplot or a density plot, whichever way was deemed more appropriate under the circumstances. In order to improve the clarity of these plots, we drew only the positive amplitudes in every case.

NOMENCLATURE

t, τ	:	time variables
ν, ξ	:	frequency variables
a	:	scale variable
$x(t)$ (resp. $x[n]$)	:	continuous (resp. discrete) time signal
$X(\nu)$	:	frequency representation of $x(t)$ or $x[n]$
$L^2(\mathbb{R})$	:	space of finite energy signals
$\langle .,. \rangle$	:	scalar product of $L^2(\mathbb{R})$
E_x	:	energy of signal x ($= \langle x, x \rangle$)
i	:	$\sqrt{-1}$
$.^*$	:	complex conjugate
Re $\{.\}$, Im $\{.\}$, $|.|$	:	real part, imaginary part, modulus
H $\{.\}$	:	Hilbert transform
pv	:	Cauchy principal value
$\widehat{A}$	:	operator associated with a function A
$\widehat{I}$	:	identity operator
$\langle \widehat{A} \rangle_x$	:	expectation value of $\widehat{A}$ ($= \langle \widehat{A}x, x \rangle$)
$\mathbf{E}\{.\}$	:	stochastic expectation value
$r_x(t,s)$	:	autocovariance function ($= \mathbf{E}\{x(t)x^*(s)\}$)
$\gamma_x(\tau)$	:	autocorrelation function (stationary covariance)
$\Gamma_x(\nu)$	:	power spectrum density
$F_x(t,\nu)$ or $F_x(t,\nu;h)$	:	short-time Fourier transform (with window h)
$T_x(t,a)$ or $T_x(t,a;h)$	:	wavelet transform (with wavelet h)
$W_x(t,\nu)$	:	Wigner-Ville distribution
$A_x(\xi,\tau)$	:	(narrowband) ambiguity function
$C_x(t,\nu;f)$	:	Cohen's class, parameter function f
$\Omega_x(t,a;f)$	:	affine class, parameter function f
$\rho_x[t,\nu)$	:	time-frequency distribution with discrete time
$U(.)$	:	Heaviside unit step function
$\delta(.)$	:	Dirac distribution
δ_{nm}	:	Kronecker symbol ($= 1$ if $n = m$, 0 otherwise)
$\mathbf{1}_I(.)$	:	characteristic function of the interval I

Chapter 1
The Time-Frequency Problem

This chapter describes some of the general concepts associated with notions of time and frequency. We dwell on the relations between these two variables and encompass the problems resulting from their combined use.

In the first part (Subsection 1.1.1) we explain why purely frequential representations (based on the Fourier transform, which erases all time dependence) prove to be insufficient from a physical point of view, as one cannot dispense with the time for describing a signal. Then we discuss two other consequences, which result from employing the Fourier transform. They are both related to mathematical impossibilities. The first is expressed by the Heisenberg-Gabor uncertainty principle (Subsection 1.1.2), which stipulates that a signal cannot be concentrated on arbitrarily small time-frequency regions. The second result (theory of Slepian-Pollak-Landau, Subsection 1.1.3) shows that a signal cannot confine its total energy to finite intervals in the time and frequency domain, no matter how large these intervals might be.

Without being able to overcome the inherent limitations of the Fourier transform, Section 1.2 holds out a prospect of possible substitutes for the Fourier frequency, which are better suited for nonstationary situations. In Subsection 1.2.1 we introduce some local concepts such as the instantaneous frequency, thus giving a meaning to the intuitive notion of a temporal evolution of a (deterministic) spectral property. Then we investigate the problem in Subsection 1.2.2 of how locality can be introduced to the representation of a nonstationary signal. Here we work in a probabilistic setting. However, the given solutions provide only partial answers to the posed problem (or answers that are difficult to interpret). This motivates the study of more general approaches in an explicitly joint time-frequency framework.

An introduction to such approaches is the subject of Section 1.3. We sketch briefly a classification of possible solutions, focusing on those that will be studied in more detail in Chapter 2.

§1.1. The Time-Frequency Duality and Its Bars

1.1.1. Fourier Analysis

The Fourier analysis [1] is one of the major accomplishments of physics and mathematics. It is indispensable to signal theory and signal processing for several reasons. Certainly, the first among them is the universal concept of *frequency* in which it is rooted: a frequential description can often be the basis of a better comprehension of the underlying phenomena, because it supplies an essential complement to the exclusively temporal description (receiver output or sequence of events) that is usually taken first for analysis. This occurs in several areas of interest, be it in physical waves (acoustics, vibrations, geophysics, optics, etc.) or in some other framework where periodic processes play an important role (economy, biology, astronomy, etc.). The second reason comes from the mathematical structure of the Fourier transform itself, as it is naturally suited to common transform methods (such as the linear filtering) by its ability to render them in a particularly simple form. Finally, we mention as a third and more pragmatic reason, that the collection of these advantages has led to the development of a large number of algorithms, programs, processors, and machines for frequency analysis, all of which contribute to its good reputation for practical use.

Limitations. Important though it is from a mathematical point of view, the Fourier analysis possesses several restrictions concerning its physical interpretation and its range of applicability. In order to explain this, let us look at the usual definition of the Fourier transform

$$X(\nu) = \int_{-\infty}^{+\infty} x(t) e^{-i2\pi\nu t} dt . \qquad (1.1)$$

Evidently, the computation of *one* frequency value $X(\nu)$ necessitates the knowledge of *the complete history* of the signal ranging from $-\infty$ to $+\infty$. Conversely, the inverse Fourier transform is given by

$$x(t) = \int_{-\infty}^{+\infty} X(\nu) e^{i2\pi\nu t} d\nu . \qquad (1.2)$$

Hence, any value $x(t)$ of the signal at *one* instant t can be regarded as an infinite superposition of complex exponentials, or *everlasting* and completely nonlocal waves. Even if this mathematical point of view may reveal the true properties of a signal in certain cases ("quasi-monochromatic" situations, steady state, etc.), it can also distort the physical reality. This happens, for example, with *transient* signals, which vanish outside a certain time

Chapter 1 The Time-Frequency Problem

interval (e.g., by switching a machine on and off). Although the nullity of its values is reflected by the Fourier analysis, this occurs only in an artificial manner: it results from an infinite superposition of virtual waves that interfere such that they annihilate each other. Hence the situation on the domain, where the signal vanishes, can be described as a "dynamic" zero (there exist waves whose resulting contribution is zero by interference). This contradicts any proper understanding of the real physical situation as a "static" zero (the signal does not exist).

Citations. In this regard we wish to give prominence to three particularly appropriate citations. (These are translations of the original texts in French.) The first, by Ville, is from 1948. In his fundamental article, which appeared in *Câbles et Transmissions*, he defined both the notion of analytic signals and the time-frequency representation that is named after him today. We can read the following in the Introduction:

> Indeed, if we consider a piece [of music] which contains several bars (which is the least we can demand), and if one note, *la* for example, appears once in the piece, the harmonic [Fourier] analysis will present the corresponding frequency with a certain amplitude and a certain phase, without locating the *la* in time. It is evident, however, that there are moments in the course of the piece where the *la* cannot be heard. Nevertheless, the representation is mathematically correct, because the phases of the notes near *la* are arranged in such a way, that they destroy this note by interference when it is not heard, and reinforce it, again by interference, when it is heard; but even if there is a versatility in this concept honoring the mathematical analysis, one should not conceal that there is also a distortion of the reality: indeed, when one does not hear the *la*, the true reason is that the *la* is not emitted.

The second citation, though more recent as it dates from 1966, is taken from an equally illustrious source. It is due to de Broglie, who asserts in his "Certitudes et Incertitudes de la Science":

> If we consider a quantity which can be represented in a Fourier manner, i.e., by a superposition of monochromatic components, then it is the superposition which has a physical meaning, but not the isolated Fourier components. If we deal, for example, with a swinging chord whose motion can be described by a series of harmonics, a movie of this motion would reveal that the chord has a very complicated form at each moment, and that it varies incessantly according to a complex rule. Nothing in this motion allows us to distinguish the various monochromatic components: these

components exist only in the minds of theorists who endeavor an abstract analysis of this motion. They would only come into physical existence if one could achieve their isolation by an operation, which, in turn, would break up the superposition. Besides, the whole theory of interferences would be inexact if this were not true. The idea that the monochromatic components have a real existence in the physical process, which comes from their superposition, seems false to me, as it vitiates parts of the theoretical reasoning which is actually common in Quantum Physics.

Finally, the third quotation is found in the work of the ineffable Bouasse, who makes a straightforward statement in his *Acoustique Générale* in 1926, saying that "[...] unless one has lost the most elementary common sense, it is impossible to attribute an *objective* existence to the harmonic oscillations which emerge in the Fourier series."

Even though the Fourier analysis has such restrictions regarding availability of interpretations or adequacy for certain types of signals, it still remains true that it possesses an immense utility, be it alone or as a computational tool. Moreover, we will see that numerous time-frequency methods, though deviating from the spirit of a Fourier analysis *stricto sensu*, stay close to it in their definition and exploit its wealth of mathematical structures.

1.1.2. Heisenberg-Gabor Uncertainty Principle

Let us first consider the case of a bandlimited signal. As its support in the frequency domain is compact, its (inverse) Fourier transform must be analytic. Therefore, the signal cannot vanish on a set of positive measure or be strictly confined to a finite duration: indeed, its analyticity would imply that it vanishes everywhere by analytic continuation. Another way to prove this fact is to start from the converse hypothesis of bounded supports in time (with duration T) and frequency (with bandwidth B). Any nonzero signal with these properties would satisfy the relation

$$x(t) = \int_{-B/2}^{+B/2} X(\nu) \, e^{i2\pi\nu t} \, d\nu = 0 \,, \quad |t| > T/2 \,.$$

As $x(t)$ is of bounded duration, the same is true for its nth derivatives. Furthermore, as $X(\nu)$ is supposed to vanish outside the interval $[-B/2, +B/2]$, we would also have

$$\frac{d^n x}{dt^n}(t) = \int_{-B/2}^{+B/2} (i2\pi\nu)^n \, X(\nu) \, e^{i2\pi\nu t} \, d\nu = 0 \,, \quad |t| > T/2 \,,$$

Chapter 1 The Time-Frequency Problem

for all $n \geq 0$. The value of the signal in a point s, which belongs to the support of $x(t)$ (so $|s| < T/2$), can now be written as

$$x(s) = \int_{-B/2}^{+B/2} X(\nu)\, e^{i2\pi\nu(s-t)}\, e^{i2\pi\nu t}\, d\nu\,, \qquad |s| < T/2\,, \quad |t| > T/2\,.$$

By replacing the first complex exponential with its power series

$$e^{i2\pi\nu(s-t)} = \sum_{n=0}^{\infty} \frac{[i2\pi(s-t)]^n}{n!} \nu^n$$

we arrive at our final relation

$$x(s) = \sum_{n=0}^{\infty} \frac{(s-t)^n}{n!} \int_{-B/2}^{+B/2} (i2\pi\nu)^n\, X(\nu)\, e^{i2\pi\nu t}\, d\nu = 0\,,$$

where $|t| > T/2$. This holds for all $|s| < T/2$ and thus contradicts our initial assumption that the signal is nonzero in the interval $[-T/2, +T/2]$.

Even if we relax the strict constraint of finite supports, it is well known that the (essential) support of a signal cannot be arbitrarily small both in time and frequency: our experience proves, for example, that a short impulse extends over a large frequency range. Vice versa, the narrower the band of a filter, the longer is its response time. This type of constraint is imposed by the Fourier duality (which exists between the time and the frequency representations of signals). It is clearly illustrated by the pair "Dirac distribution – constant function," and in a somewhat smoother way by the Gaussian functions which satisfy

$$\int_{-\infty}^{+\infty} \left(\frac{\alpha}{\pi}\right)^{1/2} e^{-\alpha t^2}\, e^{-i2\pi\nu t}\, dt = e^{-\pi^2 \nu^2/\alpha}\,.$$

If we regard the width of a Gaussian as proportional to $\alpha^{-1/2}$ in time, the preceding relation shows that the equivalent width of its Fourier transform, which is also a Gaussian, is proportional to $\alpha^{1/2}$. These magnitudes vary in opposite directions depending on the parameter α. One of them increases when the other decreases, and vice versa. The product of the two numbers remains constant. (Let us further remark that the forementioned pair "Dirac distribution – constant function" turns up when α tends to infinity.)

The time-frequency inequality. This behavior of duality is a direct consequence of the definition of the Fourier transform. It finds its simplest mathematical formulation in the so-called *Heisenberg-Gabor uncertainty principle.* [2] It is named after the "uncertainty principles" discovered by Heisenberg in the 1920s in the context of an arising quantum mechanics, and after Gabor, who performed analogous studies directly after World War II in the field of communication theory.

To establish this inequality, let us consider a signal $x(t)$ with finite energy

$$E_x = \int_{-\infty}^{+\infty} |x(t)|^2 \, dt < +\infty \ .$$

We assume for the sake of simplicity that the signal and its Fourier transform $X(\nu)$ have a vanishing center of gravity, that is

$$\int_{-\infty}^{+\infty} t \, |x(t)|^2 \, dt = 0 \quad \text{and} \quad \int_{-\infty}^{+\infty} \nu \, |X(\nu)|^2 \, d\nu = 0 \ .$$

(This can always be obtained by a suitable shift of the axes.) As measures of the time and frequency supports of the signal we introduce the respective moments of inertia

$$\Delta t^2 \equiv \frac{1}{E_x} \int_{-\infty}^{+\infty} t^2 \, |x(t)|^2 \, dt \ ; \quad \Delta \nu^2 \equiv \frac{1}{E_x} \int_{-\infty}^{+\infty} \nu^2 \, |X(\nu)|^2 \, d\nu \ . \quad (1.3)$$

Let us finally define the auxiliary quantity

$$I \equiv \int_{-\infty}^{+\infty} t \, x^*(t) \, \frac{dx}{dt}(t) \, dt \ .$$

By using Parseval's identity, we can immediately deduce that

$$(\mathrm{Re}\{I\})^2 \leq |I|^2 \leq \int_{-\infty}^{+\infty} t^2 \, |x(t)|^2 dt \cdot \int_{-\infty}^{+\infty} \left|\frac{dx}{dt}(t)\right|^2 dt = 4\pi^2 \, E_x^2 \, \Delta t^2 \, \Delta \nu^2,$$

where the first inequality holds for any complex number and the second follows from the Cauchy-Schwarz inequality. Integration by parts shows that I equals

$$I = \left[\, t \, |x(t)|^2 \,\right]_{-\infty}^{+\infty} - E_x - \int_{-\infty}^{+\infty} t \, x(t) \, \frac{dx^*}{dt}(t) \, dt = -E_x - I^* \ ,$$

hence

$$\mathrm{Re}\,\{I\} = -\frac{E_x}{2} \ ;$$

Chapter 1 The Time-Frequency Problem

this computation is justified, supposing that the squared absolute value of $x(t)$ decays fast enough such that $t|x(t)|^2$ vanishes at infinity. (This hypothesis is clearly satisfied if $x(t)$ has compact support; it also results from the finiteness of Δt.) Under this assumption the *Heisenberg-Gabor uncertainty principle* follows, which is expressed by the inequality

$$\Delta t \cdot \Delta \nu \geq \frac{1}{4\pi} . \tag{1.4}$$

Employing the preceding definitions for the duration Δt and bandwidth $\Delta \nu$ of a signal, we have thus seen that the "duration-bandwidth product" $\Delta t \cdot \Delta \nu$ of any signal is bounded from below. The lower bound is attained if we meet the conditions of equality in the Cauchy-Schwarz inequality (on which the derivation of the uncertainty principle was based). Viewed from a general geometrical point of view, this means that the two vectors in the inner product are collinear. In other words, the two signals $tx(t)$ and $(dx/dt)(t)$ must be proportional. For elements of the vector space of real-valued signals, this implies the differential equation

$$\frac{dx}{x} = kt\,dt , \quad k \in \mathbb{R} .$$

Its finite energy solutions admit the general form

$$x(t) = C\,e^{-\alpha t^2} , \quad (C, \alpha) \in \mathbb{R} \times \mathbb{R}_+ .$$

Hence the Gaussian functions are the only solutions that minimize the duration-bandwidth product in the Heisenberg-Gabor sense. This relates to our previous statement about the Gaussians and their Fourier transforms.

Remark 1. Let us note, from an angle of physical interpretation, that the given definitions of the duration and bandwidth of a signal lead to perceptibly smaller values than other measures, such as an "equivalent width" or "half-width." If we compute the quantity $\Delta t_{1/2}$ at a size of $\exp(-\pi/4)$ of the Gaussian peak value (which is rather close to the half-width, as $\exp(-\pi/4) = 0.456$), then we obtain

$$\Delta t_{1/2} = 2\sqrt{\pi}\,\Delta t .$$

This leads to a minimal duration-bandwidth product of 1.

Remark 2. One could be willing to accept that the results concerning the minimization of the duration-bandwidth product could be extended to complex-valued signals, whose absolute values $|x(t)|$ and $|X(\nu)|$ are both Gaussians. A typical instance of such signals is a "chirp" (linear frequency modulation and Gaussian envelope)

$$x_c(t) = C\,e^{-(\alpha + i\beta)t^2} , \quad (C, \alpha, \beta) \in \mathbb{R} \times \mathbb{R}_+ \times \mathbb{R} .$$

Its Fourier transform is given by

$$X_c(\nu) = C \left(\frac{\pi}{\alpha + i\beta}\right)^{1/2} e^{-\pi^2 \nu^2/(\alpha + i\beta)}.$$

Some direct computations show, however, that in this case

$$\Delta t \cdot \Delta \nu = \frac{1}{4\pi} \sqrt{1 + \gamma^2} \geq \frac{1}{4\pi} \quad \text{with} \quad \gamma \equiv \frac{\beta}{\alpha}. \tag{1.5}$$

In fact, for a given duration (fixed by α) the bandwidth is enlarged by an amount relative to the range of frequencies, which is scanned by the frequency modulation due to the quadratic phase. Consequently, although the absolute values of the signal and its Fourier transform are Gaussians, the duration-bandwidth product may exceed the lower bound in the Heisenberg-Gabor uncertainty principle. An obvious exception is the case $\beta = 0$, which reduces to the situation of a real signal without modulation. This simple example illustrates the following fact: If a signal attains the minimal value of the duration-bandwidth product, then its absolute values in time and frequency must be Gaussians. On the other hand, the converse is not true; that is, if a signal has Gaussian absolute values in time and frequency, then it need not attain the minimal value of the duration-bandwidth product. Finally, one should recall that this discussion is also valid for signals that are shifted in time and/or frequency. The only change to be made is a proper computation of the moments of inertia with respect to the centers of gravity of the shifted signals. It is important to note, however, that an extension of the same results to the case of real signals, for which only the half-spectrum of positive frequencies is considered, is not allowed. [3]

Interpretations. The Heisenberg-Gabor uncertainty principle admits different interpretations. As we have done so far, we can consider Δt and $\Delta \nu$ as extensions of the supports on which a well-defined signal exists in time and in frequency. Because $|x(t)|^2$ and $|X(\nu)|^2$ are nonnegative functions, we can also ascribe the rank of a probability density function to them. Then the forementioned quantities naturally become standard deviations. This is, in fact, the predominant approach in quantum mechanics, where the actual equivalent of the inequality (1.4) refers to the variables of position and momentum, and not to those of time and energy as in "Heisenberg's fourth uncertainty principle." [4] This latter principle can easily be derived, at least formally, from the time-frequency inequality via the relation

$$E = h\nu$$

(where h is the Planck constant). Then a simple multiplication of the two members in Eq. (1.4) by h shows that $\Delta t \cdot \Delta E \geq h/4\pi$.

Chapter 1 The Time-Frequency Problem

When we attempt to use a common formalism for quantum mechanics and signal theory, we have to pay attention to the fact that the notions of "time" in both disciplines are rather different by nature. The "time" in quantum mechanics is an evolutionary parameter relative to observable quantities that depend on the variables of the description of the system, such as the position or the momentum. With these observables we associate certain operators. (Recall that the measurements are expectation values of these operator functions over the possible states of the system.[5]) In this context, the "time" itself does not correspond to an operator, which would endow it with the status of an observable. In contrast to this situation, the "time" in the operational formalism of signal theory[6] plays a role that is analogous to the position variable in quantum mechanics. Hence it has no dynamical nature in the foregoing sense. Rather, it can be associated with an operator

$$(\hat{t}x)(t) \equiv tx(t) \ .$$

As is well known, the Fourier transform maps a multiplication by the variable in one domain into a derivative in the other domain. Hence, by duality of the Fourier transform, the frequency ν (which is the analogue of the momentum) can be associated with the operator

$$(\hat{\nu}x)(t) \equiv \frac{1}{i2\pi} \frac{dx}{dt}(t) \ .$$

One can easily see that the product of these two operators depends on their order. In fact, they satisfy the commutation relation

$$[\hat{t}, \hat{\nu}] \equiv \hat{t}\hat{\nu} - \hat{\nu}\hat{t} = \frac{i}{2\pi} \widehat{I}$$

where $\widehat{I}$ denotes the identity operator. This is traditionally expressed by saying that the time and frequency variables are *canonically conjugate*.

This operational formalism allows us to prove the Heisenberg-Gabor uncertainty principle in a different way. Let us first employ the notation of the expectation value of an operator $\widehat{A}$, which is defined by

$$\langle \widehat{A} \rangle_x \equiv \langle \widehat{A}x, x \rangle$$

where $\langle ., . \rangle$ denotes the usual inner product of $L^2(\mathbb{R})$; that is,

$$\langle x, y \rangle \equiv \int_{-\infty}^{+\infty} x(t)\, y^*(t)\, dt = \int_{-\infty}^{+\infty} X(\nu)\, Y^*(\nu)\, d\nu \ .$$

The equations

$$\Delta t^2 = \frac{1}{E_x} \langle \hat{t}^{\,2} \rangle_x \ , \qquad \Delta \nu^2 = \frac{1}{E_x} \langle \hat{\nu}^2 \rangle_x$$

hold by definition. If we introduce the operator $\hat{t} + i\lambda\hat{\nu}$, with λ being an arbitrary real number, then the positivity of the inner product yields

$$0 \leq \langle\, (\hat{t} + i\lambda\hat{\nu})x,\ (\hat{t} + i\lambda\hat{\nu})x\, \rangle = \langle\, (\hat{t} - i\lambda\hat{\nu})(\hat{t} + i\lambda\hat{\nu})\, \rangle_x \ .$$

Here the equality on the right-hand side holds, as both operators $\hat{t}$ and $\hat{\nu}$ are self-adjoint. The commutation relation for $\hat{t}$ and $\hat{\nu}$ leads further to

$$\langle \hat{t}^2 \rangle - \frac{E_x}{2\pi}\lambda + \langle \hat{\nu}^2 \rangle \lambda^2 \geq 0 \ .$$

As E_x is positive, this last inequality is satisfied for all λ, if and only if the discriminant of the quadratic polynomial (in λ) is always negative. This anew furnishes the relation

$$\Delta t \cdot \Delta \nu \geq \frac{1}{4\pi} \ .$$

1.1.3. Slepian-Pollak-Landau Theory

As fundamental as it is, the Heisenberg-Gabor uncertainty principle is not the only possible approach aiming at a mathematical description of the Fourier duality, which implies that the confinement of a signal in one domain (time or frequency) causes the loss in its confinement in the canonically conjugate domain. The Heisenberg-Gabor inequality particularly accentuates the impossibility that a signal have arbitrarily small supports both in time and frequency. But it tells nothing about the impossibility of restricting its total energy to *compact* supports in time and frequency, if arbitrarily large (but finite) intervals are allowed.

Concentrations. This impossibility is well known, however, and it calls for a quantitative explanation. Instead of characterizing the equivalence of supports in terms of measures of dispersion, it is preferable for the given task to use measures of *energy concentration*. In view of Parseval's identity we know that, for a signal $x(t)$,

$$E_x = \int_{-\infty}^{+\infty} |x(t)|^2\, dt = \int_{-\infty}^{+\infty} |X(\nu)|^2\, d\nu \ .$$

Hence, the following definitions of the energy concentrations in a time interval $[-T/2, +T/2]$ or a frequency interval $[-B/2, +B/2]$ seem to be appropriate:

$$E_x(T) \equiv \int_{-T/2}^{+T/2} |x(t)|^2\, dt \quad \text{and} \quad E_X(B) \equiv \int_{-B/2}^{+B/2} |X(\nu)|^2\, d\nu \ . \quad (1.6)$$

Chapter 1 The Time-Frequency Problem

These quantities rely on two *truncation operators*, which are defined by the restriction to the respective interval, namely

$$(\widehat{P}_T^t x)(t) = \begin{cases} x(t) & , \quad |t| \leq T/2 \, , \\ 0 & , \quad |t| > T/2 \, , \end{cases} \tag{1.7}$$

and

$$(\widehat{P}_B^f X)(\nu) = \begin{cases} X(\nu) & , \quad |\nu| \leq B/2 \, , \\ 0 & , \quad |\nu| > B/2 \, . \end{cases} \tag{1.8}$$

Equation (1.6) can now be written in the equivalent form

$$\langle \widehat{P}_T^t \rangle_x = E_x(T) \quad \text{and} \quad \langle \widehat{P}_B^f \rangle_x = E_X(B) \, . \tag{1.9}$$

The preceding operators are projections (which means that $\widehat{P}_T^t \cdot \widehat{P}_T^t = \widehat{P}_T^t$ and $\widehat{P}_B^f \cdot \widehat{P}_B^f = \widehat{P}_B^f$). Within this formalism, a perfect concentration both on the time interval $[-T/2, +T/2]$ and the frequency interval $[-B/2, +B/2]$ would result in $\langle \widehat{P}_T^t \rangle_x = \langle \widehat{P}_B^f \rangle_x = E_x$, with finite values for T and B.

Sampling. We already mentioned the impossibility of such concentrations owing to the analyticity of the Fourier transform of a function with compact support. From another point of view, we encounter one of the most common manifestations of this impossibility when we try to sample a signal. It is known as a general fact that the sampling in the time domain causes a periodization in the frequency domain. Hence, it can only be used without any loss in information (which means without changing the spectral contents of the continuous signal), if the original signal is bandlimited. Unfortunately, this forces the signal to last infinitely, which obviously never happens in practical situations. All this means that a perfect sampling is impossible in practice. However, we know from our experience that a reasonable approximation is often achievable, which is compatible with the finite duration of observations and, at the same time, with the finite frequency band of the receivers. Hence, the newly set problem of the time-frequency duality is to establish a precise mathematical framework for the empirical notion of signals that are "practically" finite in time and frequency.

A first approach consists of a qualitative analysis of the approximation error, which is induced by imperfect sampling. Let us suppose momentarily that a signal $x(t)$ is strictly limited to a frequency band $[-B/2, +B/2]$. Then it can be sampled with a sampling period $T_e \leq 1/B$ without introducing an error. Denoting by

$$x[n] \equiv x(n/B) \, , \quad n \in \mathbb{Z} \, ,$$

a sequence of sampling values at this minimal sampling rate, all values $x(t)$ of the signal can be recovered via the interpolation formula [7]

$$x(t) = \sum_{n=-\infty}^{+\infty} x[n] \, \frac{\sin \pi(Bt-n)}{\pi(Bt-n)} \, . \tag{1.10}$$

As far as the summation actually extends over the infinite sequence, this interpolation is totally exact, as one can verify that

$$\int_{-\infty}^{+\infty} \left| x(t) - \sum_{n=-\infty}^{+\infty} x[n] \, \frac{\sin \pi(Bt-n)}{\pi(Bt-n)} \right|^2 dt = 0 \, .$$

In order to work with a finite sum only, we must drop a countable number of sampling values from the sequence. Intuitively, we obtain an approximate reconstruction by restricting the summation to all sampling values inside an interval $[-T/2, +T/2]$, if $x(t)$ has negligible values outside this interval. This means that all values for n, which are used in the summation, must satisfy $|n| \leq (T/2)/(1/T_e)$, or equivalently $|n| \leq BT/2$. The resulting error of the reconstruction has the form

$$\epsilon^2 = \int_{-\infty}^{+\infty} \left| x(t) - \sum_{n=-BT/2}^{BT/2} x[n] \, \frac{\sin \pi(Bt-n)}{\pi(Bt-n)} \right|^2 dt \, .$$

This error can never be zero, yet its size is given by

$$\epsilon^2 = \frac{1}{B} \sum_{|n|>BT/2} |x[n]|^2 \approx \mathcal{O}(E_x - E_x(T)) \, ,$$

where T is fixed and B is sufficiently large.

The anticipated approximation of finite duration T of a bandlimited signal becomes better, if the temporal energy concentration is large. The problem still remains of how to find the exact bound for the largest possible energy concentration in the given time-interval. A precise answer to this question, among other related ones, was given by the theory of Slepian, Pollak, and Landau.[8] It was developed around the beginning of the 1960s and is based on the study of the eigenvalues and eigenfunctions of the forementioned projection operators (cf. Eqs. (1.7) and (1.8)).

Chapter 1 The Time-Frequency Problem

The eigenvalue equation. Given an arbitrary (nonzero) signal $x(t)$ in $L^2(\mathbb{R})$, let us first apply the truncation in time and then the truncation in frequency. The new signal must be different from the original, as its energy was reduced by at least one of the operations. The question that arises here is how the minimal decline of the energy by this double truncation can be characterized, and for which signal it is attained. A solution to this problem would enable us to define a precise notion of simultaneous concentration in time and frequency.

The operator of double truncation corresponds to the transformation

$$x(t) \longrightarrow (\widehat{P}_B^{\text{f}}\, \widehat{P}_T^{\text{t}}\, x)(t) = \int_{-B/2}^{+B/2} \left[\int_{-T/2}^{+T/2} x(s)\, e^{-i2\pi\nu s}\, ds \right] e^{i2\pi\nu t}\, d\nu\ ,$$

hence,

$$(\widehat{P}_B^{\text{f}}\, \widehat{P}_T^{\text{t}}\, x)(t) = \int_{-T/2}^{+T/2} \frac{\sin \pi B(t-s)}{\pi(t-s)}\, x(s)\, ds\ .$$

Because the truncation operators are self-adjoint, the ratio $\mu(B,T)$ of the energy of the twice-truncated signal and the original signal can be written as

$$\mu(B,T) \equiv \frac{1}{\langle x, x \rangle} \left\langle \widehat{P}_B^{\text{f}}\, \widehat{P}_T^{\text{t}} x,\ \widehat{P}_B^{\text{f}}\, \widehat{P}_T^{\text{t}} x \right\rangle$$

$$= \frac{1}{\langle x, x \rangle} \left\langle \widehat{P}_T^{\text{t}}\, \widehat{P}_B^{\text{f}}\, \widehat{P}_B^{\text{f}}\, \widehat{P}_T^{\text{t}} x,\ x \right\rangle$$

$$= \frac{1}{E_x} \left\langle \widehat{P}_T^{\text{t}}\, \widehat{P}_B^{\text{f}}\, \widehat{P}_T^{\text{t}} \right\rangle_x\ . \qquad (1.11)$$

Furthermore, the second factor in Eq. (1.11) depends only on the values of $x(t)$ inside the time interval $[-T/2, +T/2]$, hence

$$\mu(B,T) = \frac{1}{E_x} \int_{-T/2}^{+T/2} \left[\int_{-T/2}^{+T/2} \frac{\sin \pi B(t-s)}{\pi(t-s)}\, x(s)\, ds \right] x^*(t)\, dt\ .$$

Minimizing the effect of the double truncation is hereby equivalent to maximizing $\mu(B,T)$. This is obtained if $x(t)$ is an eigenfunction of the integral equation

$$\int_{-T/2}^{+T/2} \frac{\sin \pi B(t-s)}{\pi(t-s)}\, x(s)\, ds = \lambda\, x(t)\ ,\quad |t| < T/2\ , \qquad (1.12)$$

relative to the largest eigenvalue $\lambda_{\max}$. Consequently, an upper bound for the ratio of the energies (after and before the double truncation) is

$$\mu(B,T) \leq \lambda_{\max} = \max_x \frac{1}{E_x(T)} \left\langle \widehat{P}_T^{\operatorname{t}} \widehat{P}_B^{\operatorname{f}} \widehat{P}_T^{\operatorname{t}} \right\rangle_x \leq 1 \;. \tag{1.13}$$

Duration-bandwidth product. At first sight the eigenvalue equation (1.12) (and its solutions) seems to depend on B and T independently. However, it can easily be rewritten by means of a substitution of the variable $s = Tw$ and by introducing the auxiliary signal $y(u) \equiv x(Tu)$, thus rendering it in the simpler form

$$\int_{-1/2}^{+1/2} \frac{\sin \pi BT(u-w)}{\pi(u-w)} y(w)\,dw = \lambda\, y(u) \;, \quad |u| < 1/2 \;.$$

This shows that the dependence on B and T occurs only via the intermediary of the *duration-bandwidth product* BT.

Eigenvalues. Because of its structure the eigenvalue equation has a discrete spectrum of positive eigenvalues λ_n, which lie between 0 and 1 (as they provide a measure of the relative energy concentration). We can thus arrange them according to

$$0 < \ldots < \lambda_n < \ldots < \lambda_1 < \lambda_0 = \lambda_{\max} < 1 \;.$$

Each eigenvalue is to be considered as a function of the product BT.

Eigenfunctions. There is one and only one eigenfunction $\psi_n(t)$ associated with each eigenvalue λ_n. Properly normalized, this collection of eigenfunctions (which are called "prolate spheroidal wave functions" and "fonctions sphéroïdales aplaties") forms an orthonormal system on $\mathbb{R}$. We thus have that

$$\int_{-\infty}^{+\infty} \psi_n(t)\, \psi_m(t)\, dt = \delta_{nm} \;.$$

Likewise, it is a simple exercise to show that

$$\int_{-T/2}^{+T/2} \psi_n(t)\, \psi_m(t)\, dt = \lambda_n\, \delta_{nm} \;.$$

Finite bandwidth. If we rewrite the eigenvalue equation (1.12) in the form

$$\int_{-\infty}^{+\infty} \frac{\sin \pi B(t-s)}{\pi(t-s)} (\widehat{P}_T^{\operatorname{t}} \psi_n)(s)\, ds = \lambda_n\, \psi_n(t) \;,$$

it is of convolution type. We can immediately see by an application of the Fourier transform that all eigenfunctions have finite bandwidth B.

Chapter 1 The Time-Frequency Problem

Approximation of bandlimited signals. Due to the fact that the eigenfunctions form an orthonormal system, every signal with bandwidth B has a series expansion

$$x(t) = \sum_{n=0}^{+\infty} x_n \psi_n(t) \quad \text{where} \quad x_n = \int_{-\infty}^{+\infty} x(t) \psi_n(t) \, dt \ .$$

This enables us to give a definite answer to one of the previously posed problems: the bandlimited signal with bandwidth B, which maximizes the energy concentration in the time interval $[-T/2, +T/2]$, is the eigenfunction $\psi_0(t)$ relative to the maximal eigenvalue λ_0. Indeed, this result is a direct consequence of the relations

$$\int_{-\infty}^{+\infty} x^2(t) \, dt = \sum_{n=0}^{\infty} x_n^2 = E_x$$

and

$$\langle \widehat{P}_T^t \rangle_x = \int_{-T/2}^{+T/2} x^2(t) \, dt = \sum_{n=0}^{\infty} \lambda_n x_n^2 \leq \lambda_0 E_x \ , \tag{1.14}$$

with equality in the last step for $x(t) = \psi_0(t)$.

Moreover, we can quantify the approximation error, which is committed when representing a signal of bandwidth B by only finitely many eigenfunctions relative to the support T. Measuring it in terms of a mean square error as before, we obtain

$$\epsilon_N^2 \equiv \int_{-\infty}^{+\infty} \left| x(t) - \sum_{n=0}^{N} x_n \psi_n(t) \right|^2 dt = \sum_{n=N+1}^{\infty} x_n^2 \ .$$

This error is bounded by

$$\epsilon_N^2 \leq \frac{1}{1 - \lambda_{N+1}} \sum_{n=N+1}^{\infty} x_n^2 (1 - \lambda_n) \leq \frac{1}{1 - \lambda_{N+1}} \sum_{n=0}^{\infty} x_n^2 (1 - \lambda_n) \ ,$$

hence Eq. (1.14) yields

$$\epsilon_N^2 \leq \frac{1}{1 - \lambda_{N+1}} \left[E_x - E_x(T) \right] \ . \tag{1.15}$$

Approximative dimension of a signal. The approximation error in Eq. (1.15) obviously decreases when the size of the first omitted eigenvalue in the expansion gets smaller. This leads to the notion of the *approximative dimension* of a signal, which is regarded as the number of eigenfunctions that are associated with non-negligible eigenvalues. In the case of the sampling operation, the following empirical argument for a signal with a frequency band $[-B/2, +B/2]$ was raised: because it requires a minimal sampling rate of $1/B$, it should be representable by roughly $T/(1/B) = BT$ sampling values, if its (essential) temporal support is confined to $[-T/2, +T/2]$. This point of view can now be specified more accurately by the new approach using the truncation operators. Indeed, we can conclude from Mercer's theorem [9] and the value for the norm of the eigenfunctions on $[-T/2, +T/2]$, that the kernel of the integral equation has an expansion in terms of the prolate spheroidal wave functions given by

$$\frac{\sin \pi B(t-s)}{\pi(t-s)} = \sum_{n=0}^{\infty} \psi_n(t)\psi_n(s) .$$

From this relation we can deduce the fact that

$$\sum_{n=0}^{\infty} \lambda_n = \int_{-T/2}^{+T/2} \left[\frac{\sin \pi B(t-s)}{\pi(t-s)} \right]_{s=t} dt = BT .$$

This result is related to the idea that approximately BT of the dominant eigenvalues are close to 1, while the others are close to 0. An actual computation of the eigenvalues as functions of n for a fixed value of BT reveals such a behavior: one observes a fast decay of the values as soon as $n > BT$. This knowledge can be used in order to find better estimates for the error, when we want to approximate a signal with bandwidth B by a finite collection of prolate spheroidal wave functions relative to the duration T. Based on the fact that

$$\lambda_{BT+1}(BT) < 0.916 ,$$

one obtains the frequently used estimate

$$\epsilon_N^2 \leq 12 \left[E_x - E_x(T) \right] \quad \text{if} \quad N > BT . \tag{1.16}$$

Inequality of the concentrations. If we consider the time-frequency duality from an angle of the energy concentrations in time and frequency, the adopted point of view allows us to state another interesting inequality. It is closely associated with the following question: given a concentration in time (or in frequency), what is the best possible concentration in frequency

Chapter 1 The Time-Frequency Problem

(in time, respectively), which can be expected, and for which signal is it obtained? We should note, after all, that this issue arises only if the imposed concentration $E_x(T)/E_x$ (or $E_X(B)/E_x$), for a given value of BT, is strictly larger than the largest eigenvalue $\lambda_0(BT)$. Otherwise, in the case of equality, we showed before that a unique solution exists and is equal to the eigenfunction $\psi_0(t)$. Even more can be said, when we suppose that $E_x(T) < \lambda_0(BT) E_x$; then there exist infinitely many bandlimited signals with $E_X(B) = E_x$. Hence, this case implies no restriction of the frequency concentration whatsoever.

The situation is totally different for the remaining cases when $E_x(T) > \lambda_0(BT) E_x$ or $E_X(B) > \lambda_0(BT) E_x$. One can show that the individual concentrations satisfy the time-frequency inequality

$$\arccos\left((E_x(T)/E_x)^{1/2}\right) + \arccos\left((E_X(B)/E_x)^{1/2}\right) \geq \arccos\left(\lambda_0(BT)^{1/2}\right), \quad (1.17)$$

and that equality is attained for the signal

$$x(t) = \left(\frac{1 - E_x(T)/E_x}{1 - \lambda_0(BT)}\right)^{1/2} \psi_0(t) \;+$$

$$\left[\left(\frac{E_x(T)/E_x}{\lambda_0(BT)}\right)^{1/2} - \left(\frac{1 - E_x(T)/E_x}{1 - \lambda_0(BT)}\right)^{1/2}\right] (\widehat{P}_T^{\mathrm{t}} \psi_0)(t) \;.$$

As an illustration of this result, we depict several curves in Fig. 1.1 that are associated with different values of BT. Each curve forms the boundary of the domain of jointly admissible values of the energy fractions on the supports B and T. For increasing values of BT, the upper borderline of these domains stretches out towards the point $(1,1)$, which corresponds to the extreme case of total energy concentrations in time and frequency.

Remark. It is also interesting to comment on the other extreme case associated with $BT = 0$, which corresponds to the antidiagonal in the diagram. If we suppose, in fact, that the energy fractions $E_x(T)$ and $E_X(B)$ are such that $E_x(T) + E_X(B) \leq E_x$, then we have

$$\arccos\left((E_x(T)/E_x)^{1/2}\right) + \arccos\left((E_X(B)/E_x)^{1/2}\right) \geq \frac{\pi}{2} \;.$$

This guarantees that the inequality of the concentrations is verified for any value of BT, including 0, because the relation $0 < \lambda_0(BT) < 1$ is always true. Hence, there is no constraint at all in this case, and the given concentrations can be achieved on arbitrarily small supports in the time and frequency domains.

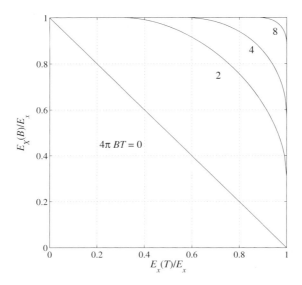

Figure 1.1. Energy concentrations in time and frequency. Borderlines of the admissible domains of energy concentrations in time and frequency, for different values of the duration-bandwidth product BT (cf. Eq. (1.17)).

§1.2. Leaving Fourier?

All forementioned limitations center around the difficulty caused by the use of global descriptions "à la Fourier" (and thus having a time *or* frequency nature) for the apprehension of a reality that exists jointly in time *and* frequency. Indeed, most time-frequency problems can be specified in terms of *local* quantities, which can either be joint (in terms of time and frequency) or not. Such specifications arise through the adoption of definitions that incorporate certain nonstationary properties, or the search for possible interpretations of these definitions. First of all they require some deeper insight into both the motivation and tools that exist for such local objects.

1.2.1. Local Quantities

When we want to describe a signal both in time and frequency, the most desirable and most natural local quantity is one that gives a meaning to an "instantaneous" spectral content. This terminology seems to be based on an inner contradiction, as a Fourier frequency in the mathematical sense

Chapter 1 The Time-Frequency Problem

is associated with a global behavior. Yet our experience (and especially that of our auditory system) suggests that one can imbue such a local quantity with a physical meaning. Only the "frequency" brought into play in this context should be defined in a different way than the usual Fourier frequency.

Instantaneous frequency. [10] In order to define the notion of an "instantaneous frequency," it is appropriate to revisit the prototype of a signal associated with the concept of steady state and stability in time: the *monochromatic wave*. It can be unambiguously represented (apart from a pure phase) by

$$x(t) = a \cos 2\pi \nu_0 t \, ,$$

where the constants a and ν_0 are to be read as the *amplitude* and the *frequency*, respectively. The latter measures the rate of change of the argument of the cosine, or its derivative with respect to the time variable (except for a factor 2π).

It is quite tempting to extend this point of view to evolutionary situations, simply by letting the constant a vary in time and by introducing an argument of the cosine with a time-varying derivative. This would lead to definitions of the form

$$x(t) = a(t) \cos \varphi(t) \, .$$

Unfortunately, this expression is not unique. In contrast to the ideal monochromatic case, there are infinitely many pairs $(a(t), \varphi(t))$ for the representation of a given signal $x(t)$. This can be seen by choosing any function $b(t)$ with $0 < b(t) < 1$. Then the equation

$$x(t) = a(t) \cos \varphi(t) = \frac{a(t)}{b(t)} b(t) \cos \varphi(t)$$

shows that $x(t)$ can be written in another form as

$$x(t) = a'(t) \cos \varphi'(t)$$

with $a'(t) = a(t)/b(t)$ and $\varphi'(t) = \arccos(b(t) \cos \varphi(t))$.

A proper solution to this problem can be found, when we first reconsider the monochromatic case prior to its generalization. A real monochromatic signal can certainly be regarded as the real part of a complex exponential

$$a \cos 2\pi \nu_0 t = \mathrm{Re} \left\{ a \, e^{i 2\pi \nu_0 t} \right\} \, . \tag{1.18}$$

The amplitude and frequency of the monochromatic signal are the modulus and phase (except for the factor 2π) of this exponential, respectively. Its imaginary part, which is $a \sin 2\pi\nu_0 t$, is derived from the real part by a phase shift of $\pi/2$: We say that the real and imaginary parts are in *quadrature*. The mathematical operation behind this transformation can be described in the frequency domain most easily. It maps the Fourier transform of a cosine function

$$\frac{1}{2}\left[\delta(\nu-\nu_0)+\delta(\nu+\nu_0)\right]$$

to the transform of the corresponding sine function

$$\frac{1}{2i}\left[\delta(\nu-\nu_0)-\delta(\nu+\nu_0)\right] .$$

This operation is a linear filtering whose frequency response is $-i \, \text{sgn}\, \nu$ (so its impulse response is $\text{pv}(1/\pi t)$ where "pv" denotes the Cauchy principal value). It is called the *Hilbert transform*. Consequently, if we associate with each real signal $x(t)$ a complex signal

$$z_x(t) \equiv x(t) + i \, \text{H}\left\{x(t)\right\} = x(t) + \frac{i}{\pi} \, \text{pv} \int_{-\infty}^{+\infty} \frac{x(s)}{t-s} \, ds , \qquad (1.19)$$

where H denotes the Hilbert transform, we obtain a "modulus-phase pair" in an equivalent (and unambiguous) way. So we are able to define an *instantaneous amplitude* $a_x(t)$ and an *instantaneous frequency* $\nu_x(t)$ (by reference to the monochromatic case of Eq. (1.18) and as well unique) by

$$a_x(t) \equiv |z_x(t)| , \qquad (1.20)$$

$$\nu_x(t) \equiv \frac{1}{2\pi} \frac{d \arg z_x}{dt}(t) . \qquad (1.21)$$

The "complexified" signal $z_x(t)$ is called an *analytic signal*. Using polar coordinates the analytic signal describes a turning vector whose length and angular velocity are time-dependent. This differentiates it from the monochromatic case, which is associated with the distinctive picture of a vector that spins round the origin at a constant angular velocity and whose trajectory is a circle (cf. Fig. 1.2).

The analytic signal admits a simple interpretation in the frequency domain. By definition, we know that

$$Z_x(\nu) = X(\nu) + i\left(-i \, \text{sgn}\, \nu\right) X(\nu) = 2 \, U(\nu) \, X(\nu) \qquad (1.22)$$

where $U(\nu)$ denotes the normalized Heaviside step function. This shows that the analytic signal is obtained from the real signal by removing the negative frequencies from its spectrum. This does not change the information

Chapter 1 The Time-Frequency Problem 29

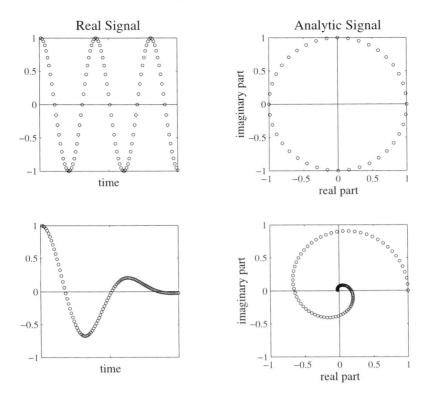

Figure 1.2. Representation of a signal by a turning vector.

Top: a pure monochromatic signal; the representation depicts a vector of constant modulus (associated with the amplitude) and constant angular velocity (associated with the frequency). Bottom: a signal with modulated amplitude and frequency; the modulus and angular velocity vary in time and define the quantities of *instantaneous amplitude* and *instantaneous frequency*.

contents of the signal, because $x(t)$ was assumed to be real, and, therefore, $X(-\nu) = X^*(\nu)$. The truncation of the negative frequencies has the effect of "complexifying" the original signal, which can be interpreted as a reallocation of the redundancy; dividing the frequency band of a signal by two allows us to sample the signal at half the sampling rate in the time domain. For a fixed period the analytic signal requires only half the number of sampling values as compared to the real signal; however, we have to compute two values at each sampling point, namely the real and imaginary part of the complex signal. Hence, the "dimension" of the signal, real or complex, globally remains the same.

Remark. If we consider only the positive frequency axis, the mean frequency $\bar{\nu}$ admits two equivalent definitions

$$\bar{\nu} = \frac{1}{E_x/2} \int_0^{+\infty} \nu |X(\nu)|^2 \, d\nu = \frac{1}{E_z} \int_{-\infty}^{+\infty} \nu_x(t) |z_x(t)|^2 \, dt \ . \quad (1.23)$$

Indeed, by using the fact that $Z_x(\nu) = 2U(\nu)X(\nu)$, and, therefore, $E_z = 2E_x$, we can write

$$\bar{\nu} = \frac{1}{E_z} \int_0^{+\infty} \nu |Z_x(\nu)|^2 \, d\nu \ .$$

An application of Parseval's identity leads to

$$\bar{\nu} = \frac{1}{i2\pi E_z} \int_{-\infty}^{+\infty} \frac{dz_x}{dt}(t) \, z_x^*(t) \, dt$$

$$= \frac{1}{i2\pi E_z} \int_{-\infty}^{+\infty} \left(\frac{d|z_x|}{dt}(t) + i |z_x(t)| \frac{d \arg z_x}{dt}(t) \right) |z_x(t)| \, dt \ .$$

The announced result follows, if we assume that $|z_x(-\infty)|^2 = |z_x(+\infty)|^2$, and this is true both for signals of finite or quasi-finite duration (whose envelope vanishes at infinity) and for signals with a constant envelope. Therefore, we can think of the mean frequency as a weighted mean value of the instantaneous frequency, with the weight given by the squared instantaneous amplitude.

The real part of an analytic signal

$$z_x(t) = a_x(t) \, e^{i\varphi_x(t)}$$

has the form

$$\text{Re}\{z_x(t)\} = a_x(t) \cos \varphi_x(t) \ .$$

It is important to keep in mind, however, that we cannot draw any conclusion in the opposite direction. This means that the analytic signal, which is associated with the real signal $a_x(t) \cos \varphi_x(t)$, need not have the same expressions $a_x(t)$ and $\varphi_x(t)$ as its modulus and phase, respectively. Motivated by a simple physical interpretation, one is at least close to such a situation, if the effects of the modulations are small.

In particular, if $a_x(t)$ has the form of a lowpass with its spectrum restricted to the interval $[-B, +B]$, and if $\cos \varphi_x(t)$ has the form of a bandpass with its spectrum in $]-\infty, -B'] \cup [+B', +\infty[$, and if $B' > B$, then (*Theorem of Bedrosian* [11])

$$\text{H}\{a_x(t) \cos \varphi_x(t)\} = a_x(t) \, \text{H}\{\cos \varphi_x(t)\} \ . \quad (1.24)$$

These considerations still do not justify the exponential expression for the analytic signal, unless the additional requirement

$$\mathcal{H}\{\cos\varphi_x(t)\} = \sin\varphi_x(t)$$

is satisfied. This is not valid in general; but it is approximately verified in the quasi-monochromatic case, where the width of the frequency band is small as compared to the central frequency of the spectrum.

Group delay. The instantaneous frequency characterizes a local frequency behavior as a function of time. In a dual manner, one can also investigate the local time behavior as a function of the frequency. The corresponding magnitude, which measures the instant of the appearance of a certain frequency, is called the *group delay*: it is defined by

$$t_x(\nu) \equiv -\frac{1}{2\pi}\frac{d\arg Z_x}{d\nu}(\nu). \quad (1.25)$$

Interpretative remarks. The local approach that we just adopted calls for several remarks:

(i) In the case of deterministic signals our intuition usually associates the notion of a stationary signal with the idea of a *steady state* governing both amplitude and frequency. The previously introduced concepts admit a more formalized point of view: it is appropriate to call a signal a *stationary deterministic signal*, if it is the sum of several components each having a constant instantaneous amplitude and frequency.

(ii) Although the instantaneous frequency and the group delay are local quantities with respect to the time or the frequency variable, they are still averages relative to the dual variable. In fact, a simple interpretation can only be given for the case of single-component signals. This denotes all situations where at a fixed instant (or frequency) the signal exists only near one frequency (or one instant, respectively). In the opposite case (multicomponent signal) the local quantities defined in Eqs. (1.20), (1.21), and (1.25) account for the local characteristics of all single components taken individually, together with their interferences. For example, the simple situation of the superposition of two monochromatic waves with the same constant amplitude a and different frequencies ν_1 and ν_2 leads to

$$z_x(t) = a\,e^{i2\pi\nu_1 t} + a\,e^{i2\pi\nu_2 t}$$
$$= 2a\,|\cos\pi(\nu_1-\nu_2)t|\,\exp\left\{i2\pi\left(\frac{\nu_1+\nu_2}{2}\right)t + i\pi\,\mathrm{sgn}(\cos\pi(\nu_1-\nu_2)t)\right\}.$$

In this special case the physical interpretation of $x(t)$ as the sum of two pure sinusoidal waves disappears where there is an average frequency with a modulated amplitude. This corresponds nicely to the notion of a *beat frequency* when ν_1 and ν_2 are very close. However, it strays from the most natural interpretation for a large frequential gap, although the characterization offered by the analytic signal remains formally valid.

It is clear that the same restrictions apply to the group delay, *mutatis mutandis*, when it is used for a signal whose components lie in the same frequency band, are localized in time, but have a time offset.

(*iii*) Let us finally remark that the notions of instantaneous amplitude and frequency, as given in Eqs. (1.20) and (1.21), are certainly *local* with respect to time. But they are obtained by the intermediate *global* knowledge of the signal that enters through the Hilbert transform, which itself has an infinite impulse response (cf. Eq. (1.19)).

An example. Let us consider the common idealization of a linear chirp; hereby we denote a signal with a linear frequency modulation and a Gaussian envelope. Its frequently adopted exponential form is given by [12]

$$x(t) = e^{-\pi(\gamma - i\beta)t^2} e^{i2\pi\nu_0 t} . \qquad (1.26)$$

Note that

$$\mathrm{Re}\,\{x(t)\} = e^{-\pi\gamma t^2} \cos 2\pi \left(\nu_0 t + \frac{\beta}{2} t^2\right) .$$

However, this model does not define an analytic signal in the strict sense, as its Fourier transform

$$X(\nu) = (\gamma - i\beta)^{-1/2} \exp\left\{-\pi \frac{\gamma + i\beta}{\gamma^2 + \beta^2} (\nu - \nu_0)^2\right\} \qquad (1.27)$$

is nonzero for $\nu < 0$.

Nevertheless, we can regard its spectrum as almost equivalent to the spectrum of an analytic signal, if the contribution at the zero frequency (and *a fortiori* at all negative frequencies) is negligible (cf. Fig. 1.3). If we measure the width of a Gaussian at a value of $\exp(-\pi/4) = 0.456$ of its peak (which almost coincides with the half-width), we find that the duration T and the bandwidth B of the signal $x(t)$ are given by

$$T = \frac{1}{\sqrt{\gamma}} \; ; \; B = \sqrt{\gamma\left(1 + \frac{\beta^2}{\gamma^2}\right)} \quad \Longrightarrow \quad BT = \sqrt{1 + \frac{\beta^2}{\gamma^2}} ,$$

according to Eqs. (1.26) and (1.27). The result of Eq. (1.27) implies that the condition of quasi-analyticity can be expressed as the *narrowband condition*

$$\frac{B}{\nu_0} \ll \frac{1}{2} . \qquad (1.28)$$

Chapter 1 The Time-Frequency Problem

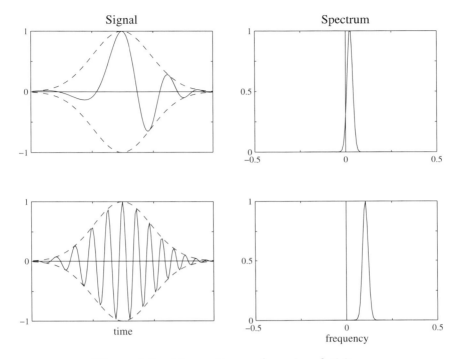

Figure 1.3. Linear chirps and quasi-analyticity.

The signals in the time domain (obtained as real parts of the exponential model in Eq. (1.26)) are shown on the left, and their spectra are shown on the right. For a given bandwidth, one verifies that a higher central frequency (i.e., a smaller relative bandwidth) leads to a better satisfaction of the condition of quasi-analyticity (negligible contribution of negative frequencies).

Under this hypothesis the instantaneous frequency and the group delay turn out to be

$$\nu_x(t) = \nu_0 + \beta t \, , \tag{1.29}$$

$$t_x(\nu) = \frac{\beta}{\gamma^2 + \beta^2} (\nu - \nu_0) \, . \tag{1.30}$$

This result enlightens the fact that the instantaneous frequency and the group delay define two curves in the time-frequency plane, which are usually distinct. One could be willing to accept, intuitively, that these two quantities are inverses of each other for frequency-modulated signals, as they provide two different readings of the same physical reality (frequency as a function of time and time as a function of frequency). However, this only becomes true if an asymptotic condition of the type "large duration-bandwidth product" holds: indeed, we obtain that

$$t_x(\nu_x(t)) = \frac{\beta^2}{\gamma^2 + \beta^2} t = \left(1 - \frac{1}{(BT)^2}\right) t,$$

which yields

$$t_x(\nu_x(t)) \approx t \quad \Longleftrightarrow \quad BT \gg 1. \qquad (1.31)$$

Remark. One can show that this asymptotic condition is stronger than the simple assumption of fast oscillations compared with the evolution of the envelope. This latter situation can persist, for example, for a narrowband signal without any frequency modulation. In our previous notations this corresponds to $\beta = 0$ and $BT = 1$. In this extreme case the instantaneous frequency is given by $\nu_x(t) = \nu_0$ for all t, while the group delay is $t_x(\nu) = 0$ for all ν. The corresponding curves in the time-frequency plane are certainly distinct.

The asymptotic reciprocity of the instantaneous frequency and the group delay can be analyzed by means of certain arguments that appeal to the method of *stationary phase*.[13] This can be carried out for signals with a monotone frequency modulation, as we will see in what follows. Note that we can actually think of the Fourier transform of an analytic signal

$$Z_x(\nu) = \int_{-\infty}^{+\infty} a_x(t)\, e^{i[\varphi_x(t) - 2\pi\nu t]}\, dt$$

as an *oscillatory integral*. Let us further assume that the variations of the phase are fast in comparison with the variation of the envelope, which means that

$$\left|\frac{da_x}{a_x}\right| \ll |d\varphi_x|.$$

Then the essential contribution to the Fourier integral is supplied by the neighborhood of the point t_s (called stationary point) for which the derivative of the phase vanishes, that is, where we have

$$\nu_x(t_s) = \nu.$$

Let us highlight this claim by giving the following qualitative interpretation: The integral over the fast oscillations yields a negligible contribution to the total integral, because the positive and negative arches of the oscillatory function are sufficiently symmetric, at least locally, in order to cancel each other. The only spot with a relevant contribution occurs where the velocity of the oscillations slows down sufficiently in order to make these

Chapter 1 The Time-Frequency Problem

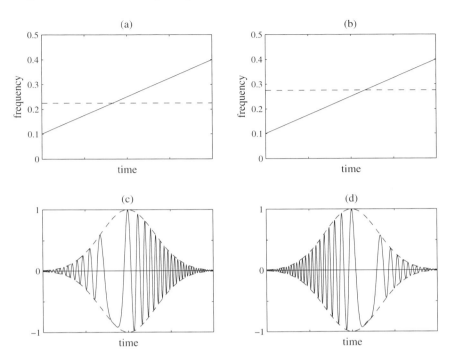

Figure 1.4. Fourier transform and stationary phase.
(a) and (b) are symbolic representations of the analysis of a signal with linear frequency modulation (solid lines) by two different reference frequencies (dashed lines). (c) and (d) are real parts of the corresponding Fourier integrals. It can clearly be seen that an essential contribution to the integral is only provided by the neighborhoods of the point (called stationary) whose time abscissa corresponds to the intersection of the frequencies.

positive and negative arches asymmetrical. In other words, the contribution to a fixed frequency of a modulated signal is a low-frequency signature of the *interference* between the signal and the reference frequency. This interference is only effective in the neighborhood of the point where the instantaneous frequency and the Fourier frequency "meet" (cf. Fig. 1.4).

Subject to the actual conditions, the approximation of $Z_x(\nu)$ by the method of stationary phase can be written as

$$Z_x(\nu) \approx \left| \frac{d\nu_x}{dt}(t_\mathrm{s}) \right|^{-1/2} a_x(t_\mathrm{s}) \exp\{i\Phi^\mathrm{s}_x(\nu)\}$$

where

$$\Phi_x^s(\nu) = \varphi_x(t_s) - 2\pi\nu t_s + \frac{\pi}{4}\operatorname{sgn}\frac{d\nu_x}{dt}(t_s) \; ; \qquad \nu_x(t_s) = \nu \; .$$

This implies that the group delay has the approximate value (again confined to the validity of this method)

$$t_x^s(\nu) = -\frac{1}{2\pi}\frac{\partial \Phi_x^s}{\partial \nu}(\nu) = t_s \; .$$

This clearly shows that the group delay is the inverse function of the instantaneous frequency. Note that the approximation by the method of stationary phase only makes sense, if the derivative of the instantaneous frequency does not vanish at the stationary point, thus excluding all cases where constant instantaneous frequencies occur.

1.2.2. Nonstationary Signals

As in the case of deterministic signals, the most natural local quantities, which intrude into the description of nonstationary *random*[14] signals, are related to the time. The "stationary" property, in its usual definition, means the "independence of statistical properties relative to an absolute time." The incorporation of a possibly nonstationary behavior forces us to reintroduce the time as a parameter needed for the description. We therefore wish to consider properties of a signal, which eventually change from one moment to the next and thus become time-varying.

Definition. Viewed from a physical point of view, the most important properties of a stationary signal are those of *second order*. This means that only the statistical properties of degrees one and two must be invariant under a translation in time. (With the exception of Gaussian distributions, for which the statistical properties of the first two degrees are known to determine those of any degree, a random signal, which is stationary in this broader sense, need not be stationary in the strict sense.)

By definition, a random signal is called *stationary in the wide sense* or *weakly stationary*, if

(*i*) its expectation value is independent of the time; that is, it has the form

$$\mathbf{E}\left\{x(t)\right\} = \mu_x \; , \qquad (1.32)$$

where μ_x is a constant (which we assume to be zero without loss of generality), and

(ii) its autocovariance function, which is the same as its *autocorrelation function* in case of a zero-mean signal (i.e., its expectation value μ_x vanishes), depends only on the difference of the two considered instants; it thus has the form
$$\mathbf{E}\left\{x(t)\,x^*(s)\right\} = \gamma_x(t-s)\ . \tag{1.33}$$

In the weakly stationary (and zero-mean) case, the variance must be constant as well, as we have
$$\mathrm{var}\left\{x(t)\right\} = \mathbf{E}\left\{|x(t)|^2\right\} = \gamma_x(0)\ .$$

If the variance is finite, the signal is called of *second order*. (One should note that the usual idealization of *white noise* in continuous time does not fall in this category, as it verifies
$$\mathbf{E}\left\{x(t)\,x^*(s)\right\} = \gamma_0\,\delta(t-s) \tag{1.34}$$
with a constant γ_0, and consequently $\mathbf{E}\left\{|x(t)|^2\right\}$ is not defined.)

It will be useful to include a brief sketch of some important properties of the autocorrelation function or the autocovariance function of a stationary signal. First of all, it possesses a *Hermitian symmetry*
$$\gamma_x(-\tau) = \gamma_x^*(\tau)\ , \tag{1.35}$$
and its maximal modulus is attained at the origin
$$|\gamma_x(\tau)| \leq \gamma_x(0)\ . \tag{1.36}$$

Moreover, it is *positive definite*; this means that for all pairs of instants t_n and t_m and all collections of complex numbers λ_n the inequality
$$\sum_{n=-\infty}^{\infty}\sum_{m=-\infty}^{\infty} \lambda_n\,\lambda_m^*\,\gamma_x(t_n - t_m) \geq 0$$
holds. This implies that the Fourier transform of the autocorrelation function of a stationary signal is *nonnegative* (Theorem of Wiener-Khinchin),
$$\Gamma_x(\nu) \equiv \int_{-\infty}^{+\infty} \gamma_x(\tau)\,e^{-i2\pi\nu\tau}\,d\tau \geq 0\ . \tag{1.37}$$

The function $\Gamma_x(\nu)$ is called the *power spectrum density* due to the identity
$$\mathbf{E}\left\{|x(t)|^2\right\} = \int_{-\infty}^{+\infty} \Gamma_x(\nu)\,d\nu\ . \tag{1.38}$$

This entity is independent of the time, and this suggests that a stationary random signal has some frequency contents that are constant in time (assuming we know how to make sense of this statement). The characterization of a signal by the mere knowledge of its power spectrum relies on a *Spectral Decomposition Theorem*. It tells that every signal $x(t)$, which is stationary in the wide sense, admits a harmonic decomposition (named after Cramér) in the form

$$x(t) = \int_{-\infty}^{+\infty} e^{i2\pi\nu t} \, dX(\nu) \; . \tag{1.39}$$

The forementioned integral is of Fourier-Stieltjes type, and the equality holds in the sense of a quadratic mean. The main interest in this decomposition stems from its property of *double orthogonality*. This means that:

(*i*) the complex exponentials, serving as (deterministic) functions for the decomposition, are orthogonal with respect to the usual inner product,

$$\int_{-\infty}^{+\infty} e^{i2\pi\nu t} \left(e^{i2\pi\xi t}\right)^* dt = \delta(\xi - \nu) \; ; \tag{1.40}$$

(*ii*) the spectral increments, being tantamount to the statistical weights associated with these functions, are orthogonal with respect to the inner product defined by the expectation value over the trial space,

$$\mathbf{E}\left\{dX(\nu)\,dX^*(\xi)\right\} = \delta(\nu - \xi)\,\Gamma_x(\nu)\,d\xi\,d\nu \; . \tag{1.41}$$

In other words, the stationary signals admit a frequential decomposition into *uncorrelated* random variables. The stochastic independence between spectral increments in the stationary case means that disjoint frequency bands share no energy. This can be regarded as a consequence of the ideal frequency localization of the Fourier decomposition in conjunction with the permanence of the frequency contents of a stationary signal.

Generalizations. The class of stationary signals is too restrictive for giving account of most of the ordinarily observed real situations. One possible generalization consists in departing from the characterization of the stationary case in terms of the doubly orthogonal decomposition and relaxing at least one of these orthogonalities. Let us first choose to retain the complex exponentials as the decomposing functions. Then we obtain a new representation (due to Loève), which is quite the same as in the stationary case, except that the spectral increments are no longer uncorrelated. Rather, they fulfill the relation

$$\mathbf{E}\left\{dX(\nu)\,dX^*(\xi)\right\} = \Phi_x(\nu,\xi)\,d\xi\,d\nu \tag{1.42}$$

Chapter 1 The Time-Frequency Problem

where the *spectral distribution function* $\Phi_x(\nu,\xi)$ can have a support of positive measure; hence, it may exist on a larger set than just the main diagonal of the frequency-frequency plane. For the existence of such a decomposition we must meet the requirement (Loève's condition)

$$\iint_{\mathbb{R}^2} |\Phi_x(\nu,\xi)| \, d\xi \, d\nu < +\infty .$$

The corresponding signals are called *harmonizable*. Their (nonstationary) autocovariance function is dual to the spectral distribution function by means of the Fourier-type relation

$$\mathbf{E}\{x(t)\, x^*(s)\} = \iint_{\mathbb{R}^2} \Phi_x(\nu,\xi)\, e^{i2\pi(\nu t - \xi s)} \, d\xi \, d\nu .$$

This equation generalizes the Wiener-Khinchin relation of Eq. (1.37) between the autocorrelation function and the power spectrum by allowing nonstationary signals. It naturally reduces to the earlier equation in the borderline case of a stationary signal, for which $\Phi_x(\nu,\xi) = \delta(\nu-\xi)\Gamma_x(\nu)$.

Another conceivable possibility of generalizing the stationary case is by retaining the double orthogonality, but replacing the complex exponentials with other functions for the decomposition. This leads to the construction of representations of the form

$$x(t) = \int_{-\infty}^{+\infty} \psi(t,\nu) \, dX(\nu) ,$$

with a comparable orthogonality relation

$$\int_{-\infty}^{+\infty} \psi(t,\nu)\, \psi^*(t,\xi) \, dt = \delta(\xi - \nu) .$$

Such decompositions (named after Karhunen) are possible, indeed. They reduce essentially to taking the eigenfunctions of the autocovariance kernel as functions for the decomposition. From this point of view, the requirements of the stationary case endow the autocovariance kernel with a convolutive structure, making it act like a linear filter. Because the eigenfunctions of the linear filter are the complex exponentials, the stationary case can again be revealed as a borderline case in this larger class.

While one can certainly find generalizations of the stationary case, it also becomes clear that they bring about certain difficulties regarding their interpretation. In the first case (harmonizable signals) we formally adhered to the concept of frequency, at the expense of introducing some correlation between spectral increments. This forces us to digress from their usual spectral interpretation in terms of selective filters, which have no interaction if they are distinct. In the second case (Karhunen) the orthogonality of the increments is assured. Their spectral interpretation, however, is questionable insofar as giving up the complex exponentials causes the (dummy) variable ν to lose its physical meaning of a frequency.

Variations. As the nonstationary behavior is a "non property," it is difficult to discuss it in general terms. Many different situations can be imagined that refer in some sense or other to the case of stationary signals. Thus we can consider signals that are *locally stationary*, which means that their autocovariance function has the form (assuming zero mean signals)

$$\mathbf{E}\left\{x(t)\,x^*(s)\right\} = m_x\left(\frac{t+s}{2}\right)\gamma_x(t-s)\,. \tag{1.43}$$

Here $m_x(t)$ is a nonnegative function and $\gamma_x(t)$ is positive definite. The "local" character of this stationarity results from the fact that the autocovariance function is not obliged to be just a function of the difference of the two instants (as in the stationary case), but that it can vary with respect to the midpoint of the two moments under consideration. In other words, the "local" symmetric autocovariance function

$$\mathbf{E}\left\{x\left(t+\frac{\tau}{2}\right)x^*\left(t-\frac{\tau}{2}\right)\right\} = m_x(t)\,\gamma_x(\tau) \tag{1.44}$$

appears as the result of a modulation in time of a stationary autocorrelation function. It is clear that the transition to the stationary case occurs when the function $m_x(t)$ tends to be constant.

Remark 1.[15] If an autocovariance function takes the form of Eq. (1.43), then $m_x(t)$ must be nonnegative and $\gamma_x(t)$ must be positive definite. The first step in the proof of this assertion is to let $s = t$, which gives

$$\mathbf{E}\left\{x(t)\,x^*(t)\right\} = m_x(t)\,\gamma_x(0) \geq 0\,.$$

Hence, we find that $m_x(t)$ has a constant sign, which is equal to the sign of $\gamma_x(0)$. There is no restriction in assuming that this sign is positive. In the second step we use the positive definiteness of any autocovariance function in order to write

$$\iint_{\mathbf{R}^2} y(t)\,y^*(s)\,\mathbf{E}\left\{x(t)\,x^*(s)\right\}\,dt\,ds \geq 0\,,$$

Chapter 1 The Time-Frequency Problem

which is true for all signals $y(t)$. In particular, if we let $y(t)$ be a complex exponential with frequency ν and insert the actual form of the autocovariance function, then we obtain

$$\int_{-\infty}^{+\infty} m_x(t)\, dt \cdot \Gamma_x(\nu) \geq 0 \; .$$

Here $\Gamma_x(\nu)$ denotes the Fourier transform of $\gamma_x(\tau)$. The integral is non-negative owing to $m_x(t) \geq 0$. We can therefore conclude that $\Gamma_x(\nu) \geq 0$. Hence, the positive definiteness of $\gamma_x(\tau)$ has been proved.

Remark 2. In order that a locally stationary signal be harmonizable, its local autocorrelation function $\gamma_x(\tau)$ must be associated with a stationary signal of second order. Moreover, the Fourier transform $M_x(\nu)$ of its modulation function $m_x(t)$ must be absolutely integrable. Indeed, as an immediate consequence of the definition of locally stationary signals, we find that the spectral distribution function has the form

$$\Phi_x(\nu,\xi) = M_x(\nu - \xi)\, \Gamma_x\!\left(\frac{\nu+\xi}{2}\right) \; .$$

We can therefore write Loève's condition as

$$\iint_{\mathbf{R}^2} |\Phi_x(\nu,\xi)|\, d\xi\, d\nu = \gamma_x(0) \int_{-\infty}^{+\infty} |M_x(\nu)|\, d\nu < +\infty \; ,$$

which is the announced result.

The notion of locally stationary signals, as already introduced here, can be expressed by a modulation in time of a stationary autocorrelation function. Another conceivable way is to impose a modulation directly on a stationary signal. Then the resulting signal, called *uniformly modulated*, has the form

$$y(t) = c(t)\, x(t) \; . \tag{1.45}$$

Here the deterministic function $c(t)$ expresses the effects of the modulation, and $x(t)$ is assumed to be weakly stationary. We can immediately conclude that its "local" autocovariance function equals

$$\mathbf{E}\left\{ y\!\left(t+\frac{\tau}{2}\right) y^*\!\left(t-\frac{\tau}{2}\right) \right\} = c\!\left(t+\frac{\tau}{2}\right) c\!\left(t-\frac{\tau}{2}\right) \gamma_x(\tau) \; .$$

If the modulation is "slow," that is, if the variation of $c(t)$ is small in comparison with the statistical memory of $x(t)$ (which can be quantified by its *radius of correlation* τ_c, or the half-width of $\gamma_x(\tau)$), then we can write

$$c\!\left(t \pm \frac{\tau}{2}\right) \approx c(t) \pm \frac{\tau}{2}\frac{dc}{dt}\!\left(t \pm \frac{\tau}{2}\right) \approx c(t) \; , \qquad |\tau| \leq \tau_\mathrm{c} \; .$$

Consequently, we obtain as a first approximation

$$c\left(t+\frac{\tau}{2}\right)c\left(t-\frac{\tau}{2}\right) \approx c^2(t)$$

whenever the values of the local correlation are (essentially) nonzero. We can thus see, given this level of exactness, that the uniformly modulated signals are constructed along the same lines as the locally stationary signals. The two definitions actually coincide with each other in the extreme case of a white noise signal (thus having a microscopic correlation) whose amplitude is modulated.

This situation of a slow evolution of a stationary autocorrelation function (slow compared with its radius) is usually called *quasi-stationary*.

§1.3. Towards Time-Frequency: Several Approaches

The Fourier duality makes both descriptions of a signal, in time and frequency, necessary and insufficient at the same time. Even though they carry the *same* information, they are both necessary, because they represent it in two complementary ways. Even though they carry *all* information, they are both insufficient because they present it in a form that is often too far from the physical reality, so they cannot be exploited conveniently. However, we have seen before that there exist tools for breaking out of the rigid framework of a Fourier analysis in the strict sense: yet most often they go only halfway. We are thus encouraged to perform a more important step, namely the search for genuinely mixed descriptions together in time and frequency. Insofar as these descriptions should be developed from the signal, they can certainly not supply any gain in information by going beyond the time axis into the time-frequency plane. The truly anticipated gain consists of a better *intelligibility*. This means that the change in the representational space corresponds to a better structuring of the information, at the eventual expense of an increased redundancy.

As there are many nonstationary situations and some inviolable theoretical bars, the issue of describing a signal simultaneously in time and frequency does not permit a unique and unanimously satisfactory answer. It suffices to glance through the literature on this subject, in order to be convinced that it offers a bestiary of methods, which is respectable by its number and its variety, combining domestic and wildlife animals, beasts of burden and racing horses, unicorns and raccoons. Before entering a detailed discussion of the possible solutions, it might be useful to close this chapter with an inventory of some guiding principles that preside over both the choice and the elaboration of a time-frequency representation. This also allows us to fix some further notations.

Chapter 1 The Time-Frequency Problem

1.3.1. The Time-Frequency Plane and Its Three Readings

The wealth of the time-frequency plane, which serves as the space for the transformed representations, roots in the possibility of different complementary readings. Having two variables for the description at our disposal, we can consider them together with their cross-relations. More generally, we can even envisage some global quantities that are inaccessible by approaches with respect to only one of the variables.

Frequency (time). The first interpretation of the plane is obtained by regarding the frequency as a function of time. This is connected with the idea of an *evolutionary spectral analysis*. As we emphasized before, the tool of the instantaneous frequency has some natural limitations concerning its interpretation when the analyzed signals are of multicomponent type: By letting the signal "burst" into the time-frequency plane, one can *a priori* overcome this difficulty, because at any moment there is a whole range of frequency values available (and not just an average).

Time (frequency). A second interpretation, which is dual to the first, considers the time as a function of the frequency. This corresponds to the idea of a sequential monitoring of the output of different frequency channels. Hence a complete history is offered for each frequency, and this gives us access to events located in time in a frequency-by-frequency manner.

These first two readings have one point in common, as they prepare the ground for some natural approaches to applications such as matched filtering or the separation of overlapping signals. As an example let us consider two signals that overlap in time and in frequency, such as parallel chirps, but which have disjoint descriptions in the time-frequency plane. In principle, the two dimensions of the time-frequency plane allow us to draw a separating curve between the two components, while this cannot be achieved in a purely temporal or frequential setting.

Time-frequency. The time-frequency plane admits a third interpretation that is more general and global in the sense that it deals with truly *joint* objects in time and frequency. The time-frequency duality, which certainly underlies every relevant description of a nonstationary signal, is thus brought to the surface — not as a concept of "twice in one dimension," but rather "once in two dimensions." This is clearly the most instructive point of view, but it also requires an intrinsically joint imagination.

1.3.2. Decompositions, Distributions, Models

Whichever reading of the plane is chosen, we can conceive several ways of associating a time-frequency function with a given signal. Different approaches can be distinguished concerning the nature of this association.

On the one hand, they relate to the physical interpretation of the representative time-frequency features in a signal, and, on the other hand, to the degree of one's *a priori* knowledge about the signal.

Decompositions. As a signal cannot be arbitrarily concentrated in time and frequency, it is tempting to regard the most concentrated signals (to be made more precise later on) as the elementary parts of every signal, the "building blocks" of an arbitrary waveform. The simplest and most natural rule for the construction process is the *linear* superposition. Here the elementary signals, also called "time-frequency atoms," play the role of a basis of the decomposition. Within the picture of a linear decomposition, the time-frequency representation is given by a (discrete or continuous) set of weights, each being associated with one atom. Moreover, these weights should be accessible by a projection of the signal onto the elements of this "basis."

Distributions. The preceding way of decomposing a signal gains some information about its energy allocation in time and frequency as a by-product. We can also pay particular attention to such energetic characterizations from an angle of energy (or power) *distributions* directly. From this viewpoint it is natural to consider a *quadratic* rule, which associates the signal with its (bilinear or sesquilinear) representation, and thus generalizes the notions of correlation or power spectrum known from the stationary case. Let us remark that it is not necessary to apply a quadratic rule for the association in order to obtain an energy distribution. Other approaches of "higher" order are also conceivable, though leading in general to more complications.

Models. Finally, if the structure of the analyzed signals is available *a priori*, it is interesting to incorporate this knowledge into the modeling by means of, for example, a parameterization. In a nonstationary context we call the model *time-dependent*, if its structural parameters may vary in time. While every decomposition of a signal by means of an *a priori* fixed "basis" can formally be considered as a modeling, we will reserve this notion to those cases where the number of parameters is noticeably smaller than the size of the observation. Understood as a savings, in this sense, the aim of the modeling is somehow opposite to the aim of a decomposition or a distribution: While the latter are mainly directed at a better representation of the data, accepting enlargement in order to gain better analysis, the modeling pays attention to a best possible reduction of the redundancy (e.g., for the purpose of coding). This shows once more that the modeling and the (nonparametric) representation without an *a priori* knowledge are complementary issues: the assumed neutrality of the representation can assist the process of finding a model, which in return will furnish an even more precise representation, because it will be well adapted.

Chapter 1 The Time-Frequency Problem

1.3.3. Moving and Joint, Adaptive and Evolutionary Methods

In order to take the nonstationary nature of the analyzed signals into account, one proceeds by reintroducing the time as a necessary parameter for the description. This can be done with or without reference to some stationary methods, thus giving rise to another classification of the possible approaches.

Moving and joint. The first approach, and no doubt the most natural and practical one, consists in using a local time window, which follows the moment of the analysis and has a limited horizon of observations ("short-time") centered around this moment: methods of this type are called *moving* methods. In the quasi-stationary case (or one that is supposed to be) the moving methods are most often some locally applied stationary methods. This evidently leads to the imposition of restraints, which may concern, for instance, the kind of "short-time" horizon or the validation of the assumptions of the quasi-stationary behavior, *a priori* and *a posteriori*. We will see, however, that the moving methods can also include some "nonstationary" approaches, thus relaxing the concept of quasi-stationarity, which is most naturally associated with a given window of short duration.

In contrast to this situation, and rendering a more general setting feasible, the "nonstationary" methods for a time-frequency analysis consider a nonstationary situation as it is, thus avoiding any *a priori* reference to the stationary or quasi-stationary case. They can be called *joint* methods, as they treat the time and the frequency symmetrically, although they are explicitly based on a time-dependence.

Adaptive and evolutionary. [16] In the setting of a parametric modeling we can adopt two viewpoints for the introduction of a time-dependence. The first, which can be called *adaptive*, consists of using a stationary model (with constant coefficients) and adjusting its estimation at each instant. Such an adaptive approach contains the time *in the algorithm* for the identification. The second point of view, which is called *evolutionary*, uses an explicitly nonstationary model, which means that its coefficients depend on the time. Hence, an evolutionary method contains the time *in the modeling* (no matter if the algorithm for the identification itself is adaptive or not).

Certainly the approaches that have been introduced herein are not mutually exclusive. One can imagine moving decompositions or joint distributions as well as joint decompositions or moving models. In particular, the large variety of approaches reflects the protean nature of the time-frequency problem. One can expect that this large variety is accompanied by a not lesser multitude of solutions: this is true, indeed, and is the subject of the next chapter.

Chapter 1 Notes

1.1.1.

¹ The Fourier analysis gave rise to considerable literature. One might look at Bracewell (1978) for a "signal" point of view, and Gasquet and Witomski (1990), Dym and McKean (1972), and Körner (1988) for a more mathematical point of view.

1.1.2.

² The Heisenberg-Gabor uncertainty principle probably appeared first in work by Weyl (1928) (in its mathematical form, which is connected with the Fourier duality, rather than in its qualitative physical interpretation). It was proved in work by both Gabor (1946) and Brillouin (1959) where further aspects were discussed, especially those concerning its links with quantum mechanics. Numerous generalizations of the inequality have been proposed, such as the extension to discrete sequences (Pearl, 1973) or to measures of concentration associated with L^p-norms (Cowling and Price, 1984). Surveys on this subject are offered by Benedetto (1990) and Folland and Sitaram (1997).

³ The case of real signals is especially treated in work by Kay and Silverman (1957) and Borchi and Pelosi (1980).

⁴ As far as quantum mechanics is concerned, a profitable source is Lévy-Leblond (1973) where a critical discussion of the notion of "uncertainty" is given.

⁵ A classical reference for a detailed discussion of quantum mechanics and its operational formalism is Cohen-Tannoudji, Diu, and Laloé (1973).

⁶ A presentation of the operational formalism of signal theory can be found in work by Bonnet (1968). The use of this formalism in the time-frequency context is contained in Flandrin (1982). We will return to this issue in Subsection 3.1.2.

1.1.3.

⁷ The sampling theorem is proved in almost every book on signal processing. One can find it, for example, in Duvaut (1991), Section 2.31.

⁸ The fundamental articles dealing with the simultaneous energy concentration in time and frequency are by Slepian and Pollak (1961) and Landau and Pollak (1961). Another very interesting presentation (which in fact inspired us) was done by Dym and McKean (1972) and Papoulis (1977).

Chapter 1 The Time-Frequency Problem 47

Among the precursors of the formalized approach by Slepian, Pollak, and Landau, one should mention Ville and Bouzitat (cf. their works from 1955 and 1957) and Fuchs (1954). We also recommend the reference by Donoho and Stark (1989), which generalizes the notion of simultaneous concentration to other domains than intervals (and exploits it for problems of signal reconstruction).

[9] See, for example, Riesz and Sz.-Nagy (1955).

1.2.1.

[10] The first attempts to give a definition of an instantaneous frequency go back to Carson and Fry (1937) as well as Van der Pol (1946). Nowadays the classical definition is the one by Ville (1948). It relies on the concept of analytic signals whose seeds can be found in Gabor (1946). One can also consult Boashash (1992a,b) for a more complete treatment of this issue.

[11] A detailed study of the conditions of the analyticity of a complex signal and the problems in connection with the Theorem of Bedrosian (1963) is given by Picinbono and Martin (1983). It also contains a bibliography on this issue. A more recent paper is by Picinbono (1997).

[12] The example of the linear chirp with a Gaussian envelope is broadly discussed in Kodera, Gendrin, and de Villedary (1978) and Gendrin and Robert (1982).

[13] An introduction to the method of stationary phase can be found in Papoulis (1977). For a more mathematical presentation one should look at the book by Copson (1967).

1.2.2.

[14] The general notions related to random signals, whether stationary or not, are discussed in a large number of books. For example, one can find them in Blanc-Lapierre and Picinbono (1981) or Priestley (1981; 1988). More precisely, the concept of harmonizable signals is due to Loève (1962), the locally stationary signals were introduced by Silverman (1957), and the uniformly modulated signals by Priestley (1965).

[15] This remark is due to Janssen (private communication).

1.3.3.

[16] The distinction "adaptive / evolutionary" is borrowed from Grenier (1984), to whom we refer for a deeper study of the parametric time-frequency aspect, as well as to his work in 1987. This issue is not discussed in this book.

Chapter 2
Classes of Solutions

We have emphasized in the first chapter a number of problems that arise in conjunction with the search for a time-frequency description of signals. This second chapter is devoted to the discussion of some classes of possible solutions and an inventory of their properties. In a certain sense the following presentation can be regarded as a "multiresolution" advance. At first, we draw the main features of a general panorama from an historical perspective (Section 2.1). Then we introduce more and more details of the principal approaches. These can be grouped into three large sets according to (linear) "atomic" decompositions (Section 2.2), (bilinear) energy distributions (Section 2.3) and power distribution functions (Section 2.4). In each case we make every effort to begin with general principles for constructing classes of such solutions. Only afterwards do we specify more details concerning the form or the properties resulting from some additional constraints.

More precisely, Section 2.2 deals with the question of decomposing a signal with respect to a family of elementary signals, which are well localized both in time and frequency. This point of view (called "atomic") is discussed in a general form in Subsection 2.2.1. Then we present two particular approaches, namely the short-time Fourier transform or Gabor decomposition (in Subsection 2.2.2) and the (time-scale) theory of wavelets (in Subsection 2.2.3). An interpretation as a "matched filtering" is discussed in Subsection 2.2.4.

By giving up the linearity of the representation, we concentrate on the construction of classes of *bilinear* solutions in Section 2.3. Here our approach relies on elementary principles of covariance. In particular, we show that simple constraints relative to translations and/or dilations yield an infinite number of (parameterized) representations in the time-frequency and the time-scale plane. These representations are divided into two large classes, which are called Cohen's class and the affine class, respectively.

Starting from this general framework, we systematically explore in Subsection 2.3.2 the properties of the resulting distributions (among which the spectrogram, the scalogram, the Wigner-Ville, Bertrand and Choï-Williams distributions come first). These properties can be expressed explicitly in terms of conditions on the parameter function, which defines the representation. This paves the way for the design of particular representations satisfying a catalogue of specifications. But it also makes the exclusive character of certain desirable properties evident: these issues, which are discussed in Subsection 2.3.3, justify the central role of the Wigner-Ville distribution in the time-frequency theory of signals.

Finally, we change our viewpoint and turn to a stochastic setting in Section 2.4. Here we itemize the principal paths that can be followed in order to gain a time-frequency description of nonstationary signals and/or their power distribution. Anticipating some profit from our investigations of the deterministic case, we recall in Subsection 2.4.1 the two main strategies (orthogonality *or* frequency in the Cramér decomposition), which are induced by the presence of nonstationary features in the signal. This leads to two different families of approaches. One of them gives priority to the double orthogonality (evolutionary spectrum "à la Priestley" in Subsection 2.4.2), and the other one supports the physical interpretation of the spectrum (Subsection 2.4.3). This latter approach emphasizes once more, but in a different light, the role of the Wigner-Ville transform, whose special properties in the stochastic case will be explained. Finally, Subsection 2.4.4 describes some links and some intersections of the two different approaches.

§2.1. An Introduction with Historical Landmarks

Before we begin a more systematic construction of the large classes of solutions for the time-frequency description of signals available today, it might be useful first to explore the pioneer work in this field and to consider it in its historical context. Since history (and especially the development of ideas) is only linear in a rather superficial way, one should not be astonished if some of the advances cited in this short overview look chaotic from time to time.

2.1.1. Short-Time Fourier and Instantaneous Spectrum [1]

It was recognized a long time ago that stationary frequency analysis is often insufficient; one can even say that this is a recurrent theme in the scientific literature, both in theoretical and applied work. According to Pimonow, it was Sommerfeld in his doctoral thesis in 1890, who used the "instantaneous spectra" for the first time. There one can also find the premises of a great deal of work that today seem so natural that no special patronymic is given to the corresponding methods. Their common denominator is the replacement of the *global* Fourier analysis, which loses all chronological information, with a sequence of *local* analyses with respect to a moving observation window. However, the actual boom of this type of approach had to wait until the middle of the 1940s and the invention of the *sonagraph*. This marked a considerable breakthrough, as it created the concept of an "instantaneous spectrum," which was still questionable theoretically, in an operational and evident form. It also left a deep impression on the following generations and their way of apprehending the time-frequency problems. (This mark is still very strong today, when users fighting for a "good" time-frequency representation want most of all a "super-sonagram.")

Sonagram and spectrogram. By construction the sonagraph operated in the frequency domain. By blackening a sensitive paper in the course of time, it recorded the output power of a bank of bandpass filters working in parallel. The filter bank was virtual, insofar as the signal under consideration was recorded on a rotating magnetic drum: only one filter was needed for scanning the useful frequency range due to the synchronous heterodyning of the rotation. The corresponding output could be written, frequency by frequency, on the sensitive paper stuck to the drum. Such an operation can be modeled as

$$S_x(t,\nu) = \left| \int_{-\infty}^{+\infty} X(n) \, H^*(n-\nu) \, e^{i2\pi nt} \, dn \right|^2 , \qquad (2.1)$$

where $X(\nu)$ denotes the Fourier transform of the analyzed signal $x(t)$ and $H(\nu)$ is the frequency response of the analyzing filter with no heterodyning.

Certainly such an analysis lends itself to several variants. For example, the way of scanning the frequency range can be varied. The forementioned scheme is classically understood as a sweep by shifting a lowpass filter. Formally, it corresponds to a uniform filter bank. However, one can also imagine more general situations, where the impulse response of each filter may depend on the analyzed frequency. Classical examples are the constant-Q filters (the quality factor Q of a filter being defined as the inverse relative bandwidth, i.e., the ratio of its central frequency and its passband). They can be described by

$$S_x(t,\nu) \; \rightarrow \; (\nu_0/\nu) \left| \int_{-\infty}^{+\infty} X(n) \, H^*(n(\nu_0/\nu)) \, e^{i2\pi nt} \, dn \right|^2 ,$$

where $H(\nu)$ is a bandpass filter with central frequency ν_0. (Although this is an old principle, it is readily connected with the concept of *wavelets*, as we will see later.)

Let us mention that some intermediate situations were also proposed in addition to these two concepts. One of them is the Frequency Time ANalyzer method (FTAN) by Levshin et al., which is defined by

$$S_x(t,\nu) \rightarrow (\nu_0/\nu)^2 \left| \int_{-\infty}^{+\infty} X(n)\, H^*((n-\nu)(\nu_0/\nu))\, e^{i2\pi nt}\, dn \right|^2.$$

For each of these cases there exists a dual interpretation in the time domain, which introduces the impulse response of the filters. It is this interpretation that imbues the notion of local analysis (in time) with a proper meaning. The locality is linked to a temporal horizon, fixed by the duration of the impulse response of the filter. In this regard we can write the sonagram equation (2.1) in an equivalent form as

$$S_x(t,\nu) = \left| \int_{-\infty}^{+\infty} x(s)\, h^*(s-t)\, e^{-i2\pi\nu s}\, ds \right|^2. \qquad (2.2)$$

This exhibits its computational structure as a "short-time observation of a signal through a window + local frequency analysis," and it is the reason for calling it a *spectrogram*. Note that improvements to the technical environment have led to a preference of the temporal point of view in most modern introductions of this method (after the development of calculators and algorithms for the fast Fourier transform took place in the middle of the 1960s).

Restrictions. The time-frequency analysis by the spectrogram/sonagram is indispensable and should not be underestimated. It has served, and is of current use today, as a basis for a considerable amount of investigations of natural signals, especially sonic and ultrasound (speech, music, animal communication, echolocation, etc.[2]). However, it is worthwhile to mention that its structure carries essential restrictions that are typical for all Fourier-type methods, if they are used in a nonstationary context. As it operates locally by employing filters (in time or in frequency, depending on the interpretation), it necessarily faces a trade-off between the temporal and the frequential localization. We can interpret those two as antagonistic *resolutions*: the time-resolution of a window analysis, such as the spectrogram, gets better when its window becomes shorter; however, the frequency-resolution degrades at the same rate, because the Fourier analysis is confined to the same short time-window. Conversely, analysis by a filter bank with more selective filters has a better frequency-resolution; but this implies a lesser time-resolution because the impulse responses of the filters have a longer duration.

This type of restriction was already known, when the sonagraph was invented. Originally it offered two possible selections of filters: one for a "wideband" (300 Hz) and another one for a narrowband (50 Hz) analysis. These special values were chosen in accordance with their main application of speech signals. In the wideband position, the fine temporal structure of the speech was retained (as manifested, for example, by the temporal periodicity of the pitch); this could only be gained by sacrificing the localization of its formantic structures. The narrowband position improved the latter, but deleted the notion of a temporal appearance of the pitch, replacing it with the corresponding harmonic structure.

The sonagram/spectrogram soon acquired a standard form and became a classical tool for the interpretation of signals. (Certain experts were even able to read messages from the sonagram.[3]) Yet there appeared different attempts to improve it, while maintaining its main features. We already mentioned the constant-Q variants or more general ones with a varying frequency-resolution. Their advantage (and also their inconvenience regarding applications) was to get away from the use of a unique filter and replace it with a bank of filters of different resolutions. Then one could hope to match each characteristic of the signal with at least one of the filters. This point of view is better adapted to the perceptual interpretation of speech signals: We know that the frequency response of the inner ear operates like a uniform filter bank at low frequencies and a bank of constant-Q filters at high frequencies, as a first approximation.[4] However, in other situations where a good resolution at all frequencies is desired, the constant-Q structure has no advantage (e.g., for the case of rotating machines with a rich harmonic structure).

Some other ways of improving the spectrogram have been explored as well (and will be discussed in Subsection 3.2.1). One tries, for instance, to improve the joint resolution of a spectrogram by including the phase information, which is usually ignored. Nevertheless, all such methods belong to the same paradigm of an "instantaneous spectrum," which is obtained via some window analysis.

Two major theoretical contributions appeared almost simultaneously with the invention of the sonagraph. The first dates from 1946 and is due to Gabor. He introduced the notion of the decomposition of signals into minimal grains of information (or time-frequency "atoms"). The second dates from 1948 and is due to Ville. It is concerned with the time-frequency decomposition (or distribution) of the *energy* of the signal, not the signal itself. As a matter of fact, these works laid out two important directions for approaching the time-frequency problem. They have remained the two reference directions for most work on this subject until today.

2.1.2. Atomic Decompositions

It is a rather natural idea to decompose a signal into a family of elementary signals and regard it as their linear superposition. For example, this is the underlying principle of every discretization of a signal. The choice of the elementary signals for the decomposition may depend on the a priori knowledge about the constitution of the signal (examples are the cardinal sine function for sampling of bandlimited signals, complex exponentials for the Fourier expansion of periodic or periodized signals, or eigenfunctions of the autocovariance kernel for Karhunen-Loève representations, etc.). In a time-frequency context, we can look upon a joint decomposition as a sampling (combined in time and frequency), which is associated with a certain paving of the plane.

Gabor. The route followed by Gabor was the following. As we have seen in Subsection 1.1.2, no signal can be arbitrarily localized both in time and frequency. Therefore, we should consider the most concentrated signal (which is the Gaussian) as an elementary signal, which carries a minimal amount of information. Such a time-frequency "atom" should be the ultimate particle, the "building block," of each signal, which can thus be written in the form

$$x(t) = \sum_{n=-\infty}^{+\infty} \sum_{m=-\infty}^{+\infty} G_x[n,m] \, g_{nm}(t) \; .$$

In this expression the $g_{nm}(t)$ represent the different time-frequency atoms. They are obtained from a single Gaussian (with widths δt and $\delta \nu$ and normalized, so that $\delta t \, \delta \nu = 1$), whose center is shifted to the point $(nt_0, m\nu_0)$. The $G_x[n,m]$ are the sampling values in the time-frequency plane. Hence, each coefficient $G_x[n,m]$ carries information, which is localized to a neighborhood of a time-frequency cell of unit area. The basic idea consists in paving the plane by means of such information cells, and this should be done in a complete and nonredundant way. Hereby we associate with each elementary signal a natural surface element of minimal area in the time-frequency plane, which represents an information "quantum," also called a *logon* by Gabor.

Intuitively, some information will be lost if the mesh is too loose ($t_0\nu_0 \gg 1$). Conversely, a mesh that is too tight ($t_0\nu_0 \ll 1$) induces redundancy and an unnecessary large number of coefficients. A choice $t_0\nu_0 \approx 1$ should represent a good compromise, as it yields a number of useful coefficients close to the "dimension" of the signal. Driving our intuitive arguments even further, the actual form of the mesh does not matter a priori, if only the correct area of the logons is imposed. This allows us qualitatively to regard the sampling "à la Gabor" as an intermediate realization between the extreme cases of "Shannon sampling" and "Fourier

Chapter 2 Classes of Solutions

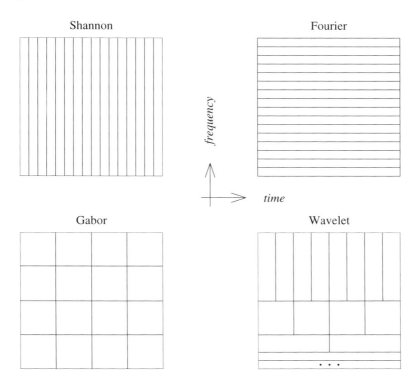

Figure 2.1. Time-frequency discretizations.

Symbolic representation of different discretizations of a signal by paving the time-frequency plane: sampling in time (Shannon) or frequency (Fourier); rectangular (Gabor) or dyadic mesh (wavelets). For all types the area of the different cells (logons) is the same.

sampling," which either advantage the temporal or the frequential aspects (cf. Fig. 2.1).

Variations. [5] As appealing as Gabor's initial idea was, it initiated only a few applications. The main reason was that, in its original form (Gaussian functions on a rectangular mesh of the plane with a minimal density of $t_0 \nu_0 = 1$), the anticipated decomposition was not coupled with a nice mathematical structure. It failed to provide a basis; in other words, the coefficients could not be obtained by a simple projection, that is,

$$G_x[n,m] \neq \int_{-\infty}^{+\infty} x(t)\, g_{nm}^*(t)\, dt \ .$$

Quite rapidly the need was felt to augment the redundancy of the representation by taking a denser mesh, and the extreme consequence was to pass to a continuum. This was performed by Helström in 1966, and it was further generalized in the following year by Montgomery and Reed. They even gave up the idea of a Gaussian for the decomposition. In this way, the initially discrete approach by Gabor lost its specificity and returned to the bosom of the short-time Fourier analyses ... .

Although some isolated attempts followed using variations of Gabor's idea (for example, the Gaussians were replaced with prolate spheroidal wave functions), atomic decompositions only reappeared on the stage of signal theory at the beginning of the 1980s. Respecting the nonorthogonality of the decomposition, it was Bastiaans in 1980 who first succeeded in showing that the Gabor coefficients can (formally) be computed by means of a dual basis. He also emphasized the bad numerical condition of the solution. Balian discovered a Gabor-type basis shortly afterwards (rectangular mesh with minimal density), but the construction led to bad localization properties in time and/or frequency. This, of course, was just the opposite of the initial plan. The "coup de grâce" was delivered by a theorem (Balian-Low), which states that there is no way out: There exists no basis associated with a rectangular mesh of minimal density, which consists of functions that are well localized in time and in frequency.

Wavelets. [6] Meantime another issue came into existence. It is closely associated with the concept of *wavelets* (or *ondelettes* in French) and was introduced by Morlet and Grossmann around 1983. The original idea was specified in terms of a continuous representation

$$T_x(t,a) = \int_{-\infty}^{+\infty} x(s)\, h_{ta}^*(s)\, ds \ .$$

Hence, it is founded on a principle similar to that of Helström's approach. Differing from that, however, it uses a projection of the signal onto a family of functions with vanishing mean value (the wavelets). These are derived from one elementary function by translations and dilations, that is,

$$h_{ta}(s) = |a|^{-1/2}\, h\left(\frac{s-t}{a}\right).$$

By construction, the name of a *time-scale* representation is better suited for such a wavelet transform. Nevertheless, one can find a time-frequency interpretation of the transform for wavelets, whose frequency response is localized to a small neighborhood of a nonzero frequency ν_0. Then the relation $\nu = \nu_0/a$ yields a suitable identification of the scale parameter a and the frequency variable ν.

The family of wavelets acts like a continuous basis. This means that there exists an exact inversion formula (also called Calderon's formula and known to mathematicians in another context since the 1960s). But it also has certain advantages as far as its discretization is concerned. If we stick to the idea of information cells of the logon type, we will find out that employing a *dyadic* paving of the plane instead of the rectangular form renders the existence of orthonormal bases possible. Then the corresponding decomposition has the form

$$x(t) = \sum_{n=-\infty}^{+\infty} \sum_{m=-\infty}^{+\infty} d_x[n,m]\, 2^{m/2} \psi(2^m t - n) ,$$

where the wavelet $\psi(t)$ may be a function that is well localized in time and frequency, and where the collection $\{\psi_{nm}(t) = 2^{m/2}\psi(2^m t - n);\ n, m \in \mathbb{Z}\}$ constitutes an orthonormal basis of $L^2(\mathbb{R})$; that is,

$$\int_{-\infty}^{+\infty} \psi_{nm}(t)\, \psi_{n'm'}(t)\, dt = \delta_{nn'}\, \delta_{mm'} .$$

The discovery of such bases (by Meyer, Daubechies, and Mallat around 1985), which enjoy many more desirable properties and come with efficient computational schemes, marked the beginning of intense activity in this field.

2.1.3. Pseudo-Densities

We already mentioned that almost in parallel with Gabor's work there was an article by Ville in 1948 that prepared the ground for another approach to the time-frequency problem: Following this route one deals with energy distributions of a quadratic nature (just like a spectral energy density or power spectrum). Thus there is an intrinsic difference to the linear decompositions of a signal into elementary components.

The energy E_x of a deterministic signal $x(t) \in L^2(\mathbb{R})$ can be written in two equivalent ways, namely

$$E_x = \int_{-\infty}^{+\infty} |x(t)|^2\, dt = \int_{-\infty}^{+\infty} |X(\nu)|^2\, d\nu .$$

This confers the rank of *energy densities* upon the quantities $|x(t)|^2$ and $|X(\nu)|^2$ (regarded as temporal or spectral densities, respectively). Therefore, it is quite natural to search for a mixed quantity $\rho_x(t,\nu)$, which allows of an intermediate point of view by providing a *joint* density, such that

$$E_x = \iint_{\mathbb{R}^2} \rho_x(t,\nu)\, dt\, d\nu .$$

The instantaneous spectra, which were briefly discussed in Subsection 2.1.1, are possible candidates for such distributions, provided that their windows satisfy some mild normalization conditions. However, they represent very particular cases, and there is enough room for defining more general bilinear representations, which are not obtained as by-products of linear transforms. Moreover, it can be easily verified that they do not satisfy a natural constraint that seems justifiable. We require from a time-frequency representation that it contain the energy densities (in time and frequency) as its marginals, that is,

$$\int_{-\infty}^{+\infty} \rho_x(t,\nu)\,dt = |X(\nu)|^2 \;, \qquad \int_{-\infty}^{+\infty} \rho_x(t,\nu)\,d\nu = |x(t)|^2 \;.$$

Assuming that the instantaneous spectrum $\rho_x(t,\nu)$ "deploys" the energy of the signal into the two-dimensional time-frequency plane, the first constraint expresses the following: An integration parallel to the time axis leads to an accumulation of the complete history of the signal, so it should yield the global spectral information. The second constraint is dual and can be interpreted accordingly.

Wigner-Ville. [7] Ville proposed to define an "instantaneous spectrum" by

$$W_x(t,\nu) = \int_{-\infty}^{+\infty} x\left(t + \frac{\tau}{2}\right) x^*\left(t - \frac{\tau}{2}\right) e^{-i2\pi\nu\tau}\,d\tau \;. \qquad (2.3)$$

This quantity satisfies the constraints of the marginal distributions. Here the signal $x(t)$ is supposed to be analytic. The motivation for this definition is its analogy with a probability density function. The preceding expression is nothing but the Fourier transform of an "acceptable form" of the characteristic function for the redistribution of energy.

We mention as a curious fact that the method employed by Ville seems to be inspired by the operational formalism of quantum mechanics. However, there is no reference or remark concerning the equivalence to the (position-momentum) function that was earlier proposed by Wigner in 1932. (Let us note in passing that Wigner said almost nothing about his motivation for giving this definition, either. It seems that it was proposed in an *ad hoc* manner, as only a footnote contains a hint that it was "found by L. Szilard and the present author some years ago for another purpose" ...).

Ville's definition exhibits some theoretical advantages. Let us once more consider the academic example of a linear chirp. One can show that the Wigner-Ville distribution ideally localizes to the instantaneous frequency. More precisely, it verifies the relation

$$x(t) = \exp\{i2\pi(\nu_0 t + (\beta/2)t^2)\} \quad \Longrightarrow \quad W_x(t,\nu) = \delta(\nu - (\nu_0 + \beta t)) \;.$$

This property is achieved regardless of the slope β of the modulation. It establishes a definite advantage over the spectrogram. Meantime, Ville's article has another surprise in store. The issue of energy distributions in the time-frequency plane fills only a very small portion of the text. As its title tells, the paper is devoted mainly to the theory and applications of the new notion of analytic signals. Surprisingly, the connection between these two concepts is not even mentioned, and nothing is said about the localization of the joint representation for signals, whose frequency modulation stays close to the rule of their instantaneous frequency. This point had already been clarified by Bass in 1945, but it was given for Wigner's function and in a context of statistical mechanics: Pursuing the analogy with a probability density function, he proposed to interpret a "local momentum" by means of a conditional expectation value. In the terminology of signal theory, this reduces to identifying the instantaneous frequency with the center of gravity of the Ville distribution

$$\nu_x(t) = \left(\int_{-\infty}^{+\infty} \nu\, W_x(t, \nu)\, d\nu \right) \Big/ \left(\int_{-\infty}^{+\infty} W_x(t, \nu)\, d\nu \right) \ .$$

However, there exists a natural limit insofar as the interpretation of the Wigner-Ville distribution as a probability density function is concerned. In fact, it turns out that it can attain negative values. This evidently bans it from being used as a probability density, and it contradicts Ville's expectations. There is no doubt that this was the main conceptual obstacle against Ville's definition for more than twenty-five years. Another reason for this "forgetting" can certainly be found in the difficulty of putting the method to work before the general advent of calculators. When this bar failed and the Wigner-Ville distribution reappeared at the end of the 1970s, some of the theoretical reservations also faded away, thanks to the offered possibilities. In 1980 there appeared a series of three articles by Claasen and Mecklenbräuker, which marked a turning point and the beginning of a true renaissance of the Wigner-Ville distribution.[8] In the meantime, one had realized by the facility of simulations that it obeys a principle of quadratic (and nonlinear) superposition according to the relation

$$W_{x+y}(t, \nu) = W_x(t, \nu) + W_y(t, \nu) + 2\, \mathrm{Re}\, \{W_{xy}(t, \nu)\}$$

where

$$W_{xy}(t, \nu) = \int_{-\infty}^{+\infty} x\left(t + \frac{\tau}{2}\right) y^*\left(t - \frac{\tau}{2}\right) e^{-i2\pi\nu\tau}\, d\tau \ .$$

Hence, a combinatorial proliferation of cross-terms, or *interferences*, is implied, which often scrambles the readability of a time-frequency diagram.

All through the 1980s a large amount of work was devoted to modifications of the initial definition in order to improve the readability of a time-frequency representation, without giving up much of its good theoretical properties. We will present a more detailed discussion of this issue in Section 3.2.

Page. [9] Ville's endeavor of defining a time-frequency energy distribution was not the only one. Contemplating the fact that the notion of a global spectral density erases all time dependence, Page proposed in 1952 to look at a *causal* spectral density; this means that at each instant it can alone be computed based on the past of the signal. He derived a definition of an instantaneous spectrum, which relies on the temporal variation of this quantity, that is,

$$P_x(t,\nu) = \frac{\partial}{\partial t} \left| \int_{-\infty}^{t} x(s) e^{-i2\pi\nu s} \, ds \right|^2. \tag{2.4}$$

It satisfies the condition of a *causal* marginal distribution

$$\int_{-\infty}^{t} P_x(s,\nu) \, ds = \left| \int_{-\infty}^{t} x(s) \, e^{-i2\pi\nu s} \, ds \right|^2.$$

In 1955, Blanc-Lapierre and Picinbono suggested some variations on this theme. They made use of "truncation operators" in time (so that causality is ensured) and certain "bandpass filters," which were defined analogously by interchanging the time and frequency variables. (In all cases, the obtained quantities were only pseudo-densities, just like the Ville distribution, as they could attain negative values.) In particular, the authors achieved an anticausal version of the Page distribution (which is usually called Levin distribution, although he only proposed it in 1967). More interestingly, they showed that the average of these two distributions, causal and anticausal, is equal to

$$\text{Re}\left\{ x(t)\, X^*(\nu)\, e^{-i2\pi\nu t} \right\}.$$

Rihaczek. [10] This last expression actually turns out to be the real part of what is nowadays called Rihaczek distribution. It was defined in 1968 by Rihaczek and is based on a completely different argument, which might be of interest to the reader and will be reproduced next. Let us consider two signals, $x_1(t)$ and $x_2(t)$. We define their (complex) energy of interaction by the inner product

$$E_{12} = \langle x_1, x_2 \rangle = \int_{-\infty}^{+\infty} x_1(s)\, x_2^*(s)\, ds.$$

Chapter 2 Classes of Solutions

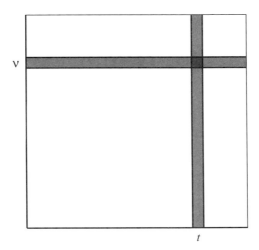

Figure 2.2. Rihaczek distribution.

The "complex energy density" of a signal x at a point (t, ν) of the time-frequency plane is defined as the energy of interaction between the restriction of x to an infinitesimal interval centered at t and its filtered version through an infinitesimal band centered at ν, normalized by the (infinitesimal) area of the corresponding time-frequency cell.

More specifically, if we let x_1 be the restriction of a given signal x to an infinitesimal interval δT about an instant t, and x_2 be the same signal x filtered through an infinitesimal band δB about a frequency ν, then we obtain (by approximating the integrals by Riemann sums)

$$E_{12} = \int_{t-\delta T/2}^{t+\delta T/2} x(s) \left[\int_{\nu-\delta B/2}^{\nu+\delta B/2} X(\xi)\, e^{i2\pi \xi s}\, d\xi \right]^* ds$$

$$\approx \int_{t-\delta T/2}^{t+\delta T/2} x(s)\, \delta B\, X^*(\nu)\, e^{-i2\pi \nu s}\, ds$$

$$\approx \delta T\, \delta B \left[x(t) X^*(\nu) e^{-i2\pi \nu t} \right].$$

This leads to an interpretation of the expression

$$R_x(t, \nu) = x(t)\, X^*(\nu)\, e^{-i2\pi \nu t} \tag{2.5}$$

as a complex energy "density" at the point (t, ν) (see Fig. 2.2). The preceding quantity is called *Rihaczek distribution*.

Extensions. Although the forementioned pseudo-densities were initially defined for deterministic finite-energy signals, they were used (at least formally) for larger classes of signals, such as distributions or random signals with eventually finite mean power. However, we should say at this place that some special approaches for random signals exist, which define a (pseudo-)power spectrum.[11] (Among others there is the "evolutionary spectrum" of Priestley in the middle of the 1960s, which is based on the Cramér decomposition of a nonstationary process.) We will come back to this issue toward the end of this chapter.

Unification.[12] Because of the different origins of the preceding definitions of time-frequency representations, they look very distinct *a priori*. Moreover, none of them is outstanding in a particular way. It is therefore desirable to find a consistent framework to enable us to compare them with each other. The first attempt aiming in this direction was made in the previously mentioned work by Blanc-Lapierre and Picinbono. But the veritable synthesis was managed by Cohen in 1966. He proposed a general form, which covers all the forementioned definitions (spectrogram/sonagram, Wigner-Ville, Page, Rihaczek, etc.) and which still gives free rein to other proposals. The definition is specified by means of an arbitrary parameter function $f(\xi, \tau)$, thus taking the form

$$C_x(t, \nu; f) = \iiint_{\mathbb{R}^3} e^{i2\pi\xi(s-t)} f(\xi, \tau)\, x\left(s + \frac{\tau}{2}\right) x^*\left(s - \frac{\tau}{2}\right) e^{-i2\pi\nu\tau} \, d\xi\, ds\, d\tau.$$

(2.6)

The original objective of the so-called "Cohen's class" was to define joint representations of position-momentum space in quantum mechanics. But Escudié and Gréa in 1976, and later Claasen and Mecklenbräuker in 1980, were responsible for its reappearance in the time-frequency context of signal theory. By the facility of its parameterization and the large number of possible interpretations, it provides the theoretical reference framework within which most of the work in the 1980s was performed.

A major exception is given by the classes of representations, which have been suggested by both Bertrands since 1983. They can be thought of as extensions of the Wigner-Ville distribution to wideband signals. They received a growing interest in parallel with the development of the wavelet methods because they are bilinear generalizations of those.

One possible justification of their definition can be given by a construction in tomography, which also puts a new light on the Wigner-Ville distribution. Let us assume that we wish to construct a time-frequency distribution $\rho_x(t, \nu)$, which has the "correct" marginal distributions, that is, such that

$$\int_{-\infty}^{+\infty} \rho_x(t,\nu)\,dt = |X(\nu)|^2, \qquad \int_{-\infty}^{+\infty} \rho_x(t,\nu)\,d\nu = |x(t)|^2.$$

These entities can be viewed as the squared absolute values of the inner products

$$X(\nu) = \int_{-\infty}^{+\infty} x(s) e^{-i2\pi\nu s}\,ds, \qquad x(t) = \int_{-\infty}^{+\infty} x(s)\,\delta(s-t)\,ds,$$

in which the analyzed signal is multiplied by the ideal analyzing signals of one pure frequency and one impulse, respectively.

As a generalization, it is conceivable to use any linear chirp as the analyzing signal. This corresponds to a straight line $\nu = \nu_0 + \beta t$ pointing in an arbitrary direction of the time-frequency plane. An appropriate extension of the constraint of a correct marginal distribution reads as

$$\int_{-\infty}^{+\infty} \rho_x(t, \nu_0 + \beta t)\,dt = |\langle x, x_{\nu_0,\beta}\rangle|^2,$$

where the member on the right is a measure for the interaction of the signal and the analyzing chirp $x_{\nu_0,\beta}(t)$, expressed by the complex inner product. The posed problem thus reduces to the inversion of a Radon transform. It turns out that its precise solution is the Wigner-Ville distribution. Next we can retain the elegant idea behind this tomographic construction, and change the underlying geometry. We do this by replacing the straight lines of integration with other curves in the time-frequency plane. In particular, when we choose *hyperbolas*, which play a central role in connection with scale changes, an analogous construction (to the preceding inversion of a Radon transform) leads to the definition of *Bertrand's unitary distribution*

$$B_x(t,\nu) = \nu \int_{-\infty}^{+\infty} \frac{(\gamma/2)}{\sinh(\gamma/2)} X\left(\nu \frac{(\gamma/2) e^{-(\gamma/2)}}{\sinh(\gamma/2)}\right)$$
$$\times\ X^*\left(\nu \frac{(\gamma/2) e^{+(\gamma/2)}}{\sinh(\gamma/2)}\right) e^{-i2\pi\nu\gamma t}\,d\gamma.$$

This is only one of the possible definitions in a huge class of affine distributions, which is comparable to Cohen's class regarding its richness and diversity.

2.1.4. The Parallel to Quantum Mechanics

We repeatedly mentioned the formalism that brings together signal theory and quantum mechanics. It is therefore not astonishing that the same mathematical objects appear in both fields, even though not only their interpretation, but also the reasons for their definitions may be completely different.

Different concerns. [13] As an example we consider a special distribution of Cohen's class, by choosing the parameter function

$$f(\xi,\tau) = \frac{\sin \pi\xi\tau}{\pi\xi\tau}.$$

We will see in Subsection 3.2.3, that this choice, in a time-frequency context, is motivated by a concern for readability of the representation in the plane. It only became known in the mid-1980s. Nevertheless, the related distribution carries the name *Born-Jordan*. This tradition goes back to Cohen's article in 1966, where reference was made to an article by the two physicists Born and Jordan, which appeared in 1925. In this inaugurating article of quantum mechanics the issue of joint representations was not raised directly. But among other problems, and in a much more general way, the authors dealt with the following question: How can one associate with a *function*, which represents a physical quantity, an *operator* whose expectation value describes all possible measurements?

From this viewpoint, the introduction of joint representations corresponds to the goal of attaining a different description of a measurement — not as an operator expectation value, taken over the possible states, but as an "ensemble average" of the classical function with respect to a "probability density" of these states. When we transfer this into a time-frequency terminology, it reduces to an implicit definition of a joint representation by the identity

$$\langle \hat{g}(\hat{t},\hat{\nu}) \rangle_x = \iint_{\mathbb{R}^2} g(t,\nu)\, \rho_x(t,\nu)\, dt\, d\nu\ .$$

Here the function $g(t,\nu)$ is associated with an operator $\hat{g}(\hat{t},\hat{\nu})$ composed of the elementary time-frequency operators $\hat{t}$ and $\hat{\nu}$ (defined in Subsection 1.1.2). It turns out that such an association, or "correspondence rule," cannot be unique in general. The nonuniqueness stems from the fact that there cannot exist an unambiguous and well-defined joint distribution that is built on two canonically conjugate variables (position-momentum or time-frequency), because their associated operators do not commute.

Indeed, as a simple example we can look at the three functions $t\nu$, νt, and $(t\nu + \nu t)/2$, which are identical, of course. However, the operators $\hat{t}\hat{\nu}$,

$\hat{\nu}\hat{t}$, and $(\hat{t}\hat{\nu}+\hat{\nu}\hat{t})/2$, which are obtained by simple substitutions, are not the same. This is true owing to the commutation relation

$$[\hat{t},\hat{\nu}] \equiv \hat{t}\hat{\nu} - \hat{\nu}\hat{t} = \frac{i}{2\pi}\widehat{I}.$$

On the other hand, it is clear that the arbitrariness in writing the operator $\hat{g}(\hat{t},\hat{\nu})$ is associated directly with the chosen definition for $\rho_x(t,\nu)$: The parameter function $f(\xi,\tau)$ of a joint distribution is, therefore, indicative of the choice of a correspondence rule. It is under this form that the time-frequency distribution associated with the forementioned choice of f is rooted in the proposed rule by Born and Jordan in their 1925 article. This point will be further discussed in Subsection 3.1.2.

Some intersections. [14] A number of definitions, properties or results that can be useful in signal theory were developed in the literature on theoretical physics and can be looked up there. The converse is true, but to a smaller extent. One is urged to draw the conclusion that there were very few links between these two areas until recently. As we mentioned before, the Ville distribution was given without reference to the one by Wigner. The same thing happened regarding the Rihaczek distribution, which was preceded by a suggestion by Margenau and Hill in 1961. Conversely, definitions from Page or Levin did not resonate, it seems, in quantum mechanics.

Finally, as another amusing observation, the pursued objectives in each of these areas and their own culture qualifies the notions of a "natural" or an "intuitive" point of view. Looking first at the community of practitioners of signal processing, one can assert that the time-frequency paradigm has its roots in the "intuitive" notion of an evolutionary spectrum (viewed as a short-time spectral analysis). It was a slow process to move it closer to the more fundamental concepts such as energy distributions, the Wigner-Ville distribution being the prototype. In quantum mechanics, one can observe a completely opposite situation: The primal and "natural" object was the Wigner distribution (this lasted until the 1970s) and thus it was quite some time beyond the trivialization of the spectrogram (and once more without any reference to it), before some position-momentum distributions "à la spectrogram" were proposed. In addition, some new difficulties arose in this context. The necessary adoption of an external quantity for the state (the equivalent of the short-time window) needs to be endowed with physical rank and interpretation.

§2.2. Atomic Decompositions

The main principle underlying every atomic decomposition is the consideration of an arbitrary signal as a linear superposition of elementary signals ("atoms"). In the time-frequency context, one requires that these atoms be "well" localized in time and frequency, so that each of them is an indivisible entity in the sense of Gabor's notion of a logon. Moreover, one generally demands that all atoms can be derived by the action of a transformation group, acting on a fixed reference element, which itself can be any member taken from the family of atoms.

It is important to emphasize, however, that the atoms appear only as a tool for the representation, just like the exponentials in Fourier analysis. They generally "have no more physical existence than the number system used to multiply the mass of the earth by that of the moon" (Meyer, 1993a).

2.2.1. Projections and Bases — General Principles

Discrete bases. Let $h_{nm}(t) \in L^2(\mathbb{R})$ be a time-frequency atom, which is localized to a neighborhood of a point of the plane indexed by n and m. The collection $\{h_{nm}(t); n, m \in \mathbb{Z}\}$ forms an orthonormal basis of $L^2(\mathbb{R})$ in the classical sense, if

$$\int_{-\infty}^{+\infty} h_{nm}(t)\, h_{n'm'}^*(t)\, dt = \delta_{nn'}\, \delta_{mm'} \tag{2.7}$$

and if every signal $x(t)$ with finite energy E_x can be written as

$$x(t) = \sum_{n=-\infty}^{+\infty} \sum_{m=-\infty}^{+\infty} L_x[n, m]\, h_{nm}(t) \ . \tag{2.8}$$

In this case the coefficients of the decomposition constitute the joint *representation* of the signal. They are simply given by the projections

$$L_x[n, m] = \int_{-\infty}^{+\infty} x(t)\, h_{nm}^*(t)\, dt \ . \tag{2.9}$$

They actually define a measure for the energy of interaction of the analyzed signal with each of the analyzing signals, the various atoms. More generally, the energy of interaction of two arbitrary signals $x(t)$ and $y(t)$ is given by

$$E_{xy} = \int_{-\infty}^{+\infty} x(t)\, y^*(t)\, dt = \sum_{n=-\infty}^{+\infty} \sum_{m=-\infty}^{+\infty} L_x[n, m]\, L_y^*[n, m] \ .$$

Chapter 2 Classes of Solutions 67

Hence, the identity of energy conservation

$$E_x = \int_{-\infty}^{+\infty} |x(t)|^2 \, dt = \sum_{n=-\infty}^{+\infty} \sum_{m=-\infty}^{+\infty} |L_x[n,m]|^2 \qquad (2.10)$$

follows. A time-frequency basis thus allows decomposition of a signal and representation of it as a linear superposition of atoms. As a by-product, it also yields an energy distribution of this signal.

Continuous bases. [15] In the discrete case, the representation is supplied by a countable family of coefficients. They are associated with a discrete lattice of points in the time-frequency plane. One can vividly look on a continuous representation as the limit of discrete decompositions (with or without respect to a basis), when the density of the lattice tends to infinity. Then the atoms of the decomposition formally behave as if they constitute a continuous basis.

A joint continuous representation thus keeps to the principle of being the projection of the signal onto an (uncountable) family of atoms indexed by $\mathbb{R}^2$. Hence, the form

$$L_x(t,\lambda) = \int_{-\infty}^{+\infty} x(s) \, h_{t\lambda}^*(s) \, ds \qquad (2.11)$$

emerges, where $\lambda \in \mathbb{R}$ is an auxiliary variable that is (directly or indirectly) connected with the frequency. Its choice is related to the group G of transformations

$$h \;\to\; h_{t\lambda} \;,$$

which generate all atoms from one single element. Let us denote the Haar measure of this group by μ_G. Then the interpretation of the collection $\{h_{s\lambda}(t); \, s, \lambda \in \mathbb{R}\}$ as a continuous basis amounts to an "inversion formula"

$$x(t) = \iint_{\mathbb{R}^2} L_x(s,\lambda) \, h_{s\lambda}(t) \, d\mu_G(s,\lambda) \qquad (2.12)$$

provided that this family is admissible. The inversion formula can be read as a continuous decomposition of the signal.

In order for such a relation to hold, we must require that

$$x(t) = \iint_{\mathbb{R}^2} \left[\int_{-\infty}^{+\infty} x(t') \, h_{s\lambda}^*(t') \, dt' \right] h_{s\lambda}(t) \, d\mu_G(s,\lambda) \;.$$

This, in turn, yields the *closure* relation, or the *resolution of the identity*,

$$\iint_{\mathbb{R}^2} h_{s\lambda}(t) h_{s\lambda}^*(t') \, d\mu_G(s,\lambda) = \delta(t-t') \ . \tag{2.13}$$

A multiplication of both sides of Eq. (2.12) by $h_{s'\lambda'}^*(t)$, followed by an integration gives

$$\int_{-\infty}^{+\infty} x(t) h_{s'\lambda'}^*(t) \, dt = \iiint_{\mathbb{R}^3} L_x(s,\lambda) h_{s\lambda}(t) h_{s'\lambda'}^*(t) \, dt \, d\mu_G(s,\lambda) \ .$$

Hence, we find

$$L_x(s',\lambda') = \iint_{\mathbb{R}^2} K(s,\lambda; s',\lambda') L_x(s,\lambda) \, d\mu_G(s,\lambda)$$

where we have introduced the so-called *reproducing kernel* of analysis

$$K(s,\lambda; s',\lambda') = \int_{-\infty}^{+\infty} h_{s\lambda}(t) h_{s'\lambda'}^*(t) \, dt \ . \tag{2.14}$$

This kernel is nothing but the representation of one atom by means of all others. It thus provides a measure for the energy shared by different atoms. At the same time, it plays the role of an evaluation functional associated with an arbitrary point in the plane.

As any continuous representation is infinitely redundant, it cannot be associated with orthogonal atoms (i.e., atoms with vanishing inner product or zero energy of interaction). Nevertheless, the reproducing kernel allows us to perform similar operations as in the case of an ordinary basis. Especially, one can derive the isometric relation

$$E_{xy} = \iint_{\mathbb{R}^2} L_x(s,\lambda) L_y^*(s,\lambda) \, d\mu_G(s,\lambda) \tag{2.15}$$

and the energy conservation

$$E_x = \iint_{\mathbb{R}^2} |L_x(s,\lambda)|^2 \, d\mu_G(s,\lambda) \ . \tag{2.16}$$

Furthermore, the squared modulus of a linear representation is an energy distribution by analogy with the case of discrete bases.

Frames. [16] The discrete decompositions, for which the family of atoms is *over-complete* and *nonorthogonal*, lie in between the continuous representations and those obtained by the projection onto an orthonormal basis. One typically runs into this situation either by discretizing a continuous representation by means of sampling the time-frequency plane, or when the atoms are given *a priori* rather than being constructed under an orthogonality constraint. In both cases, there is no obvious reason why these procedures should produce a basis. Nonetheless, it would be nice to be able to work in such a context in a controlled way. This can be achieved if one weakens the notion of orthonormal bases and uses that of *frames* (or "structures obliques" in the French terminology).

By definition, a family of atoms $\{h_{nm}(t); n, m \in \mathbb{Z}\}$ constitutes a frame if there exist two constants A and B such that $0 < A \leq B < \infty$, and so that the set of projections $L_x[n,m] = \langle x, h_{nm} \rangle$ satisfies the inequalities

$$A E_x \leq \sum_{n=-\infty}^{+\infty} \sum_{m=-\infty}^{+\infty} |L_x[n,m]|^2 \leq B E_x .$$

The interesting feature of such a structure is its similar behavior as a basis (see Eq. (2.10)). This can be observed particularly well when A is close to B. For this purpose let us consider the quantity

$$\tilde{x}(t) = \frac{2}{A+B} \sum_{n=-\infty}^{+\infty} \sum_{m=-\infty}^{+\infty} L_x[n,m] h_{nm}(t) ,$$

which, apart from a constant, would yield the discrete representation of the signal itself, if the $h_{nm}(t)$ were a basis (cf. Eq. (2.8)). In the present situation one can show that

$$\int_{-\infty}^{+\infty} |x(t) - \tilde{x}(t)|^2 \, dt \leq \mathcal{O}\left(\frac{B/A - 1}{B/A + 1}\right) E_x .$$

This implies, in particular, that the ideal case, when $A = B$ ("tight frame"), gives rise to the exact inversion formula

$$x(t) = \frac{1}{A} \sum_{n=-\infty}^{+\infty} \sum_{m=-\infty}^{+\infty} L_x[n,m] h_{nm}(t)$$

for the frame decomposition. Since $r = (B/A - 1)/(B/A + 1) < 1$ still holds in any case, it is sufficient to iterate the preceding construction in order to reduce the reconstruction error as much as desired. This reduction of the error is all the faster, as the ratio B/A is closer to 1. Indeed, it suffices to start from

$$x(t) = \tilde{x}(t) + e_1(t) ,$$

where $e_1(t)$ is the initial reconstruction error; so its energy E_1 is bounded by rE_x. A further decomposition of this error leads to

$$e_1(t) = \tilde{e}_1(t) + e_2(t)$$

with $E_2 \leq r^2 E_x$, and so forth. After k iterations we obtain $E_k \leq r^k E_x$, and hence $E_k \to 0$ as k tends to infinity. It follows that the exact reconstruction is obtained as the limit

$$x(t) = \tilde{x}(t) + \sum_{k=1}^{+\infty} e_k(t) .$$

2.2.2. Time-Frequency Examples

Short-time Fourier. [17] The most obvious example of a continuous time-frequency decomposition ($\lambda = \nu$) is given by the *short-time Fourier transform*. Recall that the classical Fourier analysis erases all chronological notion and projects the analyzed signal onto a family of everlasting monochromatic waves. By way of contrast, the short-time Fourier analysis, as its name already indicates, introduces a temporal dependence by replacing the pure waves with localized "wave packets" of the form

$$h_{t\nu}(s) = h(s-t)\, e^{i2\pi\nu s} . \tag{2.17}$$

A simple illustration is given in Fig. 2.3.

One can properly say that at each instant the "window" $h(t)$ selects only a segment of the signal by its restricted horizon, prior to the Fourier analysis. Hence we obtain a mixed representation, joint in time and frequency, which can be written as

$$F_x(t,\nu) = \langle x, h_{t\nu}\rangle = \int_{-\infty}^{+\infty} x(s)\, h^*(s-t)\, e^{-i2\pi\nu s}\, ds . \tag{2.18}$$

This representation bears the typical restrictions of the Fourier transform described in Chapter 1: An improvement of the time localization can only be obtained by shortening the window $h(t)$, which in return diminishes the frequential localization. The converse is also true, and can be rendered

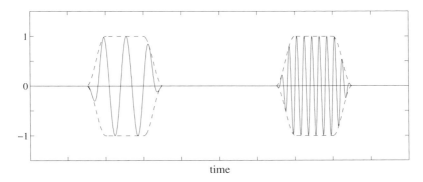

Figure 2.3. Short-time Fourier transform: time-frequency atoms.

The analysis by the short-time Fourier transform (Eq. (2.18)) can be regarded as a projection of the analyzed signal onto "time-frequency atoms" of type Eq. (2.17). Each of these atoms is obtained from a unique window $h(t)$ by a translation in time and a frequency modulation. The figure depicts two examples of such atoms (the solid line shows the real part and the dashed line shows the modulus).

even more apparent by writing out the dual representation of the short-time Fourier transform

$$F_x(t,\nu) = e^{-i2\pi\nu t} \int_{-\infty}^{+\infty} X(\xi) \, H^*(\xi - \nu) \, e^{i2\pi\xi t} \, d\xi \; . \qquad (2.19)$$

Here the Fourier transform $H(\nu)$ of $h(t)$ plays the role of a spectral "window" gliding over the spectrum $X(\nu)$ of the signal $x(t)$. This second reading (time as a function of frequency) makes the short-time Fourier transform appear as an analysis by a continuous bank of uniform filters with constant bandwidth (cf. Fig. 2.4). One verifies without difficulty that it reproduces the spectrum of the signal in the extreme case of an infinitely selective filter (i.e., if $H(\nu) = \delta(\nu)$).

Besides these two interpretations, the short-time Fourier transform is just the projection of the analyzed signal onto a family of atoms, which are all derived from one unique element (the window function $h(t)$) by time- and/or frequency-shifts. It thus relies in a fundamental way on the action of the corresponding transformation group, which is called the *Weyl-Heisenberg group*. This group is unimodular, and its natural measure is $d\mu_{\text{WH}}(t,\nu) = dt \, d\nu$. The closure condition of Eq. (2.13) shows that the short-time Fourier transform is an admissible representation if the analysis

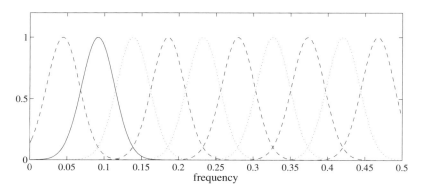

Figure 2.4. Uniform filter bank.

In its representation (Eq. (2.19)) the short-time Fourier transform can be regarded as an analysis by a uniform filter bank. In this bank, each filter is derived from a unique template by a frequency shift. The figure symbolically depicts some filters of such a bank.

window $h(t)$ has unit energy. In other words, we have

$$\int_{-\infty}^{+\infty} |h(t)|^2 \, dt = 1 \quad \Longrightarrow \quad x(t) = \iint_{\mathbb{R}^2} F_x(s,\xi) \, h_{s\xi}(t) \, ds \, d\xi \; .$$

More generally, let us consider two short-time Fourier transforms $F_x(t,\nu;h)$ and $F_x(t,\nu;g)$ based on windows $h(t)$ and $g(t)$, respectively. Then the condition [18]

$$\int_{-\infty}^{+\infty} h(t) \, g^*(t) \, dt = 1 \qquad (2.20)$$

ensures that the identities of "mixed" reconstruction

$$x(t) = \iint_{\mathbb{R}^2} F_x(s,\xi;h) \, g_{s\xi}(t) \, ds \, d\xi = \iint_{\mathbb{R}^2} F_x(s,\xi;g) \, h_{s\xi}(t) \, ds \, d\xi$$

hold as well as the generalized "energy conservation"

$$\iint_{\mathbb{R}^2} F_x(t,\nu;h) \, F_x^*(t,\nu;g) \, dt \, d\nu = E_x \; . \qquad (2.21)$$

Chapter 2   Classes of Solutions

Obstruction established by Balian-Low Theorem. As a continuous representation, the short-time Fourier transform carries the information of the signal in an extremely redundant form. This is very nicely expressed by the identity

$$F_x(t', \nu') = \iint_{\mathbb{R}^2} K(t, \nu; t', \nu') \, F_x(t, \nu) \, dt \, d\nu \;,$$

where the reproducing kernel

$$K(t, \nu; t', \nu') = e^{i2\pi(\nu - \nu')t} \, F_h(t' - t, \nu' - \nu; h)$$

is defined according to Eqs. (2.14) and (2.17). Therefore, the evaluation of the representation in one point of the time-frequency plane sets forth all values of this representation in a neighborhood of this point. The size of this neighborhood cannot be reduced arbitrarily, as it is linked to the temporal extension of $h(t)$ and the frequential extension of $H(\nu)$. This suggests that it should be possible to reduce the redundancy, without any loss in information, by a time-frequency discretization in accordance with the extension of the reproducing kernel.

Discretizing a short-time Fourier transform results in the calculation of the values

$$F_x[n, m] = F_x(nt_0, m\nu_0) = \int_{-\infty}^{+\infty} x(t) \, h^*(t - nt_0) \, e^{-i2\pi m\nu_0 t} \, dt \qquad (2.22)$$

with $n, m \in \mathbb{Z}$, where t_0 and ν_0 represent the mesh sizes in time and frequency, respectively. These latter values define a rectangular lattice of points, to which the coefficients of the representation are attached. Intuitively, a mesh that is too tight introduces some useless redundancy. Conversely, a mesh that is too loose is likely to disregard some information. It is possible to give a precise formulation of this intuitive argument. In fact, one can show [19] that a necessary condition on the lattice must be satisfied in order that the family of atoms

$$\{ h_{nm}(t) = h(t - nt_0) \, e^{i2\pi m\nu_0 t} \;; \quad n, m \in \mathbb{Z} \, \}$$

constitutes a frame. This condition has the simple form

$$t_0 \, \nu_0 \leq 1$$

and it can thus be written as a constraint of a minimal *density* of the lattice in the time-frequency plane. Taking up the preceding heuristic argument again, it is tempting to use a discretization of the short-time Fourier transform at the *critical* density $t_0\nu_0 = 1$. This is particularly interesting, as one wishes to obtain an orthonormal basis rather than just a frame. Unfortunately, there appears an obstruction right at this point, which is established by the so-called Balian-Low Theorem.[20] It affirms that, *at the critical density, there exists no frame (and hence, a fortiori, no orthonormal basis) which defines a discrete short-time Fourier transform, and whose atoms are well localized both in time and frequency.*

More precisely, if the family of atoms $h(t - nt_0) e^{i2\pi mt/t_0}$ constitutes a frame of $L^2(\mathbb{R})$, then we either have

$$\int_{-\infty}^{+\infty} t^2 |h(t)|^2 \, dt = +\infty \quad \text{or} \quad \int_{-\infty}^{+\infty} \nu^2 |H(\nu)|^2 \, d\nu = +\infty \ .$$

Note that the Balian-Low Theorem does not exclude the theoretical possibility of an economical discretization of a short-time Fourier transform. But it does rule out the possibility of gaining an atomic decomposition by sampling the time-frequency plane at the minimal rate; that is, we cannot retain the spirit of a local time-frequency analysis if such a discretization is used.

A simple example for this situation is furnished by the pair "characteristic function of an interval/cardinal sine function." An orthonormal basis can either be obtained by the choice

$$h(t) = t_0^{-1/2} \mathbf{1}_{[0,t_0]}(t)$$

(where $\mathbf{1}_I(t)$ is the characteristic function of the interval I), or by employing its Fourier transform

$$h(t) = t_0^{1/2} \frac{\sin \pi t t_0}{\pi t}$$

instead. We obtain that

$$\int_{-\infty}^{+\infty} \nu^2 |H(\nu)|^2 \, d\nu = +\infty$$

in the first case, which is the price to pay for the very good localization in time (compactly supported window). Likewise, by the Fourier duality, we have

$$\int_{-\infty}^{+\infty} t^2 |h(t)|^2 \, dt = +\infty$$

in the second case. It is surely possible to find other solutions that in a certain sense represent a more balanced compromise between time and frequency localization. None of them, however, can truly reconcile the orthogonality with the localization at the critical density.[21]

Gabor and variants. The fact that there exists no frame of localized atoms at the critical density is the major drawback that rules out any numerically stable reconstruction in this situation. This happens, in particular, if one uses a Gaussian

$$g(t) = \left(\pi t_0^2\right)^{-1/4} \exp\left(-\frac{1}{2}(t/t_0)^2\right)$$

as the window function, thus following Gabor's original idea. This choice looks quite natural because we know that it attains the minimal product of the time and frequency widths among all signals, if these values are measured by the second-order moments (cf. Eq. (1.3)). However, precisely because these two moments are both finite, the Gaussian cannot produce an orthonormal basis at the critical density (which Gabor certainly knew), or even a frame. It follows that the evaluation of the coefficients $G_x[n,m]$ in a Gabor decomposition cannot be realized by a simple projection onto the family of atoms $g_{nm}(t)$, which are defined by

$$\{\, g_{nm}(t) = g(t - nt_0)\, e^{i2\pi mt/t_0} \;;\quad n, m \in \mathbb{Z} \,\}\,.$$

Likewise the reconstruction of $x(t)$ by a combination of the $g_{nm}(t)$ cannot be done using these projections.

However, we can circumvent this problem from a purely theoretical point of view[22]: The solution relies on the idea of a *dual basis*. Given the family of functions $g_{nm}(t)$, the dual basis is defined as another family $\gamma_{nm}(t)$ such that

$$\int_{-\infty}^{+\infty} g_{nm}(t)\, \gamma_{n'm'}^*(t)\, dt = \delta_{nn'}\, \delta_{mm'}\,.$$

By employing this dual basis (which has an explicit form in terms of the Zak transform [23]), both relations

$$G_x[n, m] = \langle x, \gamma_{nm} \rangle$$

and

$$x(t) = \sum_{n=-\infty}^{+\infty} \sum_{m=-\infty}^{+\infty} \langle x, g_{nm}\rangle\, \gamma_{nm}(t)$$

are verified. From a practical point of view, however, the explicit computation of $\gamma_{nm}(t)$ reveals the limitations of this procedure. In fact, Bastiaans showed that the function $\gamma(t)$, from which the $\gamma_{nm}(t)$ are constructed, has infinite energy and possesses a "spiky" structure, which makes the numerical computations very unstable.

There is no opportunity to use this exact, but unstable approach in practice. However, one can develop approximate and more stable solutions, if one is willing to give up the idea of a minimal density and accept a certain amount of redundancy of the representation. All this is well understood nowadays and quantified in terms of the associated *frame bounds*. Daubechies has actually computed the constants A and B (hence the ratio B/A, which determines the stability of the reconstruction) as well as the dual window functions $\gamma(t)$ for several sizes of the lattice beyond the critical density. She found that $\gamma(t)$ tends to the Gaussian or to Bastiaans' function, when $t_0 \nu_0$ tends to 0 or 1, respectively. Furthermore, one can describe the precise rate of growth of the ratio B/A, when $t_0 \nu_0$ approaches 1. This provides a measure for the bad numerical condition of the solution.[24]

2.2.3. Time-Scale Examples

Continuous wavelets. For the time-scale case ($\lambda = a$), the natural transformation group is no longer the Weyl-Heisenberg group of time-frequency shifts. Here we use the group of translations and dilations, also called the *affine* or 'ax+b' group. This group acts on a function $h(s)$ by generating the transformed entity

$$h_{ta}(s) = |a|^{-1/2} h\left(\frac{s-t}{a}\right) , \qquad (2.23)$$

which corresponds to a shift (and a normalization) after stretching ($|a| > 1$) or compressing ($|a| < 1$) the function. It is crucial that such a transformation leaves the shape of the function *invariant* (cf. Fig. 2.5).

Remark. The scaling parameter a is usually supposed to be strictly positive. As a generalization, one can also introduce negative scales, which play a role similar to that of the negative frequencies in Fourier analysis.

By the projection of an arbitrary signal $x(s)$ onto the family of functions $\{h_{ta}(s); t, a \in \mathbb{R}\}$, one obtains a time-scale representation of the signal, which has the form

$$T_x(t,a) = \langle x, h_{ta}\rangle = |a|^{-1/2} \int_{-\infty}^{+\infty} x(s) h^*\left(\frac{s-t}{a}\right) ds . \qquad (2.24)$$

It is called the *continuous wavelet transform*.[25] In order to define an effective representation, this quantity must be invertible so as to yield

$$x(t) = \iint_{\mathbb{R}^2} T_x(s,a) h_{sa}(t) d\mu_A(s,a) .$$

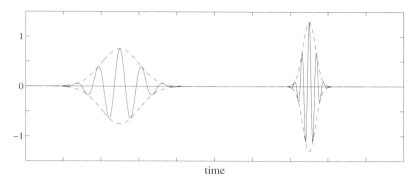

Figure 2.5. Wavelet analysis: time-scale atoms.

The wavelet analysis (Eq. (2.24)) can be regarded as the projection of the signal onto "time-scale atoms" of the form of Eq. (2.23). Each atom is obtained from one wavelet $h(t)$ by a translation in time and a dilation. The figure depicts two examples of such atoms (solid line for the wavelet and dashed line for its envelope).

Here μ_A is the canonical measure of the affine group, which is not unimodular. Rather, its left-invariant measure is defined by [26]

$$d\mu_A(t,a) = \frac{dt\,da}{a^2} \ .$$

Hence, we can rewrite the closure relation of Eq. (2.13), which is directly associated with the inversion formula, as

$$\delta(t - t') = \iint_{\mathbb{R}^2} |a|^{-1} h\left(\frac{t-s}{a}\right) h^*\left(\frac{t'-s}{a}\right) \frac{ds\,da}{a^2}$$

$$= \int_{-\infty}^{+\infty} \left[\int_{-\infty}^{+\infty} h(s)\, h^*\left(s - \frac{t-t'}{a}\right) ds\right] \frac{da}{a^2} \ .$$

By taking Fourier transforms of both sides of this equation, we further derive that

$$\int_{-\infty}^{+\infty} |H(\nu)|^2 \frac{d\nu}{|\nu|} = 1 \ . \tag{2.25}$$

This constitutes the *admissibility condition* of the wavelet transform. Its physical interpretation is rather simple. In fact, in order for the foregoing integral to be finite (which is the important part, while the value

1 is only related to the normalization) one needs to check its convergence both at infinity and at the origin. The first condition is very mild, as it reduces to the requirement that the spectrum of $h(t)$ decrease at least as fast as $|\nu|^{-1/2}$. The second condition is more severe, as it imposes a suitable annihilation on the spectrum of $h(t)$ at the origin, in order to obviate the possible divergence stemming from the measure $|\nu|^{-1}$. This second condition implies, in particular, that the mean value of $h(t)$ vanishes; that is,

$$\int_{-\infty}^{+\infty} h(t)\, dt = H(0) = 0 \ . \tag{2.26}$$

Hence, the function shows at least some oscillations, and this is the reason it is called a *wavelet*.

As in the case of the short-time Fourier transform, the admissibility condition of Eq. (2.25) can be generalized to a mixed condition

$$\int_{-\infty}^{+\infty} H(\nu)\, G^*(\nu)\, \frac{d\nu}{|\nu|} = 1 \ , \tag{2.27}$$

which furnishes the identities of the "mixed reconstruction"

$$x(t) = \iint_{\mathbb{R}^2} T_x(s, a; g)\, h_{sa}(t)\, \frac{ds\, da}{a^2} = \iint_{\mathbb{R}^2} T_x(s, a; h)\, g_{sa}(t)\, \frac{ds\, da}{a^2}$$

on the one hand, and the generalized "energy conservation"

$$E_x = \iint_{\mathbb{R}^2} T_x(s, a; g)\, T_x^*(s, a; h)\, \frac{ds\, da}{a^2} \tag{2.28}$$

on the other hand.

By its necessary extinction at the origin of the frequency domain and its decay at infinity, any admissible wavelet has the character of a bandpass filter (in a broader sense). We can thus regard the wavelet transform as a continuous bank of *constant-Q filters*. This can be better explained, perhaps, by rewriting the definition Eq. (2.24) of the wavelet transform in its equivalent form

$$T_x(t, a) = |a|^{1/2} \int_{-\infty}^{+\infty} X(\nu)\, H^*(a\nu)\, e^{i2\pi\nu t}\, d\nu \ , \tag{2.29}$$

which operates in the frequency domain. Obviously, the spectrum $X(\nu)$ is multiplied by the scaled Fourier transform $|a|^{1/2} H^*(a\nu)$ of the wavelet.

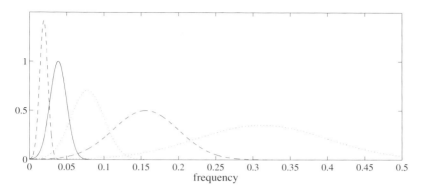

Figure 2.6. Filter bank with constant quality factor.

In its form Eq. (2.29), the wavelet transform can be regarded as an analysis by a filter bank with constant quality factor, or constant-Q filter bank. In this bank each (bandpass) filter is derived from one model by a frequency dilation or compression. The figure symbolically depicts several filters of such a bank.

Suppose $H(\nu)$ possesses a central frequency ν_0 and an equivalent band $[\nu_0 - B/2, \nu_0 + B/2]$. Then any value $a \neq 0$ of the scaling parameter defines a filter with an equivalent band $[(\nu_0 - B/2)/a, (\nu_0 + B/2)/a]$. Hence, $H(\nu)$ is a template filter, whose central frequency and bandwidth are modified by the action of the affine group, while its quality factor Q (the inverse of its relative bandwidth)

$$Q = \frac{\nu_0}{B} = \frac{\nu_0/a}{B/a}$$

remains constant.

Although the wavelet transform is a time-scale representation in the first place, it also admits a time-frequency interpretation by considering the variation of the scaling parameter a as an exploration of the frequency axis. This interpretation particularly matches the situation, where the analyzing wavelet is unimodal and localized to a neighborhood of a frequency ν_0, which can be used as a reference for the "natural" scale $a = 1$ (cf. Fig. 2.6). The previously evolved argument allows us to look on the associated wavelet transform as a function of time and frequency by means of the formal identification $\nu = \nu_0/a$.

Viewed from this perspective, the resolution of the wavelet transform depends on the point of the evaluation and varies as a function of the frequency (cf. Fig. 2.1). This is opposite to the short-time Fourier transform, which offers an identical resolution at each point of the plane. Indeed, an analysis with constant-Q filters has a fine frequency-resolution at low frequencies (i.e., for big a), at the expense of its temporal localization due to the big dilation of the analyzing wavelet. Conversely, at high frequencies (for small a) the compression of the wavelet is in favor of the temporal resolution, but it reduces the frequency-resolution. Whichever point in the time-frequency plane is considered, a trade-off of Heisenberg type (cf. Eq. (1.4)) between time- and frequency-resolution always persists. Here it takes a local form, which functionally depends on the evaluated frequency according to

$$\Delta t_{h_{t\nu}}(\nu)\,\Delta\nu_{h_{t\nu}}(\nu) \geq \frac{1}{4\pi} \ .$$

More specifically, the constant-Q property of the wavelet analysis implies that

$$\Delta t_{h_{t\nu}}(\nu) = \frac{\Delta t_h}{\nu}\nu_0 \ , \qquad \Delta\nu_{h_{t\nu}}(\nu) = \frac{1}{\nu_0}\nu\,\Delta\nu_h \ . \qquad (2.30)$$

Let us finally note that a wavelet transform and a short-time Fourier transform contain the same information, as both represent the signal in a one-to-one correspondence. We can thus pass from one transform to the other without losing information. A straightforward calculation shows that they are connected via the pair of relations

$$T_x(t, a; \psi) = \iint_{\mathbb{R}^2} F_x(s, \xi; \varphi)\,\langle\varphi_{s\xi}, \psi_{ta}\rangle\,ds\,d\xi \ ,$$

$$F_x(t, \nu; \varphi) = \iint_{\mathbb{R}^2} T_x(s, a; \psi)\,\langle\psi_{sa}, \varphi_{t\nu}\rangle\,\frac{ds\,da}{a^2} \ ,$$

where $\varphi(t)$ (and $\psi(t)$, respectively) denotes the window (the wavelet) of the short-time Fourier transform (the wavelet transform, respectively). The atoms $\varphi_{s\xi}(t)$ and $\varphi_{t\nu}(t)$ (or $\psi_{ta}(t)$ and $\psi_{sa}(t)$, respectively) emerge from Eq. (2.17) (Eq. (2.23), respectively).

Discrete wavelets. The forementioned equivalence between the *continuous* short-time Fourier and wavelet transforms becomes dissymmetrical when we attempt to discretize the representations. A natural way of paving the plane relative to a wavelet transform (cf. Fig. 2.1) is to use a discretization that is denser in time (and therefore wider in frequency) for all points at a higher frequency. This type of nonuniform mesh

$$\{\,(t, a) = (nt_0 a_0^{-m}, a_0^{-m}) \ ; \quad n, m \in \mathbb{Z}\,\} \ , \qquad t_0 > 0,\ a_0 > 0,$$

Chapter 2 Classes of Solutions

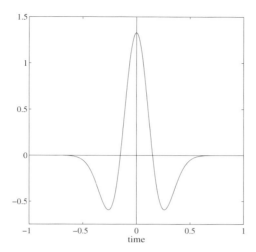

Figure 2.7. Mexican hat function.

A simple example of a wavelet is the "Mexican hat," the second derivative of a Gaussian.

leads to the definition of the *discrete* wavelet transform

$$T_x[n,m] = a_0^{m/2} \int_{-\infty}^{+\infty} x(s)\, h^*(a_0^m s - nt_0)\, ds\ ,\quad n,m \in \mathbb{Z}\ . \qquad (2.31)$$

The fundamental point to be made here is that such a construction does not suffer from an obstruction of Balian-Low type. This means that there exist "good" frames (i.e., the condition number B/A of the frame is close to 1) associated with the wavelet transform, that are well localized in time and frequency. [27]

One possible example is supplied by the function

$$h(t) = \left(2\pi\, \delta t^2\right)^{-1/2} \left(1 - (t/\delta t)^2\right) \exp\left(-\frac{1}{2}(t/\delta t)^2\right),$$

called the "Mexican hat" (second derivative of a Gaussian, cf. Fig. 2.7). It verifies the relation $\Delta t \cdot \Delta \nu = 5/4\pi$. Moreover, the choices of $a_0 = 2$ and $t_0 = \delta t$ in Eq. (2.31) amount to a ratio $B/A = 1.116$.

The choice of $a_0 = 2$ corresponds to a decomposition into *dyadic* scales (only one sequence of coefficients per octave). Combined with the chosen sampling of the time domain, this defines a rather thin lattice in the plane (though it is difficult to make sense of the notion of density in the time-scale

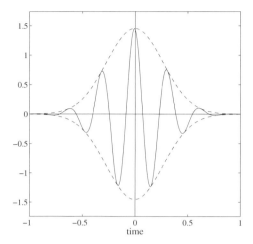

Figure 2.8. Morlet wavelet.

The Morlet wavelet is obtained by a modulation of a Gaussian. The real part of such a wavelet is drawn as a solid line and its modulus as a dashed line. The number of oscillations of the Morlet wavelet is usually chosen, so that a good compromise between the quality factor and the respect of the admissibility condition is realized (see the text).

case). One can show that a small degree of oversampling in scale and in time improves notably the quality of the decomposition. So by introducing two voices per octave (i.e., two sequences of coefficients instead of one, which is realized by intertwining two different dyadic decompositions) and taking $t_0 = \delta t/2$ (which means doubling the temporal sampling rate on each scale), the condition number B/A drops down to 1.0002. This results in a practically negligible error, if one uses the frame just like an orthonormal basis. [28]

Remark. This behavior is not a specialty of the "Mexican hat." It is caused by the way of paving the plane, which is associated with a wavelet decomposition. It was foreseen, and even experimentally verified, by Morlet at the beginning of the 1980s. [29] His own definition of a wavelet (today called *Morlet wavelet*, cf. Fig. 2.8) was inspired by the basic Gabor signal. It had the form of a modulated Gaussian

$$h(t) = \left(\pi t_0^2\right)^{-1/4} \exp\left(-\frac{1}{2}(t/t_0)^2 + i2\pi\nu_0 t\right).$$

Strictly speaking, such a wavelet is not even admissible (in the sense of

Eq. (2.26)), because

$$H(0) = \sqrt{2} \left(\pi t_0^2\right)^{1/4} \exp\left(-\frac{1}{2}(2\pi t_0 \nu_0)^2\right) \neq 0.$$

For fixed t_0, however, it becomes asymptotically admissible when ν_0 tends to infinity. But in this case the function $h(t)$ tends to look more like a wave than a wavelet. The intermediate solution adopted by Morlet consisted of imposing a small (absolute) value on $H(0)$, while only allowing a small number of oscillations of $h(t)$. Hereby, the constraint $|H(0)|/\max|H(\nu)| = 10^{-6}$ led to a relation of the form $2\pi t_0 \nu_0 = 5.4285$. A classical choice is to use a value between 5 and 6.

Multiresolution analyses and orthonormal bases. The discretization of a wavelet transform can do even better than providing good frames: In certain cases it can be associated with truly *orthonormal bases*, composed of time-scale atoms, which are well localized in time and frequency. This point is of great importance and definitely deserves a more comprehensive explanation. As it is somewhat beyond the scope of this work, we sketch only its principles here and refer to other books for a complete treatment of this subject.[30]

Let us restrict our attention to the case of decompositions of real finite energy signals into *dyadic* scales ($a_0 = 2$). Then we set the problem of finding a wavelet $\psi(t) \in \mathbb{R}$, such that the family

$$\{ \psi_{nm}(t) = 2^{m/2} \psi(2^m t - n) \; ; \quad n, m \in \mathbb{Z} \}$$

is an orthonormal basis of $L^2(\mathbb{R})$. We thus require that

$$\int_{-\infty}^{+\infty} \psi_{nm}(t) \psi_{n'm'}(t) \, dt = \delta_{nn'} \delta_{mm'}$$

and

$$x(t) = \sum_{n=-\infty}^{+\infty} \sum_{m=-\infty}^{+\infty} \left(\int_{-\infty}^{+\infty} x(s) \psi_{nm}(s) \, ds \right) \psi_{nm}(t) .$$

This problem admits at least two solutions. The first is provided by the *Haar system*

$$\psi(t) = \begin{cases} +1 & , \text{ if } 0 \leq t < 1/2, \\ -1 & , \text{ if } 1/2 \leq t < 1, \\ 0 & , \text{ otherwise,} \end{cases} \tag{2.32}$$

and the second by the *Littlewood-Paley decomposition*

$$\Psi(\nu) = \begin{cases} 1 & , \text{ if } 1/2 \leq |\nu| < 1, \\ 0 & , \text{ otherwise.} \end{cases}$$

However, in both cases the basic wavelet is badly localized in one of the domains and has a very low regularity (is in fact discontinuous) in the dual domain. (These two properties are of course linked to each other, as a function is smooth of order r, i.e., it has $r-1$ continuous derivatives, if its spectrum decreases like $|\nu|^{-r}$ at infinity.) It was one of the major accomplishments in wavelet theory to show that a construction of bases with better localization and regularity properties is possible, and that one can control either of these features to a certain extent.

Multiresolution analysis. The construction of orthonormal wavelet bases relies on the notion of *multiresolution analysis* in a central way.[31] This notion gives a formal description of the intuitive idea that every signal can be constructed by a successive refinement, which means by adding *details* to an *approximation*, and by iterating this process. More precisely, a multiresolution analysis of $L^2(\mathbb{R})$ is defined to be a sequence of nested subspaces

$$\ldots \supset V_1 \supset V_0 \supset V_{-1} \supset \ldots \;,$$

so that

(i) $$\bigcap_{m=-\infty}^{+\infty} V_m = \{0\} \;;$$

(ii) $$\bigcup_{m=-\infty}^{+\infty} V_m \text{ is dense in } L^2(\mathbb{R}) \;;$$

(iii) $$x(t) \in V_m \quad \Longleftrightarrow \quad x(2t) \in V_{m+1} \;;$$

(iv) there exists a function φ so that $\{\varphi(t-n)\,;\; n \in \mathbb{Z}\}$ is a basis of V_0.

With each V_m we can associate a time-resolution of 2^m, and the approximation of a signal $x(t)$ at this resolution level is obtained by the projection onto the corresponding subspace. Due to the properties (*iii*) and (*iv*) in the foregoing list, a basis of V_m can be derived from the basis of V_0 in the following way: Starting from the single function $\varphi(t)$, which is called the *scaling function*, the basis of V_m is obtained by taking the family of dilates and translates

$$\varphi_{nm}(t) = 2^{m/2} \varphi(2^m t - n) \;.$$

It is evident that the coefficients of the approximations

$$a_x[n,m] = \int_{-\infty}^{+\infty} x(t)\, \varphi_{nm}(t)\, dt \;,$$

Chapter 2 Classes of Solutions

which are associated with the collection of all these bases, share a large amount of information. This makes it an extremely redundant representation. Going back to the idea of describing a signal in terms of *successive approximations*, a much more economical representation consists of finding the *information difference* between two consecutive approximations. This means that we are interested in the *detail* that must be added to the coarser approximation in order to pass to the finer one. For each approximation space V_m, this amounts to saying that the details belong to a space W_m, which is the orthogonal complement of V_m in V_{m+1}. Hence, we infer the relation

$$V_{m+1} = V_m \oplus W_m ,$$

which leads to the decomposition

$$L^2(\mathbb{R}) = \bigoplus_{m=-\infty}^{+\infty} W_m .$$

Therefore, it is enough to find a function $\psi(t)$ (which is the proper wavelet), so that $\{\psi(t-n);\ n \in \mathbb{Z}\}$ is a basis of W_0. Then the set

$$\{\ \psi_{nm}(t) = 2^{m/2}\psi(2^m t - n)\ ;\quad n, m \in \mathbb{Z}\ \}$$

constitutes an orthonormal bases of $L^2(\mathbb{R})$.

Scaling function. As the scaling function $\varphi(t)$ is an element of V_0, it is also in V_1. So there exists a set of coefficients $h[n]$, which give

$$\varphi(t) = \sqrt{2} \sum_{n=-\infty}^{+\infty} h[n]\, \varphi(2t - n) , \tag{2.33}$$

and

$$h[n] = \sqrt{2} \int_{-\infty}^{+\infty} \varphi(t)\, \varphi(2t - n)\, dt ,\qquad \sum_{n=-\infty}^{+\infty} h^2[n] = 1 .$$

Because the integer translates of the scaling function $\varphi(t)$ form a basis of V_0, the discrete filter with coefficients $h[n]$ has very particular properties. The Fourier transform of the two-scale equation (2.33) for $\varphi(t)$ gives

$$\Phi(\nu) = H(\nu/2)\, \Phi(\nu/2) \tag{2.34}$$

where

$$H(\nu) = \frac{1}{\sqrt{2}} \sum_{n=-\infty}^{+\infty} h[n]\, e^{i 2\pi \nu n} .$$

This transfer function (which is periodic with period 1) cannot be arbitrary, because the orthogonality of the shifts $\varphi_{n0}(t)$ in V_0 implies that

$$\delta_{k0} = \int_{-\infty}^{+\infty} \varphi(t)\, \varphi(t-k)\, dt$$

$$= \int_{-\infty}^{+\infty} |\Phi(\nu)|^2\, e^{i2\pi\nu k}\, d\nu$$

$$= \int_0^1 \left(\sum_{n=-\infty}^{+\infty} |\Phi(\nu+n)|^2 \right) e^{i2\pi\nu k}\, d\nu,$$

and this yields the result

$$\sum_{n=-\infty}^{+\infty} |\Phi(\nu+n)|^2 = 1. \qquad (2.35)$$

Furthermore, by bringing out the transfer function (cf. Eq. (2.34)) in this equation, using its periodicity and putting $\nu = 2\zeta$, we obtain

$$1 = \sum_{n=-\infty}^{+\infty} |H(\zeta + \tfrac{n}{2})|^2\, |\Phi(\zeta + \tfrac{n}{2})|^2$$

$$= \sum_{k=-\infty}^{+\infty} |H(\zeta + k)|^2\, |\Phi(\zeta + k)|^2 + \sum_{k=-\infty}^{+\infty} |H(\zeta + \tfrac{1}{2} + k)|^2\, |\Phi(\zeta + \tfrac{1}{2} + k)|^2$$

$$= |H(\zeta)|^2 \sum_{k=-\infty}^{+\infty} |\Phi(\zeta + k)|^2 + |H(\zeta + \tfrac{1}{2})|^2 \sum_{k=-\infty}^{+\infty} |\Phi(\zeta + \tfrac{1}{2} + k)|^2.$$

Hence, the relation

$$|H(\nu)|^2 + |H(\nu + \tfrac{1}{2})|^2 = 1 \qquad (2.36)$$

follows for all ν by an application of Eq. (2.35).

Wavelet. Let us now return to the wavelet $\psi(t)$. As it is an element of V_1, it is also characterized by a discrete filter with coefficients $g[n]$, so that

$$\psi(t) = \sqrt{2} \sum_{n=-\infty}^{+\infty} g[n]\, \varphi(2t - n)$$

and
$$g[n] = \sqrt{2} \int_{-\infty}^{+\infty} \psi(t)\,\varphi(2t-n)\,dt\;,\qquad \sum_{n=-\infty}^{+\infty} g^2[n] = 1\;.$$

With the same notational conventions as before, we find that
$$\Psi(\nu) = G(\nu/2)\,\Phi(\nu/2)\;,$$
where $G(\nu)$ is the (1-periodic) transfer function of the discrete filter $g[n]$ associated with the wavelet $\psi(t)$. Also note that $\psi(t)$ is an element of W_0. Thus it is orthogonal to the space V_0, and this gives

$$\begin{aligned}
0 &= \int_{-\infty}^{+\infty} \psi(t)\,\varphi(t-k)\,dt \\
&= \int_{-\infty}^{+\infty} \Psi(\nu)\,\Phi^*(\nu)\,e^{i2\pi\nu k}\,d\nu \\
&= \int_0^1 \left(\sum_{n=-\infty}^{+\infty} \Psi(\nu+n)\,\Phi^*(\nu+n) \right) e^{i2\pi\nu k}\,d\nu\;.
\end{aligned}$$

Because this identity is true for all $k \in \mathbb{Z}$, we infer that
$$\sum_{n=-\infty}^{+\infty} \Psi(\nu+n)\,\Phi^*(\nu+n) = 0\;.$$

If we proceed as before, which is bringing out the functions H and G in this equation and using their periodicity, we finally obtain
$$G(\nu)\,H^*(\nu) + G(\nu+\tfrac{1}{2})\,H^*(\nu+\tfrac{1}{2}) = 0\;. \tag{2.37}$$

The discrete filters $h[n]$ and $g[n]$ form a pair of *quadrature mirror filters*[32], employing the nomenclature of the theory of filter banks. For given H, the solution G to the preceding equation has the form
$$G(\nu) = \lambda(\nu)\,H^*(\nu+\tfrac{1}{2})\;,$$
where $\lambda(\nu)$ is a 1-periodic function, so that
$$\lambda(\nu) + \lambda(\nu+\tfrac{1}{2}) = 0\;.$$

One possible choice is $\lambda(\nu) = -\exp(i2\pi\nu)$, and this gives
$$G(\nu) = e^{i2\pi(\nu+\frac{1}{2})}\,H^*(\nu+\tfrac{1}{2}) \quad\Longrightarrow\quad g[n] = (-1)^n h[1-n]\;. \tag{2.38}$$

It still remains to show that the integer shifts of the wavelet $\psi(t)$, which is constructed in this way, form a basis of W_0. We show in a first step that

$$\sum_{n=-\infty}^{+\infty} |\Psi(\nu+n)|^2 = \sum_{n=-\infty}^{+\infty} \left|G(\tfrac{\nu}{2}+\tfrac{n}{2})\right|^2 \left|\Phi(\tfrac{\nu}{2}+\tfrac{n}{2})\right|^2$$

$$= \sum_{n=-\infty}^{+\infty} \left|H(\tfrac{\nu+1}{2}+\tfrac{n}{2})\right|^2 \left|\Phi(\tfrac{\nu}{2}+\tfrac{n}{2})\right|^2$$

$$= \left|H(\tfrac{\nu+1}{2})\right|^2 \sum_{k=-\infty}^{+\infty} \left|\Phi(\tfrac{\nu}{2}+k)\right|^2 + \left|H(\tfrac{\nu}{2})\right|^2 \sum_{n=-\infty}^{+\infty} \left|\Phi(\tfrac{\nu+1}{2}+k)\right|^2$$

$$= \left|H(\tfrac{\nu+1}{2})\right|^2 + \left|H(\tfrac{\nu}{2})\right|^2 = 1 .$$

Hence, we can conclude that

$$\int_{-\infty}^{+\infty} \psi(t)\,\psi(t-k)\,dt = \int_{-\infty}^{+\infty} |\Psi(\nu)|^2\, e^{i2\pi\nu k}\,d\nu$$

$$= \int_0^1 \left(\sum_{n=-\infty}^{+\infty} |\Psi(\nu+n)|^2 \right) e^{i2\pi\nu k}\,d\nu$$

$$= \delta_{k0} .$$

Let us now summarize our findings: Starting from a scaling function $\varphi(t)$ (such that the family $\{\varphi(t-n)\,;\,n \in \mathbb{Z}\}$ is a basis of V_0) and coefficients $h[n]$ of the associated discrete filter, the wavelet

$$\psi(t) = \sqrt{2} \sum_{n=-\infty}^{+\infty} (-1)^n\, h[1-n]\, \varphi(2t-n) \tag{2.39}$$

defines a set $\{\psi_{nm}(t) = 2^{m/2}\psi(2^m t - n)\,;\,n, m \in \mathbb{Z}\,\}$, which constitutes a basis of $L^2(\mathbb{R})$.

Remark 1. The scaling function $\varphi(t)$ and its associated filter $h[n]$ have the characteristics of a *lowpass* filter, while the wavelet $\psi(t)$ and its related filter $g[n]$ resemble a *bandpass* filter. Indeed, we infer from Eqs. (2.35) and (2.36) that the relations $|\Phi(\nu)| \leq 1$ and $|H(\nu)| \leq 1$ hold. Moreover, we can write

$$\Phi(\nu) = H(\nu/2)\, \Phi(\nu/2)$$
$$= H(\nu/2)\, H(\nu/4)\, \Phi(\nu/4) .$$

By iterating this factorization process and passing to the limit we obtain

$$\Phi(\nu) = \Phi(0) \prod_{m=1}^{+\infty} H\left(2^{-m}\nu\right).$$

This indicates that $\Phi(0)$ must be finite and nonzero. Hence, it follows that

$$\Phi(0) = \int_{-\infty}^{+\infty} \varphi(t)\, dt$$

$$= \int_{-\infty}^{+\infty} \sqrt{2} \sum_{n=-\infty}^{+\infty} h[n]\, \varphi(2t - n)\, dt$$

$$= \Phi(0)\, \frac{1}{\sqrt{2}} \sum_{n=-\infty}^{+\infty} h[n],$$

and this gives $H(0) = 1$. In fact, it also implies $H(1/2) = 0$. Now we can easily deduce that $G(0) = 0$ (which yields $\psi(0) = 0$) and $|G(1/2)| = 1$.

Remark 2. Incidentally, the relations of the quadrature mirror filters imply that (cf. Fig. 2.9)

$$|H(\nu)|^2 + |G(\nu)|^2 = 1.$$

By using the nullity of $\Phi(\nu)$ at infinity (due to its lowpass character), this leads to

$$|\Phi(\nu)|^2 = |\Psi(2\nu)|^2 + |\Phi(2\nu)|^2 = |\Psi(2\nu)|^2 + |\Psi(4\nu)|^2 + |\Phi(4\nu)|^2.$$

Hence, again by iteration, we find that

$$|\Phi(\nu)|^2 = \sum_{m=1}^{+\infty} |\Psi(2^m \nu)|^2$$

for every nonzero frequency.

Pyramidal algorithms. The introduction of the discrete filters $h[n]$ and $g[n]$ is a major tool for the practical computation of the coefficients of the *approximation*

$$a_x[n, m] = \int_{-\infty}^{+\infty} x(t)\, \varphi_{nm}(t)\, dt \tag{2.40}$$

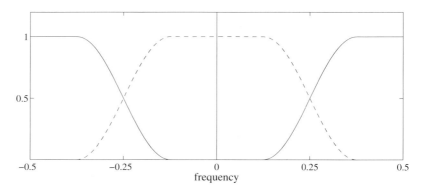

Figure 2.9. Quadrature mirror filters.

For orthonormal wavelet bases, the filter coefficients $h[n]$ and $g[n]$ of the scaling function and the wavelet, respectively, are linked to each other by the "quadrature mirror filter" relation. The figure depicts this situation symbolically. The (lowpass) pattern of the scaling filter is represented by a dashed line and the (bandpass) pattern of its related wavelet filter by a solid line. The "quadrature mirror" relation $|H(\nu)|^2 + |G(\nu)|^2 = 1$ derived from Eqs. (2.36) and (2.37) connects both.

and the *detail*

$$d_x[n, m] = \int_{-\infty}^{+\infty} x(t)\,\psi_{nm}(t)\,dt \qquad (2.41)$$

of an orthogonal wavelet decomposition.

Analysis. A straightforward computation based on the preceding relations gives

$$a_x[n, m] = \int_{-\infty}^{+\infty} x(t)\,2^{m/2}\varphi(2^m t - n)\,dt$$

$$= \int_{-\infty}^{+\infty} x(t)\,2^{m/2} \left[\sqrt{2} \sum_{k=-\infty}^{+\infty} h[k]\,\varphi\big(2(2^m t - n) - k\big)\right] dt$$

$$= \sum_{k=-\infty}^{+\infty} h[k] \int_{-\infty}^{+\infty} x(t)\,2^{(m+1)/2}\varphi\big(2^{m+1}t - (k+2n)\big)\,dt$$

$$= \sum_{k=-\infty}^{+\infty} h[k]\,a_x[k + 2n, m+1]\ .$$

We have thus proved the formula

$$a_x[n,m] = \sum_{k=-\infty}^{+\infty} h[k-2n]\, a_x[k, m+1] \ . \tag{2.42}$$

One can show in the same way that

$$d_x[n,m] = \sum_{k=-\infty}^{+\infty} g[k-2n]\, a_x[k, m+1] \ . \tag{2.43}$$

This shows that the coefficients of the approximation and the detail at a fixed resolution level can be derived by means of a *filtering*, followed by a *decimation*, from the known coefficients of the approximation at the next higher level. Operating step by step, one thus achieves a fast and recursive algorithm, which only brings two discrete filters into play for the iterative procedure. Such an algorithm is called *pyramidal*.[33] Regarding its practical aspects, the initialization of the algorithm is performed either by a projection of the analyzed signal onto V_0, if continuous data are given, or by mapping the available sequence of sampling values into this space, if only discrete values are known.

Synthesis. The *analysis* scheme can be inverted, thus leading to a dual algorithm for the *synthesis*. Here an approximation at a fixed resolution level is derived from the approximation and the detail at the next lower level. In order to establish the structure of this algorithm, we need only observe that the approximation $x_m(t)$ of a signal $x(t)$ at a fixed scaling level m is obtained by the projection of $x(t)$ onto V_m. As the family $\{\varphi_{nm}(t) = 2^{m/2}\varphi(2^m t - n);\ n \in \mathbb{Z}\}$ is a basis of this subspace of $L^2(\mathbb{R})$, we can write

$$x_m(t) = \sum_{n=-\infty}^{+\infty} a_x[n,m]\, \varphi_{nm}(t) \ .$$

Recall that $V_{m+1} = V_m \oplus W_m$ and that the orthogonal subspace W_m admits the basis $\{\psi_{nm}(t) = 2^{m/2}\psi(2^m t - n);\ n \in \mathbb{Z}\}$. So we can conclude that

$$x_{m+1}(t) = x_m(t) + \sum_{k=-\infty}^{+\infty} d_x[k,m]\, \psi_{km}(t) \ .$$

The projection of this equality onto $\varphi_{n,m+1}(t)$ gives

$$a_x[n, m+1] = \sum_{k=-\infty}^{+\infty} a_x[k,m]\, \langle \varphi_{km}, \varphi_{n,m+1}\rangle + \sum_{k=-\infty}^{+\infty} d_x[k,m]\, \langle \psi_{km}, \varphi_{n,m+1}\rangle \ .$$

Moreover, it is easy to see that

$$\langle \varphi_{km}, \varphi_{n,m+1} \rangle = \int_{-\infty}^{+\infty} 2^{m/2} \varphi(2^m t - k) \, 2^{(m+1)/2} \varphi(2^{m+1} t - n) \, dt$$

$$= \sqrt{2} \int_{-\infty}^{+\infty} \varphi(t) \, \varphi\bigl(2t - (n - 2k)\bigr) \, dt$$

$$= h[n - 2k] \, ,$$

and by an analogous reasoning

$$\langle \psi_{km}, \varphi_{n,m+1} \rangle = g[n - 2k] \, .$$

Hence, we finally conclude that the wanted reconstruction has the coefficients

$$a_x[n, m+1] = \sum_{k=-\infty}^{+\infty} h[n - 2k] \, a_x[k, m] + \sum_{k=-\infty}^{+\infty} g[n - 2k] \, d_x[k, m] \, . \quad (2.44)$$

In contrast to the algorithm for the analysis, which performs a filtering followed by a decimation, the algorithm for the synthesis operates by *interpolation* first, followed by a filtering.

Both algorithms for the analysis and the synthesis are schematically represented in Fig. 2.10. One can observe that they bring two identical cascades into play. Those related to the analysis (A) consist of the two filters H and G, followed by an operation of decimation, which only keeps every other output sample. Conversely, the cells associated with the synthesis (S) first perform an interpolation of the input (by inserting a zero between two consecutive samples), and then apply the filters H' and G', which are the transposes of H and G (in the sense that $h'[n] = h[-n]$ and $g'[n] = g[-n]$).

Some wavelet bases. Relying on the general framework for the construction of wavelet bases that we just developed, we can now turn to some special examples.

Haar. The simplest example of an orthonormal wavelet basis employs the scaling function

$$\varphi(t) = \begin{cases} 1 & , \text{ if } 0 \leq t < 1, \\ 0 & , \text{ otherwise,} \end{cases}$$

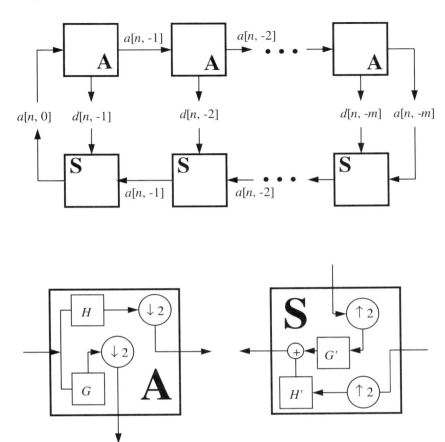

Figure 2.10. Pyramidal algorithm for analysis and synthesis by orthogonal wavelets.

The analysis and synthesis by orthogonal wavelets lend themselves to an efficient computation composed of identical units, which are used iteratively. For the analysis part, each unit A maps an approximation to a coarser approximation and a detail by the action of a scaling filter H and a wavelet filter G (with impulse responses $h[n]$ and $g[n]$, respectively), both followed by a decimation by a factor 2. For the synthesis part, each unit S operates on an approximation and a detail by producing a finer approximation. It first performs an interpolation (inserting a zero between two successive samples), then a filtering by $h'[n] = h[-n]$ and $g'[n] = g[-n]$, respectively, and finally adds the obtained output signals.

which essentially reduces to the investigation of piecewise–continuous signals. The related filter is given by

$$h[n] = \sqrt{2} \int_{-\infty}^{+\infty} \varphi(t)\,\varphi(2t - n)\,dt = \begin{cases} 1/\sqrt{2} & \text{, if } n = 0 \text{ or } 1, \\ 0 & \text{, otherwise.} \end{cases}$$

This leads further to

$$g[n] = (-1)^n h[1 - n] = \begin{cases} +1/\sqrt{2} & \text{, if } n = 0, \\ -1/\sqrt{2} & \text{, if } n = 1, \\ 0 & \text{, otherwise.} \end{cases}$$

Consequently, the associated wavelet has the form

$$\psi(t) = \varphi(2t) - \varphi(2t - 1),$$

or in explicit terms

$$\psi(t) = \begin{cases} +1 & \text{, if } 0 \leq t < 1/2, \\ -1 & \text{, if } 1/2 \leq t < 1, \\ 0 & \text{, otherwise.} \end{cases}$$

We can thus recognize the Haar wavelet of Eq. (2.32).

Battle-Lemarié. Even if the Haar system forms a basis for $L^2(\mathbb{R})$, it employs a wavelet that is not smooth (in fact discontinuous) and badly localized in frequency (i.e., its spectrum only decreases like $|\nu|^{-1}$). More regular bases can be constructed by imposing some constraints on the scaling function. For instance, we can demand that this function generate a space of *spline* functions: Then the family of *Battle-Lemarié* wavelets is obtained, which appeared as the first generalizations of the Haar wavelet.[34] Let us define the scaling function

$$\varphi(t) = \begin{cases} 1 - |t| & \text{, if } 0 \leq |t| < 1, \\ 0 & \text{, otherwise.} \end{cases}$$

It is continuous (with discontinuous derivative); however, the collection of its integer translates does not form an orthonormal family. Nevertheless, we can derive a function $\varphi_+(t)$ from $\varphi(t)$, which has this orthogonality feature. It is defined in the frequency domain by putting

$$\Phi_+(\nu) = \Phi(\nu) \left(\sum_{n=-\infty}^{+\infty} |\Phi(\nu + n)|^2 \right)^{-1/2}.$$

The foregoing series has an explicit form (as a trigonometric polynomial) [35], and this enables us to find the frequency response of its related lowpass filter and the corresponding wavelet. Note that by construction, and in contrast to the Haar wavelet, the Battle-Lemarié wavelet has infinite support.

We can follow the same procedure for scaling functions with higher regularity. They correspond to polynomial splines of higher degree. Such functions are gained from taking convolution powers of the characteristic function of the unit interval. Consequently, their spectrum satisfies

$$|\Phi(\nu)| = \left|\frac{\sin \pi\nu}{\pi\nu}\right|^{N+1} \leq C\,(1+|\nu|)^{-(N+1)}$$

if N is the degree of the splines. Because the considered scaling functions $\varphi(t)$ decay exponentially, one can show that the same feature remains true for $\varphi_+(t)$ and the wavelet $\psi(t)$, which are derived from it. By taking higher and higher degrees N of the spline we can thus construct wavelet bases, which are generated by a function that is *well localized in time* (due to its exponential decay) and has a *high degree of smoothness*.

The regularity of a wavelet is closely associated with its properties of *cancellation* (i.e., the number of vanishing moments). More precisely, $\psi(t)$ being a member of the class C^r implies that

$$\int_{-\infty}^{+\infty} t^k\,\psi(t)\,dt = 0\;,\qquad k=0,1,\ldots,r\;,$$

or equivalently

$$\frac{d^k\Psi}{d\nu^k}(0) = 0\;,\qquad k=0,1,\ldots,r\;.$$

Daubechies. We have just seen that starting from spline functions leads to wavelet bases that are well localized, yet have an unbounded support. This means that they are defined via discrete filters with an infinite number of coefficients $h[n]$. The newly set problem to be solved now is, how one can find wavelets that satisfy the stronger property of having *compact support*. Such solutions involve only a finite number of coefficients in the related filters.[36] For this reason, it is natural to begin with a scaling function that itself is compactly supported, even in its orthonormal form. An immediate consequence of this restriction is the fact that

$$H(\nu) = \frac{1}{\sqrt{2}}\sum_{n=-\infty}^{+\infty} h[n]\,e^{i2\pi\nu n}$$

must be a *trigonometric polynomial*. It is a 1-periodic function and must fulfill (according to Eq. (2.36)) the "quadrature mirror" relation

$$|H(\nu)|^2 + \left|H(\nu + \tfrac{1}{2})\right|^2 = 1 \ .$$

The problem becomes even more constrained, if we further demand that the solution have a certain regularity, because the necessary smoothness of the spectrum of the wavelet at the origin carries over to the frequency response of its (bandpass) filter. In accordance with the "quadrature mirror" relations of Eq. (2.37), one can show that the (lowpass) filter H must have a zero of multiplicity r at the frequency $\nu = 1/2$, if it corresponds to a wavelet with r vanishing moments. Therefore, it admits a factorization

$$H(\nu) = \left(\frac{1 + e^{i2\pi\nu}}{2}\right)^r L(\nu) \ ,$$

where L is a 1-periodic function and belongs to C^r, if H does. In case of a finite impulse response, L must be a trigonometric polynomial.

As the quadrature mirror relation only brings the squared modulus of H into play, we can put

$$|H(\nu)|^2 = \left(\cos^2 \pi\nu\right)^r P\left(\sin^2 \pi\nu\right) \ ,$$

where the polynomial P must satisfy the relation

$$(1-x)^r P(x) + x^r P(1-x) = 1 \ .$$

Daubechies showed that such a polynomial is given by

$$P(x) = \sum_{k=0}^{r-1} \binom{r-1+k}{k} x^k .$$

After a spectral factorization of the resulting function $|H(\nu)|^2$, this provides a solution to the posed problem: it defines a compactly supported wavelet with $2r$ nonzero filter coefficients and a regularity, which grows linearly with r.

A commonly used practice for the spectral factorization insists on the property that all zeros of the transfer function lie inside the unit circle. This corresponds to the *minimum-phase* solution. Meanwhile, other choices are possible. One can construct a filter that is closest to a symmetrical (or *linear phase*) solution. The perfect linearity of the phase, however, is incompatible with the compact support in case of a (real) orthonormal basis. [37]

Some examples of compactly supported (Daubechies) wavelets are displayed in Fig. 2.11 (right column). Their related scaling functions are also included (left column). The simplest example, the Haar wavelet of Eq. (2.32), is drawn on top. It is characterized by only two coefficients and is discontinuous. The next two examples use filters of 4 and 16 nonzero coefficients, respectively. They clearly demonstrate the influence of a growing size of the filters on the regularity of the basis functions. The shown solutions of these two examples correspond to the choice of a minimum-phase factorization. The last example displays the result for a filter with 16 coefficients and a spectral factorization, which leads to an almost symmetrical solution.

2.2.4. A "Detection-Estimation" Viewpoint

There exists a large variety of problems in signal processing (sonar, radar, ultrasound scan, nondestructive testing, telemetry, etc.), which can schematically be formulated in the following way: (i) a known signal $x(t)$ is emitted for "questioning" a system; and (ii) a "response" is received in the form of a signal $y(t)$, whose modifications relative to $x(t)$ carry some information about the system.

Although such a scheme is initially directed at problems of the "detection-estimation" type, it turns out to be useful as a formal framework for investigating atomic decompositions of a signal as well. This allows us to look upon the previously discussed solutions from a different angle.

Ambiguity functions. In order to be more precise about the "detection-estimation" point of view, let us first consider the classical problem of a statistical test with binary hypotheses

$$\begin{cases} H_0 : & y(t) = b(t) \\ H_1 : & y(t) = x_{\theta_0}(t) + b(t) \ . \end{cases}$$

The null hypothesis in this model signifies that the observation $y(t)$ contains only noise $b(t)$ (supposed to be zero-mean Gaussian white noise). The alternative claims that there is a useful signal $x_{\theta_0}(t)$ superimposed with the same type of noise. The first problem arising here is the *detection*, that is, the determination to which hypothesis the known observation should be attributed. A second problem arises when we assume that the signal $x_{\theta_0}(t)$ to be detected is equal to a reference signal $x(t)$, which is known apart from an unknown parameter vector θ_0. Then the problem of *estimating* this parameter vector is set forth, provided that the alternative H_1 has been selected.

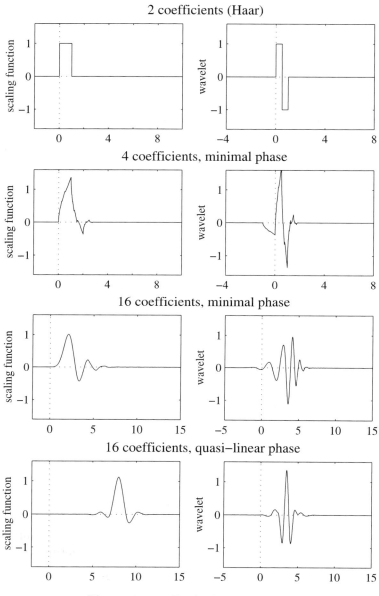

Figure 2.11. Daubechies wavelets.

The Daubechies wavelets form a family of compactly supported orthonormal wavelets. They are parameterized by an index that controls the size of the filter, the support of the wavelet, and its regularity.

Chapter 2 Classes of Solutions

In the forementioned setting, the theory of optimal decision [38] (e.g., in the form of a *maximum likelihood*) dictates that we should retain the alternative H_1 if the quantity

$$\max_{\theta}\{\lambda(\theta) = \langle y, x_\theta \rangle\}$$

exceeds a certain threshold. The strategy of an optimal detection thus comprises the following steps: one *correlates* the observation $y(t)$ with a battery of copies of the reference $x(t)$, which is modified by conceivable values of the parameter vector θ; then one decides that the searched signal is present, if the largest of the obtained values lies above a threshold. Intuitively, this maximal value should be obtained when the tested vector and its true value coincide, because then the (noise-free) correlation is maximal. One can actually show that this is true in general and that a proper estimate for θ is

$$\widehat{\theta} = \arg\max_{\theta}\{\lambda(\theta) \mid H_1\}\ .$$

Remark. As the procedure of detection-estimation compares the observed signal with a known reference signal (in the sense of their correlation and for each value of the parameter vector), it is called a *matched filtering*.

Two simple examples of this approach are furnished by *radar* and *sonar*.[39] In both cases, a known signal is emitted and its echo is recorded. Apart from the additive noise, and as a first approximation, we can consider the differences in the structure of the emitted and the received signal as caused by the existence of a propagation *delay* (related to the distance between the emitter and the target) and a *Doppler effect* (induced by the relative motion emitter-target). Therefore, the vector θ_0, which parameterizes the modifications of the emitted signal $x(t)$, is a vector of two components: one for coding the delay and a second one for the Doppler effect.

Let us momentarily suppose that the emitted signal has a narrow band and/or the relative velocity is small (which is most often the case for radar). Then the Doppler effect is manifested by a global frequency shift of the signal, called *Doppler shift*. Hence, we can use the transformation

$$x(t) \quad \to \quad x_\theta(t) = x(t-\tau)\, e^{i2\pi\xi t}$$

as a model for the underlying modifications of the emitted signal. Here τ determines the delay, and ξ denotes the Doppler shift. Certainly, this transformation can be recognized as the action (Eq. (2.17)) of the group of time-frequency shifts, which appear in the short-time Fourier transform and Gabor analysis. Indeed, if we come back to the statistical decision problem, we immediately obtain (using, e.g., the notation as in Eq. (2.21))

$$\lambda(\theta) = F_y(\tau, \xi; x) .$$

This is formally equivalent to a short-time Fourier transform of the observation with the window function being the emitted signal. Bearing in mind the nature of the observation as "signal + noise," we can finally write

$$\lambda(\theta) = A_x(\xi - \xi_0, \tau - \tau_0) e^{-i2\pi(\xi-\xi_0)\tau} + \text{fluctuations}$$

with

$$A_x(\xi, \tau) = F_x(\tau, \xi; x) = \int_{-\infty}^{+\infty} x(t) \, x^*(t - \tau) \, e^{-i2\pi\xi t} \, dt . \qquad (2.45)$$

This function is called (*narrowband*) *ambiguity function* in the literature on radar (or ambiguity function *in translation*, or again *in Woodward's sense*) [40]. It essentially measures a *time-frequency correlation*, that is, the degree of resemblance of a signal to its different translates in the plane. Playing the role of a correlation function, its maximal value appears at the origin. Hence, one can see that, apart from the additive fluctuations, the decision statistics are maximal near the true value of the delay-Doppler pair in the ambiguity plane. The better the emitted signal is jointly localized in time and frequency, the better is the estimation of this pair *a priori*. If we extend this definition to two interacting signals, the short-time Fourier transform can be regarded, in a detection-estimation context, as a *cross-ambiguity function* of the signal and the window.

This reasoning was based on the assumption that we can assimilate the Doppler effect with a simple shift of the spectrum. This can only be an approximation of the real situation, as the Doppler effect expresses itself by a *compression* or *dilation* of the frequencies. When this latter situation can actually be observed (and this especially happens with sonar), the previous modeling must be replaced with a transformation of the type

$$x(t) \quad \to \quad x_\theta(t) = \sqrt{\eta} \, x\big(\eta(t - \tau)\big) .$$

The parameter η is a *compression factor* or *Doppler rate*. Herein we can recognize the action (Eq. (2.23)) of the affine group. Hence, an analogous line of arguments as before leads to

$$\lambda(\theta) = T_y(\tau, 1/\eta; x) .$$

Hence, the wavelet transform of the observed signal naturally appears in the decision statistics, if a simple Doppler shift is not satisfactory (e.g., wideband signals). The analyzing wavelet is formally identified with the emitted signal. The corresponding ambiguity function has the form

$$A_x(\eta, \tau) = T_x(\tau, 1/\eta; x) = \sqrt{\eta} \int_{-\infty}^{+\infty} x(t)\, x^*\left(\eta(t-\tau)\right) dt\ ,\qquad (2.46)$$

and we obtain that

$$\lambda(\theta) = A_x\left(\eta/\eta_0, (\tau-\tau_0)/\eta_0\right) + \text{fluctuations}.$$

This new function is called *wideband ambiguity function* (or *ambiguity function in compression*).[41] It also has the properties of a correlation function in an *affine* sense: it measures the degree of resemblance of a signal to its dilates and translates. By an extension of this definition, the wavelet transform can be regarded as a (wideband) *cross-ambiguity function* of the analyzed signal and the analyzing wavelet.

So both cases, narrow- and wideband, lay open an equivalence of the mathematical structures between the *decomposition* tools, which are the short-time Fourier and the wavelet transform, and the *decision* tools, such as the narrow- and wideband cross-ambiguity functions. (We note in passing, and for the sake of recalling a previously used notation, that both (auto-)ambiguity functions are nothing but the *reproducing kernels* of the associated linear decomposition.) As a consequence, many properties of the considered representations can be found in the (*a priori* remote) literature on radar and sonar.

Atoms and matched filtering. Once we have recognized the equivalence between the structures of decompositions and decision statistics, we can easily find a natural interpretation. For this purpose, let us continue regarding the analyzed signal as being "received" and the analyzing signal as being "emitted." Then the decomposition of the signal into atoms amounts to identifying, at which place and by which weight, the different atoms of the signal manifest themselves. This means that we test for all hypothetical components that can be derived from the elementary analyzing signal by the action of a natural group of transformations. It thus corresponds to a problem of *detection* (are there atoms?) and *estimation* (where and with which weight?).

The use of "atoms" (in the sense of well-localized signals in time and frequency) as elements for the decomposition corresponds to the goal of emitting signals that permit a fine estimation of the transformations induced by the questioned system. We observe, by pushing the analogy further, that a "good" emitted signal should have an autoambiguity function that is maximally concentrated about the origin. In a context of signal analysis, this comes to requiring that the different atoms approximate the situation of a *basis*: their cross-ambiguity functions should be zero or negligible.

Remark. The constraint of a concentration of its ambiguity function does not necessarily imply that the signal itself is well localized in time and frequency. The chirps with large duration-bandwidth product fall in this category by an effect called "pulse compression." This is well known in detection theory. Moreover, one can recognize that, without having a large duration-bandwidth product, the Daubechies wavelets with minimal phase (and sufficiently high regularity) have such a structure of a chirp (see Fig. 2.11).

Although the mathematical structure of linear decompositions and (cross-)ambiguity functions is the same, they still feature major differences concerning the useful range of their variables. This can be well explained, in particular, for the Doppler effect. Indeed, the Doppler rate relative to a radial velocity w is expressed by

$$\eta = \frac{1+w/c}{1-w/c}$$

when c is the speed of the propagation of the waves in the medium. Hence, in the case of electromagnetic waves, for which $c = 3 \cdot 10^8$ m/s, a radial velocity of 3600 km/h leads to the small value of $\eta - 1 \approx 6.66 \cdot 10^{-6}$. This justifies perfectly the approximation $\eta \approx 1 + 2w/c$ commonly used in radar. On the other hand, if we consider the case of acoustic waves in the air, for which the speed of the propagation is $c \approx 340$ m/s, a velocity of about 110 km/h leads to $\eta \approx 1.2$. Although we cannot use the preceding approximation for this value of η, it still stays close to one. However, a wavelet transform usually analyzes the signal over several octaves (or powers of ten), which uses a scaling parameter that notably digresses from one. Expressed in a shortened form, one can thus look upon a wideband cross-ambiguity function as the fine exploration of a wavelet transform in the neighborhood of the natural scale (chosen to be 1) of the analyzing wavelet.

A final consequence of the detection-estimation point of view for linear decompositions results from their interpretation as *matched filtering*. In fact, it is intuitively clear that an atomic representation is all the more pertinent (and economic in the sense that it has fewer significant nonzero coefficients), as it brings into play atoms that preferentially exist in the analyzed signal. Precisely at this point, a philosophy of matched filtering can be discovered. This paves the way for less general and less rigid decompositions than those, which are bound to an analysis by a constant

Chapter 2 Classes of Solutions

or constant-Q frequency band. The anticipated gain in efficiency of the representation must clearly be compared with an expense, either counted as an *a priori* knowledge about the nature of the analyzed signal, or as the implementation cost for an adaptive algorithm for the selection of the most pertinent atoms. [42]

§2.3. The Energy Distributions

In contrast to the linear (atomic) representations, which decompose a signal into elementary constituents, the objective of a joint *energy* distribution is to redistribute the energy of the signal between the two variables of the description.

General setting. As before, we are going to consider the two large classes of time-frequency ($\lambda = \nu$) and time-scale ($\lambda = a$) distributions. The present goal consists in finding a distribution $\rho_x(t, \lambda)$, so that

$$E_x = \iint_{\mathbb{R}^2} \rho_x(t, \lambda) \, d\mu_G(t, \lambda) \; . \tag{2.47}$$

As the energy is a quadratic entity by nature, it suggests a search for quadratic distributions (bilinear or sesquilinear), even though this is not a necessary constraint. In other words, we are after generalizations of the well-known temporal and spectral energy densities

$$E_x = \int_{-\infty}^{+\infty} |x(t)|^2 \, dt = \int_{-\infty}^{+\infty} |X(\nu)|^2 \, d\nu \; .$$

We already noticed in Eqs. (2.21) and (2.28) that some quadratic by-products of the linear decompositions, such as the spectrogram or the scalogram, can play the role of such energy distributions. This results from the identities

$$\iint_{\mathbb{R}^2} |F_x(t, \nu)|^2 \, dt \, d\nu = \iint_{\mathbb{R}^2} |T_x(t, a)|^2 \, \frac{dt \, da}{a^2} = E_x \; ,$$

which are valid subject to adequate normalizations.

However, the squared modulus of a linear transformation is only a very particular case of a bilinear transformation. It is, therefore, important to find more general solutions to the posed problem. A conceivable approach is to start from the generic form

$$\rho_x(t, \lambda) = \iint_{\mathbb{R}^2} K(u, u'; t, \lambda) \, x(u) \, x^*(u') \, du \, du' \; , \tag{2.48}$$

which is parameterized by a kernel K. In order to satisfy the constraint of an energy distribution Eq. (2.47), the kernel K has to be chosen such that

$$\iint_{\mathbf{R}^2} K(u, u'; t, \lambda) \, d\mu_G(t, \lambda) = \delta(u - u') \,. \tag{2.49}$$

Covariance principles. After the general setting has been fixed, the actual choice of the kernel for the parameterization requires the imposition of additional constraints on the searched distribution. One way to proceed is to specify "natural" constraints for the distribution $\rho_x(t, \lambda)$, and express them as admissibility conditions for its kernel $K(u, u'; t, \lambda)$. We give precedence to those constraints, which are related to principles of covariance; this means that the effect of certain transformations of the signal can be observed on the signal itself as well as on its bilinear representation. In other words, if $\mathbf{T}$ denotes any transformation, imposing a covariance principle relative to $\mathbf{T}$ is equivalent to demanding that the diagram

$$\begin{array}{ccc} x & \rightarrow & \rho_x \\ \downarrow & & \downarrow \\ \mathbf{T}x & \rightarrow & \rho_{\mathbf{T}x} = \mathbf{T}\rho_x \end{array}$$

be commutative. (The operator acting on the distribution must be considered as a natural extension of the operator $\mathbf{T}$ acting on the signal. This will be made clear in each considered case separately.)

The first covariance principles considered here are naturally linked to the transformation groups of the time-frequency and time-scale classes themselves. We will soon see that such a point of view, as simple as it may look to us now, allows the space of solutions to be reduced considerably. On the other hand it will still comprise a wealth of possibilities. In particular, we will show how these covariance principles generate certain classes of distributions "à la Cohen." This surpasses the empirical remarks made before in connection with the issue of unification. Instead, it rather replaces them with some constructive approaches. [43]

2.3.1. Construction of the Bilinear Classes

Time-frequency. Our goal is to define a general class of bilinear time-frequency representations. Let us therefore consider the operator of time-frequency shifts, whose action is defined by

$$x(t) \quad \rightarrow \quad x_{t'\nu'}(t) = x(t - t') \, e^{i 2\pi \nu' t} \,.$$

Chapter 2 Classes of Solutions

Imposing the covariance with respect to this operator means that a time-frequency representation must "follow" the signal, when it is shifted in the plane. The constraint can thus be written as

$$\rho_{x_{t'\nu'}}(t,\nu) = \rho_x(t-t',\nu-\nu') . \qquad (2.50)$$

When we insert the parameterization equation (2.48) of the distribution by its bilinear kernel into this relation, it is easy to show that

$$\rho_{x_{t'\nu'}}(t,\nu) = \iint_{\mathbf{R}^2} K(u+t',u'+t';t,\nu)\, e^{i2\pi\nu'(u-u')}\, x(u)\, x^*(u')\, du\, du'$$

and

$$\rho_x(t-t',\nu-\nu') = \iint_{\mathbf{R}^2} K(u,u';t-t',\nu-\nu')\, x(u)\, x^*(u')\, du\, du' .$$

The equality of these two quantities for every signal implies that the kernel must verify the relation

$$K(u+t',u'+t';t,\nu)\, e^{i2\pi\nu'(u-u')} = K(u,u';t-t',\nu-\nu') .$$

If we now fix $t' = t$ and $\nu' = \nu$ and reorganize the variables, we find that

$$K(u,u';t,\nu) = K(u-t,u'-t;0,0)\, e^{-i2\pi\nu(u-u')} . \qquad (2.51)$$

The kernel, being *a priori* a function of *four* variables, therefore turns into a function of only *two* independent variables. This is an immediate consequence of the preceding constraint of covariance with respect to time-frequency shifts.

Remark. There is an analogous situation in the space of linear operators: imposing the covariance relative to the temporal shifts cuts down the class of all linear operators to the subspace of linear *filters*.[44]

If we use this form of the kernel in the generic expression of Eq. (2.48) and perform a change of variables, we obtain that

$$\rho_x(t,\nu) = \iint_{\mathbf{R}^2} K\left(s-t+\frac{\tau}{2},\ s-t-\frac{\tau}{2};0,0\right)$$
$$\times\, x\left(s+\frac{\tau}{2}\right) x^*\left(s-\frac{\tau}{2}\right) e^{-i2\pi\nu\tau}\, d\tau\, ds .$$

Now we can rewrite the integral with respect to τ in its frequency form. By letting

$$\Pi(t,\nu) = \int_{-\infty}^{+\infty} K\left(t+\frac{\tau}{2}, t-\frac{\tau}{2}; 0, 0\right) e^{i2\pi\nu\tau} d\tau$$

and

$$W_x(t,\nu) = \int_{-\infty}^{+\infty} x\left(t+\frac{\tau}{2}\right) x^*\left(t-\frac{\tau}{2}\right) e^{-i2\pi\nu\tau} d\tau$$

we end up with the final result

$$\rho_x(t,\nu) = \iint_{\mathbb{R}^2} \Pi(s-t, \xi-\nu) W_x(s,\xi) \, ds \, d\xi \ . \qquad (2.52)$$

Recall from Eq. (2.3) that $W_x(t,\nu)$ is the Wigner-Ville distribution. It is easy to see that the foregoing general form can also be written as

$$C_x(t,\nu; f) = \iiint_{\mathbb{R}^3} e^{i2\pi\xi(s-t)} f(\xi,\tau) \, x\left(s+\frac{\tau}{2}\right) x^*\left(s-\frac{\tau}{2}\right) e^{-i2\pi\nu\tau} d\xi \, ds \, d\tau$$

$$(2.53)$$

where

$$f(\xi,\tau) = \iint_{\mathbb{R}^2} \Pi(t,\nu) \, e^{-i2\pi(\nu\tau+\xi t)} dt \, d\nu \ . \qquad (2.54)$$

Once again, we recover the definition of Eq. (2.6) of Cohen's class, which was previously proposed as a means for including most time-frequency energy distributions in one common framework. *We have just shown that this class can actually be constructed, and that it comprises all bilinear representations that are covariant relative to time-frequency shifts.*

Remark. The characterization of Cohen's class given here supposes that the parameter function equation (2.54) does not depend on the signal. Then the transformation of Eq. (2.53) is bilinear. We should note, however, that this characterization is more restrictive than the definition, which was initially proposed by Cohen. The original form allows f to depend on the analyzed signal. Nevertheless, we will keep to the foregoing terminology throughout this book, sanctioned by usage, and call the family of equation (2.53) Cohen's class.

Let us recall the constraint equation (2.47), which was fixed *ab initio* and shows that the bilinear representation is an energy distribution. According to Eq. (2.49), it can be expressed as

$$\iint_{\mathbb{R}^2} K(u-t, u'-t; 0, 0) \, e^{-i2\pi\nu(u-u')} dt \, d\nu = \delta(u-u') \ .$$

In the setting of the developed parameterization, this reduces to

$$\iint_{\mathbb{R}^2} \Pi(t,\nu)\, dt\, d\nu = 1 \;,$$

or equivalently to
$$f(0,0) = 1 \;. \qquad (2.55)$$

We always take this simple normalization condition for granted. Then Cohen's class is exactly the set of all bilinear energy distributions, which are covariant with respect to translations.

Let us finally note that the definition of Cohen's class as the correlation of the Wigner-Ville distribution and an arbitrary parameter function can be expressed as a product in the conjugate Fourier domain. More specifically, we obtain that

$$\iint_{\mathbb{R}^2} C_x(t,\nu;f)\, e^{i2\pi(\xi t + \nu \tau)}\, dt\, d\nu = f(\xi,\tau)\, A_x(\xi,\tau)\;, \qquad (2.56)$$

where
$$A_x(\xi,\tau) = \int_{-\infty}^{+\infty} x\left(s + \frac{\tau}{2}\right) x^*\left(s - \frac{\tau}{2}\right) e^{i2\pi \xi s}\, ds \qquad (2.57)$$

denotes the Fourier transform of the Wigner-Ville distribution. Recall that the last quantity represents the (symmetrized) *ambiguity function* of the signal. Except for a pure phase, it is the same as the expression in Eq. (2.45).

Time-scale. The definition of a general class of bilinear time-scale representations parallels the time-frequency case. It is enough to replace the covariance relative to the shifts with the covariance relative to the affine group. Its action on an arbitrary signal $x(t)$ is given by

$$x(t) \quad \rightarrow \quad x_{t'a'}(t) = |a'|^{-1/2}\, x\left(\frac{t-t'}{a'}\right).$$

Here the constraint of covariance reads as

$$\rho_{x_{t'a'}}(t,a) = \rho_x\left(\frac{t-t'}{a'},\frac{a}{a'}\right), \qquad (2.58)$$

and this reveals the identity

$$|a'|\, K(a'u + t',\, a'u' + t';\, t, a) = K\left(u, u';\, \frac{t-t'}{a'},\, \frac{a}{a'}\right)$$

for the kernel. Because the last equation must be verified for arbitrary values of the variables, the special choice $t' = t$ and $a' = a$ leads to the simplified form

$$K(u, u'; t, a) = |a|^{-1} K\left(\frac{u-t}{a}, \frac{u'-t}{a}; 0, 1\right).$$

Again, we can see that only two degrees of freedom are left for the parameterization. We finally obtain, by inserting this reduced form into the general bilinear expression, that

$$\rho_x(t, a) = \iint_{\mathbf{R}^2} \Pi\left(\frac{s-t}{a}, a\xi\right) W_x(s, \xi) \, ds \, d\xi \,, \tag{2.59}$$

where we use the quantity

$$\Pi(t, \nu) = \int_{-\infty}^{+\infty} K\left(t + \frac{\tau}{2}, t - \frac{\tau}{2}; 0, 1\right) e^{i2\pi\nu\tau} \, d\tau \,.$$

This parameterization constitutes the "affine class" of bilinear representations, which are covariant relative to the action of translations-dilations.[45] In order to gain veritable energy distributions, we must demand that

$$E_x = \iiint_{\mathbf{R}^3} \psi(0, \nu) \, W_x(s, \xi) \, ds \, d\xi \, \frac{d\nu}{|\nu|} \,.$$

Here $\psi(\xi, \nu)$ denotes the partial Fourier transform

$$\psi(\xi, \nu) = \int_{-\infty}^{+\infty} \Pi(t, \nu) \, e^{-i2\pi\xi t} \, dt \,.$$

We can further employ the marginal property of the Wigner-Ville distribution, which asserts that

$$\int_{-\infty}^{+\infty} W_x(s, \xi) \, ds = |X(\xi)|^2 \,.$$

Then the admissibility condition takes its final form

$$\int_{-\infty}^{+\infty} \psi(0, \nu) \, \frac{d\nu}{|\nu|} = 1 \,. \tag{2.60}$$

Provided that this condition is satisfied, the affine class is the set of all bilinear energy distributions, which are covariant with respect to

Chapter 2 Classes of Solutions

translations (in time) and dilations. Just like Cohen's class, it admits several equivalent forms depending on the representational space chosen for the free parameter function. In particular, we obtain (by introducing a notation for Eq. (2.59) similar to Cohen's class equations (2.53)–(2.54)) that

$$\Omega_x(t, a; f) = \iint_{\mathbb{R}^2} \Pi\left(\frac{s-t}{a}, a\zeta\right) W_x(s, \zeta)\, ds\, d\zeta$$

$$= \iint_{\mathbb{R}^2} \Pi\left(\frac{s-t}{a}, a\zeta\right) \iint_{\mathbb{R}^2} A_x(\xi, \tau)\, e^{-i2\pi(\zeta\tau + \xi s)}\, d\xi\, d\tau\, ds\, d\zeta$$

$$= \iint_{\mathbb{R}^2} f\left(a\xi, \frac{\tau}{a}\right) A_x(\xi, \tau)\, e^{-i2\pi\xi t}\, d\xi\, d\tau \,. \tag{2.61}$$

The last expression favors the ambiguity plane. By an analogous calculation, one can also conclude that

$$\Omega_x(t, a; f) = \frac{1}{|a|} \iint_{\mathbb{R}^2} \psi(\xi, \nu)\, X\left(\frac{1}{a}\left(\nu - \frac{\xi}{2}\right)\right) \times$$
$$X^*\left(\frac{1}{a}\left(\nu + \frac{\xi}{2}\right)\right) e^{-i2\pi\xi(t/a)}\, d\xi\, d\nu \,. \tag{2.62}$$

This represents the time-scale distribution in terms of the Fourier transform of the signal.

2.3.2. The Troika of Parameterizations-Definitions-Properties

The unified formulations, which are offered by the bilinear classes (Cohen's and affine class), reflect three main interests:

(*i*) It is possible, by an appropriate specification of the parameter function, to cover most definitions of known energy distributions;

(*ii*) It is relatively easy to translate a given constraint of a joint representation into a corresponding admissibility condition for the parameter function; and

(*iii*) It is possible, by the use of such arguments of admissibility, either to specify the properties of a particular representation *a priori*, or to construct a class of solutions relative to a whole set of given specifications.

Definitions. In Table 2.1 we present several examples of definitions of time-frequency distributions, which can be recognized as special members of Cohen's class in terms of a correctly specified parameter function. The given examples do not exhaust all conceivable definitions, not even the existing ones; the table should rather be regarded as a list of representatives of most classes of solutions. This means that other definitions can be derived from these by more or less minor variations.[46]

For the complex-valued representations (such as the Rihaczek distribution), one can evidently consider the associated quantities of the complex conjugate, the real part or the imaginary part. For the example of the Rihaczek representation and its by-products, one can show that

$$f(\xi,\tau) = \cos \pi \xi \tau \iff C_x(t,\nu;f) = \mathrm{Re}\left\{x(t)X^*(\nu)e^{-i2\pi\nu t}\right\}.$$

This definition is usually attributed to Ackroyd.

We intentionally include the class of s-Wigner (or generalized Wigner-Ville) distributions in the table, as it permits a smooth transition from the Wigner-Ville distribution to the one(s) by Rihaczek by means of a single real parameter s. Its value is usually restricted to $|s| \leq 1/2$. One can readily check that the cases $s = 0, +1/2$ or $-1/2$ correspond to the Wigner-Ville distribution, the one by Rihaczek or its complex conjugate, respectively.

The Born-Jordan and Choï-Williams distributions appear in Table 2.1. It is important to note, however, that they are special examples of an important class of representations: here we refer to those representations, whose parameter functions behave like an *apodization* kernel (in the sense of spectral analysis) operating on the *product* of the variables. More precisely, we have

$$f(\xi,\tau) = \varphi(\xi\tau) \quad \text{and} \quad |\varphi(\alpha)| \leq \varphi(0) = 1.$$

This gives free rein to variations of the form of the apodization, which can be achieved by replacing the cardinal sine or the Gaussian with any other form deemed appropriate. As an example let

$$f(\xi,\tau) = \frac{1}{1 + (\pi^2 \xi^2 \tau^2 / 2\sigma^2)^N};$$

we can regard it as a Butterworth parameterization by analogy with filter theory.

Here and in the case of the Born-Jordan representation, we observe that the Wigner-Ville distribution appears as the limit, when the "equivalent width" σ tends to infinity. Otherwise, the representation related to

Table 2.1

Cohen's class: some examples of parameter
functions and the corresponding time-frequency representations

$f(\xi,\tau)$	$C_x(t,\nu;f)$	Name
1	$\int_{-\infty}^{+\infty} x\left(t+\frac{\tau}{2}\right) x^*\left(t-\frac{\tau}{2}\right) e^{-i2\pi\nu\tau}\, d\tau$	Wigner-Ville
$e^{i2\pi s\xi\tau}$	$\int_{-\infty}^{+\infty} x\left(t-(s-\tfrac{1}{2})\tau\right) x^*\left(t-(s+\tfrac{1}{2})\tau\right) e^{-i2\pi\nu\tau}\, d\tau$	s-Wigner
$e^{i\pi\xi\tau}$	$x(t)\, X^*(\nu)\, e^{-i2\pi\nu t}$	Rihaczek
$e^{\pm i\pi\xi\lvert\tau\rvert}$	$\pm\dfrac{\partial}{\partial t}\left\lvert \int_{-\infty}^{+\infty} x(s)\, U(\pm(t-s))\, e^{-i2\pi\nu s}\, ds \right\rvert^2$	Page-Levin
$\dfrac{\sin\pi\xi\tau}{\pi\xi\tau}$	$\int_{-\infty}^{+\infty}\left[\dfrac{1}{\lvert\tau\rvert}\int_{t-\lvert\tau\rvert/2}^{t+\lvert\tau\rvert/2} x\left(s+\tfrac{\tau}{2}\right) x^*\left(s-\tfrac{\tau}{2}\right) ds\right] e^{-i2\pi\nu\tau}\, d\tau$	Born-Jordan
$e^{-(\pi\xi\tau/\sigma)^2/2}$	$\iint_{\mathbb{R}^2} \dfrac{\sigma}{\lvert\tau\rvert}\, e^{-2\sigma^2(s-t)^2/\tau^2}\, x\left(s+\tfrac{\tau}{2}\right) x^*\left(s-\tfrac{\tau}{2}\right) e^{-i2\pi\nu\tau}\, ds\, d\tau$	Choï-Williams
$A_h^*(\xi,\tau)$	$\left\lvert \int_{-\infty}^{+\infty} x(s)\, h^*(s-t)\, e^{-i2\pi\nu s}\, ds \right\rvert^2$	spectrogram
$G(\xi)h(\tau)$	$\iint_{\mathbb{R}^2} h(\tau)\, g(s-t)\, x\left(s+\tfrac{\tau}{2}\right) x^*\left(s-\tfrac{\tau}{2}\right) e^{-i2\pi\nu\tau}\, ds\, d\tau$	separable

an apodization kernel of product type is a smoothed version (in time and frequency) of the Wigner-Ville distribution. The smoothing effect is more significant, *a priori*, when the regulating parameter σ is small. This interpretation by a smoothing can give rise to other variations of the definitions of Born-Jordan type. For instance, we can make its time-frequency behavior dissymmetrical. While it was governed by a single parameter σ in its original form, we can thus employ transformations of the type

$$\xi^2 \tau^2 / \sigma^2 \quad \rightarrow \quad (\xi/\sigma_\xi)^{2M} (\tau/\sigma_\tau)^{2N} \ .$$

Let us finally remark that the use of separable parameter functions, that is, such that $f(\xi,\tau) = G(\xi)h(\tau)$, offers another natural way to deal with the issue of smoothing (or at least of convolving) the Wigner-Ville distribution. It is clear that this modification can also be performed with any other distribution. This gives rise to so many modified versions of a given representation, as there are reasonable changes of the parameter function according to

$$f(\xi,\tau) \quad \rightarrow \quad G(\xi)h(\tau)f(\xi,\tau) \ .$$

The representations, which are based on so-called "cone-shaped kernels," were introduced by Zhao, Atlas, and Marks. They can be considered in this context, as their parameter functions have the form

$$f(\xi,\tau) = g_0(\tau) |\tau| \frac{\sin \pi \xi \tau}{\pi \xi \tau} \ .$$

Hence, they are formally equivalent to smoothed versions (in frequency) of the Born-Jordan distribution (and this explains why they do not appear in Table 2.1).

Table 2.2 offers a similar panorama of time-scale distributions of the affine class. As before, we only include some large classes. One of them is the *scalogram*, of course (i.e., the squared modulus of the wavelet transform). It plays a role in the time-scale context similar to that of the *spectrogram* in the time-frequency setting. One also recovers the Wigner-Ville distribution again, which is the core of the general class in itself. But its reappearance in the time-scale context necessitates the introduction of an arbitrary nonzero reference frequency ν_0. It is needed for the formal identification "scale = inverse of frequency" in terms of the relation

$$\nu = \frac{\nu_0}{a} \ .$$

This type of identification can also be extended to other distributions. In order to see how it works, let us start over from the definition of the affine

Chapter 2 Classes of Solutions

Table 2.2

Affine class: some examples of parameter
functions and the corresponding time-scale representations

$f(\xi,\tau)$	$\Omega_x(t,a;f)$	Name
$e^{-i2\pi\nu_0\tau}$	$\displaystyle\int_{-\infty}^{+\infty} x\left(t+\frac{\tau}{2}\right) x^*\left(t-\frac{\tau}{2}\right) e^{-i2\pi(\nu_0/a)\tau}\,d\tau$	Wigner-Ville
$A_h^*(\xi,\tau)$	$\displaystyle\frac{1}{a}\left\lvert\int_{-\infty}^{+\infty} x(s)\, h^*\!\left(\frac{s-t}{a}\right) ds\right\rvert^2$	scalogram
$G(\xi)h(\tau)$	$\displaystyle\frac{1}{a}\iint_{\mathbf{R}^2} h\!\left(\frac{\tau}{a}\right) g\!\left(\frac{s-t}{a}\right) x\!\left(s+\frac{\tau}{2}\right) x^*\!\left(s-\frac{\tau}{2}\right) ds\,d\tau$	separable
$\varphi(\xi\tau)\, e^{-i2\pi\nu_0\tau}$	$C_x\!\left(t,\dfrac{\nu_0}{a};\varphi\right)$	Cohen product kernel
$G(\xi)\, e^{-i2\pi H(\xi)\tau}$	$\displaystyle\int_{-\infty}^{+\infty} G(a\xi)\, X\!\left(\frac{H(a\xi)}{a}-\frac{\xi}{2}\right) X^*\!\left(\frac{H(a\xi)}{a}+\frac{\xi}{2}\right) e^{-i2\pi t\xi}\,d\xi$	affine Wigner

class in the ambiguity plane. We suppose that the parameter function has the form

$$f(\xi,\tau) = \varphi(\xi\tau)\, e^{-i2\pi\nu_0\tau} \ .$$

Then it immediately follows that

$$\Omega_x(t,a;f) = \iint_{\mathbf{R}^2} \varphi(\xi\tau)\, A_x(\xi,\tau)\, e^{-i2\pi(\xi t+(\nu_0/a)\tau)}\, d\xi\, d\tau = C_x\!\left(t,\frac{\nu_0}{a};\varphi\right).$$

This last expression is nothing but the *time-frequency* distribution in Cohen's class associated with the parameterization $\varphi(\xi\tau)$ and the identification $\nu = \nu_0/a$.

The corresponding representations lie in the intersection of both classes (Cohen's and affine class). Hence they are amenable to both the time-frequency and the time-scale interpretation.

A particularly interesting class of time-scale representations, which a priori maintain the distinction between scale and frequency, are the so-called *affine Wigner distributions*. Their parameter functions admit the generic form

$$f(\xi, \tau) = G(\xi)\, e^{-i2\pi H(\xi)\tau}, \tag{2.63}$$

where $G(\xi)$ and $H(\xi)$ are two arbitrary real functions. By means of the identification $a = \nu_0/\nu$, we can associate with every time-scale distribution of this type a time-frequency distribution

$$\widetilde{C}_x(t, \nu; f) := \Omega_x(t, \nu_0/\nu; f). \tag{2.64}$$

The details of the most current definitions that belong to this class are given in Table 2.3 together with their parameterizations. (Note that the name *affine* or *wideband* time-frequency representations is also used in the literature.) [47] Let us further assume (and this will be implicitly done in the sequel) that the signal, on which these different distributions act, is given in its *analytic* form (i.e., $X(\nu)$ is zero for all negative frequencies). Then we can verify that all these parameterizations satisfy the necessary condition

$$H(\xi) \geq \frac{|\xi|}{2}.$$

Constraints. There are many conceivable constraints that can be imposed on a joint representation. The most important ones are associated with:

(*i*) The nature of a representation and its physical meaning, which meets the concern for the *interpretation* of the mathematical object;

(*ii*) the properties of covariance or the compatibility with respect to usual or natural transformations in signal processing; this meets a concern for *coherence* with the descriptions that exist only in time or in frequency; and

(*iii*) the possibility of exposing the specifically nonstationary properties, which concerns the need of theoretical and practical *exploitation*.

We are going to investigate some of these constraints and establish the related admissibility condition in each case. We begin with a consideration of Cohen's class and then turn to the affine class. [48]

Table 2.3
Affine Wigner distributions (cf. Table 2.2) for analytic signals

$H(\xi)$	$G(\xi)$	Name
	$\widetilde{C}_x(t,\nu;f)$	
$\nu_0(\xi/2\nu_0)\coth(\xi/2\nu_0)$	$\dfrac{(\xi/2\nu_0)}{\sinh(\xi/2\nu_0)}$	Bertrand
$\nu\displaystyle\int_{-\infty}^{+\infty}\dfrac{(\gamma/2)}{\sinh(\gamma/2)}X\!\left(\nu\dfrac{(\gamma/2)e^{-(\gamma/2)}}{\sinh(\gamma/2)}\right)X^{*}\!\left(\nu\dfrac{(\gamma/2)e^{+(\gamma/2)}}{\sinh(\gamma/2)}\right)e^{-i2\pi\nu\gamma t}\,d\gamma$		
$\nu_0\left(1+(\xi/2\nu_0)^2\right)^{1/2}$	1	Unterberger (active form)
$\nu\displaystyle\int_{0}^{+\infty}\left(1+\dfrac{1}{\gamma^2}\right)X(\nu\gamma)\,X^{*}\!\left(\dfrac{\nu}{\gamma}\right)e^{i2\pi\nu(\gamma-1/\gamma)t}\,d\gamma$		
$\nu_0\left(1+(\xi/2\nu_0)^2\right)^{1/2}$	$\left(1+(\xi/2\nu_0)^2\right)^{-1/2}$	Unterberger (passive form)
$\nu\displaystyle\int_{0}^{+\infty}\dfrac{2}{\gamma}X(\nu\gamma)\,X^{*}\!\left(\dfrac{\nu}{\gamma}\right)e^{i2\pi\nu(\gamma-1/\gamma)t}\,d\gamma$		
$\nu_0\left[1+(\xi/4\nu_0)^2\right]$	$1-(\xi/4\nu_0)^2$	D-distribution
$\nu\displaystyle\int_{-4}^{+4}\left[1-\left(\dfrac{\gamma}{4}\right)^2\right]X\!\left(\nu\left(1-\dfrac{\gamma}{4}\right)^2\right)X^{*}\!\left(\nu\left(1+\dfrac{\gamma}{4}\right)^2\right)e^{-i2\pi\nu\gamma t}\,d\gamma$		

Cohen's class

Energy. In order for a time-frequency representation to be an energy distribution, we must have

$$E_x = \iint_{\mathbf{R}^2} C_x(t,\nu;f)\, dt\, d\nu \ . \tag{2.65}$$

As seen before (cf. Eq. (2.55)), this condition is equivalent to a simple constraint on the parameter function reading as

$$f(0,0) = 1 \ .$$

Another way to show this equivalence relies on a well-known property of the Fourier transform: Integrating a function over the whole space yields the value of its Fourier transform at the origin. We recall from Eq. (2.56) that the Fourier transform of $C_x(t,\nu;f)$ is given by $f(\xi,\tau)\, A_x(\xi,\tau)$, and $A_x(0,0) = E_x$ holds by definition (cf. Eq. (2.57)). Hence, the relation

$$\iint_{\mathbf{R}^2} C_x(t,\nu;f)\, dt\, d\nu = f(0,0)\, E_x$$

follows immediately, and this yields the desired result.

Marginal distributions. Let us now impose the additional constraint that the "univariate" energy distributions (spectral energy density and instantaneous power) can be obtained as the marginal distributions of the joint representation; that is,

$$\int_{-\infty}^{+\infty} C_x(t,\nu;f)\, dt = |X(\nu)|^2 \quad \text{and} \quad \int_{-\infty}^{+\infty} C_x(t,\nu;f)\, d\nu = |x(t)|^2 \ . \tag{2.66}$$

Then we can easily prove that the respective identities

$$f(0,\tau) = 1 \quad \text{and} \quad f(\xi,0) = 1 \tag{2.67}$$

must hold. Let us consider the second constraint, for example. The definition equation (2.53) of Cohen's class leads directly to

$$\int_{-\infty}^{+\infty} C_x(t,\nu;f)\, d\nu = \int_{-\infty}^{+\infty} \left[\int_{-\infty}^{+\infty} f(\xi,0)\, e^{i2\pi\xi(s-t)}\, d\xi \right] |x(s)|^2\, ds \ ,$$

and this gives the announced result.

Chapter 2 Classes of Solutions

The constraints of correct marginal distributions have several interpretations and consequences. Certainly, the first interpretation results from the analogy of energy distributions and probability density functions. If both constraints are met, we can recover the *a priori* densities from the joint density by means of its marginal distributions.

As a second interpretation we can regard the fact that a joint representation "deploys" the energy of a signal between the variables of time and frequency. As a consequence, the integration with respect to either variable must restore the global information relative to the conjugate variable. Therefore, an integration over time corresponds to collecting the complete history of the signal into one value; all chronology is thus erased and only the global aspect of the spectral density subsists. In a dual manner, integrating over the frequency axis suppresses the specialty of different spectral contributions at the considered instant, and thus (from an energetic point of view) comes back to the instantaneous power.

Let us finally note in this context that attaining at least one of the correct marginal distributions is stronger than the constraint of the (global) energy conservation; this can be expressed, for instance, by the relations

$$\int_{-\infty}^{+\infty} C_x(t,\nu; f)\, dt = |X(\nu)|^2 \quad \Longrightarrow$$

$$\iint_{\mathbb{R}^2} C_x(t,\nu; f)\, dt\, d\nu = \int_{-\infty}^{+\infty} |X(\nu)|^2\, d\nu = E_x \,,$$

or equivalently

$$f(0,\tau) = 1 \quad \Longrightarrow \quad f(0,0) = 1 \,.$$

Reality. The (local or global) integral constraints of energy distributions guarantee by no means that a representation of Cohen's class be real-valued. In order for this property to be met, we need the representation to have an Hermitian Fourier transform. Recall that the latter has the value $f(\xi,\tau)\,A_x(\xi,\tau)$, and note that by Eq. (2.57) the ambiguity function $A_x(\xi,\tau)$ itself is Hermitian. Hence, the condition of reality of the representation is equivalent to

$$f(\xi,\tau) = f^*(-\xi,-\tau) \,. \tag{2.68}$$

Positivity. If one finally wants to carry on the same interpretation, it is tempting to require an energy distribution to be nonnegative everywhere; then it would acquire the rank of an energy density function.[49] In order to investigate this property, let us introduce the change of the parameterization

$$F(s,\tau) = \int_{-\infty}^{+\infty} f(\xi,\tau)\, e^{i2\pi\xi s}\, d\xi \tag{2.69}$$

and the notation
$$y_{t\nu}(s) = x(s+t)\,e^{-i2\pi\nu s}\;.$$
Then the condition of positivity, imposed on the general form of Cohen's class, reads as
$$\iint_{\mathbf{R}^2} F(s,\tau)\,y_{t\nu}\left(s+\frac{\tau}{2}\right) y_{t\nu}^*\left(s-\frac{\tau}{2}\right) ds\,d\tau \geq 0\;.$$

Hence, $F(s,\tau)$ must be the kernel of a positive–definite operator. If we restrict ourselves to the class of square integrable parameterizations, then this kernel admits a decomposition
$$F(s,\tau) = \sum_{k=1}^{+\infty} c_k\,h_k^*\left(s+\frac{\tau}{2}\right) h_k\left(s-\frac{\tau}{2}\right)$$
with
$$c_k \geq 0\;,\qquad \sum_{k=1}^{+\infty} c_k < +\infty\;.$$

Here the $h_k(t)$ form an orthonormal family of functions in $L^2(\mathbb{R})$. By taking Fourier transforms with respect to the first variable, we can infer from the previous relation that
$$f(\xi,\tau) = \sum_{k=1}^{+\infty} c_k\,A_{h_k}^*(\xi,\tau)\;. \tag{2.70}$$

This signifies that the associated positive representation has the form
$$C_x(t,\nu;f) = \sum_{k=1}^{+\infty} c_k \left|\int_{-\infty}^{+\infty} x(s)\,h_k^*(s-t)\,e^{-i2\pi\nu s}\,ds\right|^2\;, \tag{2.71}$$

subject to the normalization condition
$$\sum_{k=1}^{+\infty} c_k = 1\;.$$

Therefore, the positivity of the representation rules out all representations of Cohen's class except for linear combinations of spectrograms with positive coefficients (at least, if we assume that the parameterization is square integrable).

Chapter 2 Classes of Solutions

Causality. The evaluation of a joint representation of Cohen's class at a fixed instant, employing its most general definition, involves the future and the past of the signal. Meanwhile, it is legitimate to raise the issue of whether there exist causal solutions in this class; this means that at each instant t they may refer only to the past of the signal given by the set of values $\{x(u);\ u \leq t\}$. An answer to this question can be given by rewriting the general form of Eq. (2.53) as

$$C_x(t,\nu;f) = \iint_{\mathbb{R}^2} F\left(\frac{u+s}{2} - t, u - s\right) x(u)\, x^*(s)\, e^{-i2\pi\nu(u-s)}\, du\, ds\,. \tag{2.72}$$

Then the causality of the representation is equivalent to the condition

$$F\left(\frac{u+s}{2} - t, u - s\right) = F\left(\frac{u+s}{2} - t, u - s\right) U(t-u)\, U(t-s)\,,$$

which must be met by the parameter function; here $U(t)$ denotes the Heaviside unit step function. This admissibility condition can further be simplified to

$$F(t,\tau) = F(t,\tau)\, U\left(-t - \frac{|\tau|}{2}\right). \tag{2.73}$$

Taking partial Fourier transforms of both sides leads to

$$f(\xi,\tau) = \frac{i}{\pi}\, \text{pv} \int_{-\infty}^{+\infty} \frac{f(\zeta,\tau)}{\xi - \zeta}\, e^{i\pi\zeta|\tau|}\, d\zeta\,. \tag{2.74}$$

The last constraint, which ensures the causality of a bilinear time-frequency representation, can be compared with the classical relation (due to Kramers-König) guaranteeing the causality of a linear filter.[50] As a corollary, the parameter function $f(\xi,\tau)$ cannot just have real or imaginary values. Indeed, by separating the real and imaginary part, the admissibility condition is equivalent to the two identities

$$\text{Re}\left\{f(\xi,\tau)\right\} = -\frac{1}{\pi}\, \text{pv} \int_{-\infty}^{+\infty} \big[\text{Im}\left\{f(\zeta,\tau)\right\} \cos(\pi\zeta\tau)$$
$$+\ \text{Re}\left\{f(\zeta,\tau)\right\} \sin(\pi\zeta|\tau|)\big]\, \frac{d\zeta}{\xi - \zeta}$$

$$\text{Im}\left\{f(\xi,\tau)\right\} = +\frac{1}{\pi}\, \text{pv} \int_{-\infty}^{+\infty} \big[\text{Re}\left\{f(\zeta,\tau)\right\} \cos(\pi\zeta\tau)$$
$$-\ \text{Im}\left\{f(\zeta,\tau)\right\} \sin(\pi\zeta|\tau|)\big]\, \frac{d\zeta}{\xi - \zeta}\,.$$

Let us suppose, for instance, that Im $\{f(\zeta,\tau)\} = 0$. We immediately conclude that the quantity $Q_\tau(\zeta) = \text{Re}\{f(\zeta,\tau)\}\cos(\pi\zeta\tau)$ has a vanishing Hilbert transform. This implies, in return, that $Q_\tau(\zeta)$ must vanish everywhere, hence the real part of $f(\zeta,\tau)$ vanishes as well. Unless $f(\zeta,\tau)$ is identically zero, it must therefore have strictly complex values.

The simplest example that is compatible with the constraint of causality is provided by the function

$$f(\xi,\tau) = e^{i\pi\xi|\tau|} .$$

It corresponds to the Page distribution (cf. Table 2.1), and we recapitulate that it was explicitly constructed (in Subsection 2.1.3) using an argument of causality.

Invertibility. The bilinear transformation, which associates a signal with an energy distribution, can define only a veritable representation, if no information is lost by this operation. Therefore it has to be invertible; in other words, there must be a one-to-one correspondence between the Fourier transform of the distribution $C_x(t,\nu;f)$ (which is $f(\xi,\tau)A_x(\xi,\tau)$) and the signal $x(t)$ itself. One easily checks that the ambiguity function $A_x(\xi,\tau)$ is invertible (apart from a pure phase), since

$$x(t)\,x^*(0) = \int_{-\infty}^{+\infty} A_x(\xi,t)\,e^{-i\pi\xi t}\,d\xi . \tag{2.75}$$

Hence, the invertibility of $C_x(t,\nu;f)$ follows, if $f(\xi,\tau)$ has no zeros. Then we can formally write

$$x(t)\,x^*(0) = \iint_{\mathbf{R}^2} F_{-1}\left(s - \frac{t}{2}, t\right) C_x(s,\nu;f)\,e^{i2\pi\nu s}\,ds\,d\nu \tag{2.76}$$

by putting

$$F_{-1}(t,\tau) = \int_{-\infty}^{+\infty} \frac{1}{f(\xi,\tau)}\,e^{i2\pi\xi t}\,d\xi .$$

Translations. The most important covariance to be called for refers to the translations in the time-frequency plane. Obviously, it does not give rise to any restrictions within Cohen's class whatsoever, as this class was precisely built on this covariance.

Dilations. While we stay inside the class of time-frequency representations, a second natural covariance can be imposed relative to dilations. Let us consider the transformation

$$x(t) \quad \to \quad x_k(t) = \sqrt{k}\,x(kt) , \qquad k > 0 .$$

Chapter 2 Classes of Solutions

We know that a dilation ($k < 1$) or compression ($k > 1$) in the time domain produces an opposite effect in frequency, because

$$X_k(\nu) = \frac{1}{\sqrt{k}}\, X\left(\frac{\nu}{k}\right) .$$

Consequently, the natural constraint of covariance relative to dilations is met if both the time and frequency behavior of the joint representation are conjugate in the sense that

$$C_{x_k}(t,\nu;f) = C_x\left(kt, \frac{\nu}{k}; f\right) . \tag{2.77}$$

We first observe that the ambiguity function equation (2.57) has this property, that is,

$$A_{x_k}(\xi,\tau) = A_x\left(\frac{\xi}{k}, k\tau\right) . \tag{2.78}$$

Hence, the wanted condition simply reads as

$$f(\xi,\tau) = f\left(\frac{\xi}{k}, k\tau\right), \qquad \forall\, k . \tag{2.79}$$

This eliminates one degree of freedom from the corresponding parameter function: It is no longer a function of two variables ξ and τ, but only a function of the product $\xi\tau$. Therefore, it must have one of the forms

$$f(\xi,\tau) = \varphi(\xi\tau), \quad \varphi(\xi|\tau|), \quad \varphi(|\xi|\tau), \quad \text{or} \quad \varphi(|\xi\tau|) .$$

It is remarkable that as soon as such parameter functions define energy distributions, that is, $\varphi(0) = 1$, the constraints of marginal distributions are automatically satisfied due to $f(\xi, 0) = f(0, \tau) = \varphi(0) = 1$.

Filtering. When we look at a linear filtering with impulse response $h(t)$ and frequency response $H(\nu)$, it is a well-known fact that the relation between the input signal $x(t)$ and the output $y(t)$ is governed by a convolution in time and a multiplication in frequency,

$$y(t) = \int_{-\infty}^{+\infty} h(t-s)\, x(s)\, ds \quad \Longleftrightarrow \quad Y(\nu) = H(\nu)\, X(\nu) . \tag{2.80}$$

Requiring a joint time-frequency representation to reflect these two points of view can formally be expressed as

$$C_y(t,\nu;f) = \int_{-\infty}^{+\infty} C_h(t-s,\nu;f)\, C_x(s,\nu;f)\, ds . \tag{2.81}$$

Starting over from the most general bilinear form (Eq. (2.48)), which was parameterized by the kernel $K(u, u'; t, \nu)$, this constraint is equivalent to

$$K(u+w, u'+w'; t, \nu) = \int_{-\infty}^{+\infty} K(u, u'; t-s, \nu) K(w, w'; t, \nu) \, ds \; .$$

It follows that

$$K_{\xi\nu}(u, u') K_{\xi\nu}(w, w') = K_{\xi\nu}(u+w, u'+w')$$

by putting

$$K_{\xi\nu}(u, u') = \int_{-\infty}^{+\infty} K(u, u'; t, \nu) e^{-i2\pi\xi t} \, dt \; .$$

If we use the explicit form (Eq. (2.51)) of $K(u, u'; t, \nu)$ inside Cohen's class and bring out the parameter function $f(\xi, \tau)$ in the latter equation, we obtain

$$K_{\xi\nu}(u, u') = f(\xi, u' - u) \exp\left\{-i2\pi\left(\xi \frac{u+u'}{2} + \nu(u - u')\right)\right\} \; .$$

This yields the admissibility condition

$$f(\xi, \tau) f(\xi, \tau') = f(\xi, \tau + \tau') \; . \tag{2.82}$$

The parameter function thus submits to the form

$$f(\xi, \tau) = e^{G(\xi)\tau} \; , \tag{2.83}$$

where $G(\xi)$ may be an arbitrary function.

By inserting $\tau = 0$ into the last expression, we confirm that $f(\xi, 0) = 1$. This shows, by itself, that the marginal property of the instantaneous power is a by-product of the compatibility with the linear filtering. Moreover, the second property of a correct marginal distribution (spectral energy distribution) is assured if and only if $G(0) = 0$.

Modulations. These are dual to the linear filtering: the modulation of the signal by means of a multiplication by $m(t)$ has the form of a product in the time-domain and a convolution in the frequency-domain; more precisely, we have

$$y(t) = m(t)x(t) \quad \Longleftrightarrow \quad Y(\nu) = \int_{-\infty}^{+\infty} M(\nu - \zeta) X(\zeta) \, d\zeta \; . \tag{2.84}$$

Chapter 2 Classes of Solutions 123

The corresponding time-frequency constraint reads as

$$C_y(t,\nu;f) = \int_{-\infty}^{+\infty} C_m(t,\nu-\zeta;f)\,C_x(t,\zeta;f)\,d\zeta \;, \tag{2.85}$$

and one can show as already mentioned that the associated admissibility condition is given by

$$f(\xi,\tau)\,f(\xi',\tau) = f(\xi+\xi',\tau) \;. \tag{2.86}$$

This requires the parameter function to be of the form

$$f(\xi,\tau) = e^{\xi\,g(\tau)} \;, \tag{2.87}$$

where $g(\tau)$ is an arbitrary function.

If we insert $\xi = 0$ into Eq. (2.87), we easily confirm that $f(0,\tau) = 1$. Hence, the property of the correct marginal spectral energy density appears as a by-product of the compatibility with modulations. The dual marginal property regarding the instantaneous power is ascertained if and only if $g(0) = 0$.

Remark. The simultaneous compatibility with filtering and modulation necessarily leads to the form

$$f(\xi,\tau) = e^{\alpha\xi\tau} \tag{2.88}$$

of the parameter function. It thus guarantees *a fortiori* the covariance with dilations and, as aftereffects, the correct marginal distributions.

Supports. It is desirable to demand from a joint representation beforehand that it should preserve the time and frequency support of a signal. The corresponding constraint can actually appear in two different forms according to the considered nature of the support.

The first form (conservation of support in a wide sense) only insists upon

$$\begin{cases} x(t) = 0\;,\; |t| > T & \Longrightarrow \quad C_x(t,\nu;f) = 0\;, |t| > T\;, \\ X(\nu) = 0\;,\; |\nu| > B & \Longrightarrow \quad C_x(t,\nu;f) = 0\;, |\nu| > B\;. \end{cases} \tag{2.89}$$

If we consider the first constraint, for instance, Eq. (2.72) implies that the parameterization must satisfy

$$\int_{-\infty}^{+\infty} F(s,\tau)\,x\!\left(s+t+\frac{\tau}{2}\right) x^*\!\left(s+t-\frac{\tau}{2}\right) ds = 0\;, \qquad |t| > T\;.$$

As the integration over s is restricted to the domain $-T - t + |\tau|/2 \leq s \leq T - t - |\tau|/2$, the values of the integral for $t > T$ (or $t < -T$, respectively) will refer to the interval $s < -|\tau|/2$ (and $s > |\tau|/2$, respectively). Hence the nullity of these values is ensured if we have

$$F(s, \tau) = 0 \quad \text{for} \quad |s| > \frac{|\tau|}{2} . \tag{2.90}$$

In the same way we can establish the admissibility condition relative to the dual constraint, which refers to the conservation of the frequency support (in the wide sense). Then we obtain the condition

$$\psi(\xi, \nu) = 0 \quad \text{for} \quad |\nu| > \frac{|\xi|}{2} , \tag{2.91}$$

where

$$\psi(\xi, \nu) = \int_{-\infty}^{+\infty} \Pi(t, \nu) e^{-i2\pi\xi t} dt = \int_{-\infty}^{+\infty} f(\xi, \tau) e^{i2\pi\nu\tau} d\tau .$$

Remark. The constraint of the conservation of the supports is responsible for fixing the usual domain of the free parameter s in the s-Wigner distribution. Indeed, one can easily compute

$$F(t, \tau) = \delta(t + s\tau) \quad \text{and} \quad \psi(\xi, \nu) = \delta(\nu + s\xi)$$

in this case. In order that the equation $t + s\tau = 0$ be unsolvable for $|t| > |\tau|/2$, we must have $|s| \leq 1/2$.

The conservation of supports in the wide sense can be rated as insufficient, when one considers signals with disconnected supports. The Wigner-Ville distribution may once more serve as an example. It satisfies the foregoing conditions of Eqs. (2.90) and (2.91). Therefore, the corresponding representation of a compactly supported signal

$$x(t) = x(t) \, \mathbf{1}_{[-T,+T]}(t)$$

identically vanishes outside the same interval (here $\mathbf{1}_I(t)$ is the characteristic function of the interval I). However, in the opposite case when the signal has the form

$$x(t) = x(t) \, \mathbf{1}_{(-\infty,-T]\cup[+T,+\infty)}(t) ,$$

the relation

$$W_x(0, \nu) = \int_{|\tau|/2 > T} x\left(\frac{\tau}{2}\right) x^*\left(-\frac{\tau}{2}\right) e^{-i2\pi\nu\tau} d\tau \neq 0$$

Chapter 2 Classes of Solutions

exemplifies that the distribution can have nonzero values at an instant where the signal vanishes. It is, henceforth, possible to envisage a stricter constraint (conservation of supports in the strict sense), which demands that

$$\begin{cases} x(t_0) = 0 & \Longrightarrow \quad C_x(t_0, \nu; f) = 0 \,, \\ X(\nu_0) = 0 & \Longrightarrow \quad C_x(t, \nu_0; f) = 0 \,. \end{cases} \qquad (2.92)$$

In this case, one forces the time-frequency representation to vanish at all (time or frequency) points where the signal itself is zero. The Rihaczek distribution of Eq. (2.5) provides a solution to this problem. However, it is not the only possible one. In order to find the general admissibility condition, let us first express the temporal constraint as

$$x(t_0) = 0 \implies \int_{-\infty}^{+\infty} F(s,\tau)\, x\left(s + t_0 + \frac{\tau}{2}\right) x^*\left(s + t_0 - \frac{\tau}{2}\right) ds = 0 \,.$$

If we now define the auxiliary signal

$$y(t) = x(t) + \delta(t - t_1) \quad \text{with} \quad t_1 \neq t_0 \,,$$

we can deduce the equivalent relation

$$\int_{-\infty}^{+\infty} F(s,\tau)\, y\left(s + t_0 + \frac{\tau}{2}\right) y^*\left(s + t_0 - \frac{\tau}{2}\right) ds = 0 \,.$$

By expanding the terms we arrive at

$$F\left(t_1 - t_0 + \frac{\tau}{2}, \tau\right) x(t_1 + \tau)$$
$$+ F\left(t_1 - t_0 - \frac{\tau}{2}, \tau\right) x^*(t_1 - \tau) + F(t_1 - t_0, 0)\, \delta(\tau) = 0 \,.$$

This can only be true for every signal $x(t)$ and every delay $\tau \neq 0$, if the time-time form of the parameter function satisfies

$$F\left(s \pm \frac{\tau}{2}, \tau\right) = 0 \,, \quad s \neq 0 \,.$$

Hence, we can write the admissibility condition as

$$F(s,\tau) = g_+(\tau)\, \delta\left(s + \frac{\tau}{2}\right) + g_-(\tau)\, \delta\left(s - \frac{\tau}{2}\right), \qquad (2.93)$$

where $g_+(\tau)$ and $g_-(\tau)$ may be chosen arbitrarily.[51] One can readily check that the Rihaczek distribution corresponds to a solution of this type, employing the functions $g_+(\tau) = 0$ and $g_-(\tau) = 1$. More generally, the set of all such solutions is composed of all linear combinations, which are formed of smoothed versions of the Rihaczek distribution and its complex conjugate, the smoothing taking place in the frequency domain.

In a dual fashion, the constraint of conservation of the frequency support in the strict sense leads to a solution of the type

$$\psi(\xi,\nu) = G_+(\xi)\,\delta\left(\nu + \frac{\xi}{2}\right) + G_-(\xi)\,\delta\left(\nu - \frac{\xi}{2}\right). \tag{2.94}$$

Unitarity. The Fourier transform is an isometry of $L^2(\mathbb{R})$. Thus the inner product of two signals remains the same, no matter if we represent it in the time or the frequency domain,

$$\int_{-\infty}^{+\infty} x(t)\,y^*(t)\,dt = \int_{-\infty}^{+\infty} X(\nu)\,Y^*(\nu)\,d\nu \ .$$

It is often desirable to ensure a relation of the same type, *mutatis mutandis*, when passing from the time or frequency axis to the time-frequency plane. In view of the bilinear character of the distributions in Cohen's class, it is natural to identify the inner product of two time-frequency distributions with the square of the usual inner product in the time domain (referring to the signals) or the frequency domain (their spectrum, respectively). Hence, the condition of unitarity takes the form

$$\iint_{\mathbb{R}^2} C_x(t,\nu;f)\,C_y^*(t,\nu;f)\,dt\,d\nu = \left|\int_{-\infty}^{+\infty} x(t)\,y^*(t)\,dt\right|^2. \tag{2.95}$$

The identity of Eq. (2.95) is usually called *Moyal's formula*.[52]

In order to investigate this constraint, let us rewrite (by Parseval's relation and Eq. (2.56)) the left-hand side of the previous equation as

$$\iint_{\mathbb{R}^2} C_x(t,\nu;f)\,C_y^*(t,\nu;f)\,dt\,d\nu = \iint_{\mathbb{R}^2} |f(\xi,\tau)|^2 A_x(\xi,\tau)\,A_y^*(\xi,\tau)\,d\xi\,d\tau \ .$$

Note that the definition equation (2.57) yields

$$\iint_{\mathbb{R}^2} A_x(\xi,\tau)\,A_y^*(\xi,\tau)\,d\xi\,d\tau = \left|\int_{-\infty}^{+\infty} x(t)\,y^*(t)\,dt\right|^2. \tag{2.96}$$

We can thus conclude that the condition of unitarity is equivalent to

$$\iint_{\mathbb{R}^2} (|f(\xi,\tau)|^2 - 1)\,A_x(\xi,\tau)\,A_y^*(\xi,\tau)\,d\xi\,d\tau = 0 \ ,$$

Chapter 2 Classes of Solutions 127

and this requires the parameter function to be unimodular, that is,

$$|f(\xi, \tau)| = 1 \ . \tag{2.97}$$

It should be noted that the fulfillment of this condition guarantees that the parameter function cannot vanish. Thus the associated representation is invertible (apart from a pure phase).

Instantaneous frequency. We were already in search of an analogy between time-frequency representations and (joint) probability distributions when we looked for an interpretation of the marginal distributions in terms of the *a priori* probability densities. Carrying this analogy farther, we are now interested in the local behavior of a time-frequency distribution. For example, we can look upon the frequential cross section at a fixed instant as a conditional probability density with respect to this instant. Consequently, the center of gravity of such a section has the same nature as a conditional expectation value. It thus provides some information about the (average) instantaneous frequential contents of the signal. A proper requirement signifies that this information matches the instantaneous frequency, as it was defined for analytic signals in Eq. (1.21). (Intuitively, in case of a single-component signal with frequency modulation, a "good" time-frequency representation should essentially live near the curve of the instantaneous frequency.)

For a signal $x(t)$ let $z_x(t)$ be the associated analytic signal. The imposed constraint can be written as

$$\frac{\int_{-\infty}^{+\infty} \nu C_{z_x}(t, \nu; f) \, d\nu}{\int_{-\infty}^{+\infty} C_{z_x}(t, \nu; f) \, d\nu} = \frac{1}{2\pi} \frac{d \arg z_x}{dt}(t) \ . \tag{2.98}$$

Expanding the numerator leads to

$$\int_{-\infty}^{+\infty} \nu C_{z_x}(t, \nu; f) \, d\nu = \frac{1}{2\pi} \int_{-\infty}^{+\infty} F(s-t, 0) \, \text{Im} \left\{ \frac{dz_x}{dt}(s) z_x^*(s) \right\} ds$$

$$+ \frac{i}{2\pi} \int_{-\infty}^{+\infty} \frac{\partial F}{\partial \tau}(s-t, 0) \, |z_x(s)|^2 \, ds \ .$$

This gives

$$\int_{-\infty}^{+\infty} \nu C_{z_x}(t, \nu; f) \, d\nu = \frac{1}{2\pi} \text{Im} \left\{ \frac{dz_x}{dt}(t) \, z_x^*(t) \right\} ,$$

on the assumption that the parameter function satisfies the following admissibility conditions (the first of which guarantees a correct marginal distribution):

$$f(\xi, 0) = 1 \quad \text{and} \quad \frac{\partial f}{\partial \tau}(\xi, 0) = 0 . \tag{2.99}$$

Bringing out the modulus and the phase of the analytic signal yields

$$\frac{dz_x}{dt}(t) z_x^*(t) = \left(\frac{d|z_x|}{dt}(t) + i|z_x(t)| \frac{d \arg z_x}{dt}(t) \right) |z_x(t)| .$$

Consequently, we arrive at

$$\frac{1}{2\pi} \operatorname{Im} \left\{ \frac{dz_x}{dt}(t) z_x^*(t) \right\} = |z_x(t)|^2 \left(\frac{1}{2\pi} \frac{d \arg z_x}{dt}(t) \right) .$$

Hence, the forementioned admissibility conditions (Eq. (2.99)) ensure the satisfaction of the given constraint: the instantaneous frequency is equal to the center of gravity (or local first-order moment) of the distribution.

Group delay. We can also consider the local first-order moment (or the conditional expectation) relative to the frequency and proceed in a similar manner. Then we find that the group delay is obtained as

$$\frac{\int_{-\infty}^{+\infty} t\, C_{z_x}(t, \nu; f)\, dt}{\int_{-\infty}^{+\infty} C_{z_x}(t, \nu; f)\, dt} = -\frac{1}{2\pi} \frac{d \arg Z_x}{d\nu}(\nu) , \tag{2.100}$$

provided that the following two conditions are satisfied:

$$f(0, \tau) = 1 \quad \text{and} \quad \frac{\partial f}{\partial \xi}(0, \tau) = 0 . \tag{2.101}$$

Stationarity. In Chapter 1 we appealed to the notion of *stationary deterministic signals*. They were defined as the sum of components with constant amplitudes and instantaneous frequencies, namely

$$x_{\rm s}(t) = \sum_n a_n e^{i 2\pi \nu_n t} .$$

This definition is connected with the idea of a steady-state or time-invariant system. It is reasonable to demand from a time-frequency representation

Chapter 2 Classes of Solutions

of a stationary deterministic signal to be independent of the time as well. This is expressed by the necessary condition

$$\frac{\partial C_{x_s}}{\partial t}(t,\nu;f) = 0 \ .$$

The definition of Cohen's class (Eq. (2.53)) and straightforward computations show that

$$C_{x_s}(t,\nu;f) = \sum_n \sum_m a_n a_m \int_{-\infty}^{+\infty} \Pi\left(s-t, \frac{\nu_n+\nu_m}{2} - \nu\right) e^{i2\pi(\nu_n-\nu_m)s} \, ds.$$

Therefore it is clear that the searched property imposes independence of the time on the parameter function $\Pi(t,\nu)$. Only the frequency variable remains at our disposal. Consequently, the parameter function $f(\xi,\tau)$ must have the form

$$f(\xi,\tau) = \delta(\xi)\, g(\tau) \ .$$

This condition, in fact, is very restrictive, as it gives up the concept of energy distributions ($f(0,0)$ is not defined) and is only formally acceptable. It defines some smoothed versions (in frequency) of the spectral energy density, that is,

$$C_x(t,\nu;f) = \int_{-\infty}^{+\infty} G(\xi-\nu) |X(\xi)|^2 \, d\xi \ .$$

Localization. For a single-component frequency-modulated signal, our intuition suggests that the time-frequency representation should be concentrated near the curve of the instantaneous frequency $\nu_x(t)$. We restrict ourselves to the case of a unimodular analytic signal (i.e., its instantaneous amplitude is constant and equal to 1). The "ideal" situation would be that

$$C_{z_x}(t,\nu;f) = \delta(\nu - \nu_x(t)) \ . \tag{2.102}$$

Let us first suppose that the analyzed signal is a linear chirp characterized by

$$\nu_x(t) = \nu_0 + \beta t \ .$$

Some direct computations show that

$$C_{z_x}(t,\nu;f) = \int_{-\infty}^{+\infty} f(\beta\tau,\tau) \, e^{-i2\pi(\nu-\nu_x(t))\tau} \, d\tau \ . \tag{2.103}$$

Hence, for a nonzero slope β of the modulation, the only solution is given by $f(\xi,\tau) = 1$, and this characterizes the Wigner-Ville distribution. On the other hand, for a pure frequency ($\beta = 0$) every definition according to a parameter function with $f(0, \tau) = 1$ is suitable. This evidently agrees with the fact that this same condition guarantees a correct marginal distribution in frequency (cf. Eq. (2.67)).

If one goes further and wishes to maintain a perfect localization to a more general curve of an instantaneous frequency, which is neither constant nor linear, there exist no solutions in Cohen's class. Nevertheless, one can find a formal solution whose parameter function may depend on the analyzed signal. (This is obviously outside the scope of bilinear distributions.) Indeed, for a (unimodular) analytic signal with arbitrary phase the condition of ideal localization reads as

$$\int_{-\infty}^{+\infty} F(t - s, \tau) \exp\left\{i \arg z_x \left(s + \frac{\tau}{2}\right) - i \arg z_x \left(s - \frac{\tau}{2}\right)\right\} ds = e^{i 2 \pi \nu_x(t) \tau}.$$

If we consider, for instance, a parabolic chirp (i.e., a signal with cubic phase), which is characterized by

$$\nu_x(t) = \nu_0 + \beta t + \gamma t^2,$$

some simple calculations lead to the condition

$$f(\xi, \tau) = \exp\left\{-i \frac{\pi \gamma}{6} \tau^3\right\}. \tag{2.104}$$

This defines a parameter function, which depends on the curvature of the instantaneous frequency.

Remark. At first sight, the representation associated with Eq. (2.104) functionally depends on the analyzed signal by an *a priori* unknown parameter (the curvature). However, this parameter can also be written in terms of the moments of the usual Wigner-Ville representation. In fact, one can show that

$$\gamma = -8\pi^2 \left(\mu_3(t) - 6\mu_1(t)\mu_2(t) + 5\mu_1^3(t)\right)$$

where

$$\mu_k(t) = \left(\int_{-\infty}^{+\infty} \nu^k W_{z_x}(t, \nu) d\nu\right) \Big/ \left(\int_{-\infty}^{+\infty} W_{z_x}(t, \nu) d\nu\right).$$

Thus the Wigner-Ville distribution provides, at least theoretically, a way to find estimates of the characteristics of a quadratic rule of the frequency modulation. Inserting those estimates into a parameterization of Cohen's class is likely to yield a better localization. [53]

The association of constraints and admissibility conditions for the representations in Cohen's class is summarized in Table 2.4. The constraints can easily be checked for the special representations in Table 2.1 by inspecting their parameter functions. For a subset of these distributions the verified conditions are marked in Table 2.5.

Chapter 2 Classes of Solutions

Table 2.4
Cohen's class: typical constraints on $C_x(t,\nu;f)$ and the corresponding admissibility conditions for the parameter function $f(\xi,\tau)$

Constraint	Admissibility Condition
Energy	$f(0,0) = 1$
Marginal in time	$f(\xi,0) = 1$
Marginal in frequency	$f(0,\tau) = 1$
Reality	$f(\xi,\tau) = f^*(-\xi,-\tau)$
Positivity	$f(\xi,\tau) = \sum_{k=1}^{+\infty} c_k\, A_{h_k}^*(\xi,\tau),\quad c_k > 0 \quad$ (if $f(\xi,\tau) \in L^2(\mathbb{R}^2)$)
Causality	$f(\xi,\tau) = \dfrac{i}{\pi}\,\mathrm{pv}\displaystyle\int_{-\infty}^{+\infty} \dfrac{f(\zeta,\tau)}{\xi-\zeta}\, e^{i\pi\zeta|\tau|}\, d\zeta$
Invertibility	$f(\xi,\tau) \neq 0$
Dilations	$f(\xi,\tau) = f\left(\dfrac{\xi}{k}, k\tau\right),\quad \forall\, k$
Convolution	$f(\xi,\tau)\, f(\xi,\tau') = f(\xi,\tau+\tau')$
Modulation	$f(\xi,\tau)\, f(\xi',\tau) = f(\xi+\xi',\tau)$
Time support (in the wide sense)	$\lvert s \rvert > \dfrac{\lvert \tau \rvert}{2} \;\Rightarrow\; \displaystyle\int_{-\infty}^{+\infty} f(\xi,\tau)\, e^{i2\pi\xi s}\, d\xi = 0$
Frequency support (in the wide sense)	$\lvert \nu \rvert > \dfrac{\lvert \xi \rvert}{2} \;\Rightarrow\; \displaystyle\int_{-\infty}^{+\infty} f(-\xi,\tau)\, e^{-i2\pi\nu\tau}\, d\tau = 0$
Unitarity	$\lvert f(\xi,\tau) \rvert = 1$
Instantaneous frequency	$f(\xi,0) = 1,\quad \dfrac{\partial f}{\partial \tau}(\xi,0) = 0$
Group delay	$f(0,\tau) = 1,\quad \dfrac{\partial f}{\partial \xi}(0,\tau) = 0$

Table 2.5

Cohen's class: a list of constraints that are verified by several representations (S = spectrogram; WV = Wigner-Ville; R = Rihaczek; P = Page; BJ = Born-Jordan; CW = Choï-Williams)

Constraint	Distributions					
	S	WV	R	P	BJ	CW
Energy	√	√	√	√	√	√
Marginal in time		√	√	√	√	√
Marginal in frequency		√	√	√	√	√
Reality	√	√		√	√	√
Positivity	√					
Causality				√		
Invertibility		√	√	√		√
Dilations		√	√	√	√	√
Convolution		√	√			
Modulation		√	√	√		
Time support (in the wide sense)		√	√	√	√	
Frequency support (in the wide sense)		√	√	√	√	
Unitarity		√	√	√		
Instantaneous frequency		√	√	√	√	√
Group delay		√	√		√	√

Affine class

The analysis of the time-scale case runs in parallel to the time-frequency case. We therefore list only some of its specific properties.

Energy. A time-scale representation is an energy distribution if

$$\iint_{\mathbb{R}^2} \Omega_x(t, a; f) \frac{dt\, da}{a^2} = E_x \; . \tag{2.105}$$

As seen before (cf. Eq. (2.60)), this corresponds to the admissibility condition

$$\int_{-\infty}^{+\infty} \psi(0, \nu) \frac{d\nu}{|\nu|} = 1 \; . \tag{2.106}$$

Chapter 2 Classes of Solutions

For the special case of the scalogram, which has a parameter function $f(\xi, \tau) = A_h^*(\xi, \tau)$, we compute that

$$\psi(\xi, \nu) = H\left(\nu + \frac{\xi}{2}\right) H^*\left(\nu - \frac{\xi}{2}\right) \implies \psi(0, \nu) = |H(\nu)|^2.$$

Hence, the simplified condition

$$\int_{-\infty}^{+\infty} |H(\nu)|^2 \frac{d\nu}{|\nu|} = 1 \qquad (2.107)$$

follows, and this is the usual admissibility condition (cf. Eq. (2.25)) of a wavelet analysis.

The constraint equation (2.106) must be slightly modified, when we turn back to the time-frequency interpretation in Eq. (2.64), where

$$\iint_{\mathbb{R}^2} \widetilde{C}_x(t, \nu; f) \, dt \, d\nu = E_x. \qquad (2.108)$$

In order for this condition to be met by the parameter function we must have

$$\int_{-\infty}^{+\infty} \psi(0, \nu) \frac{d\nu}{|\nu|} = \frac{1}{\nu_0}. \qquad (2.109)$$

For the affine Wigner distributions this leads to the constraint

$$|H(0)| = \nu_0 \, G(0). \qquad (2.110)$$

One easily checks that this identity is verified by all distributions in Table 2.3.

Marginal distributions. Let us first look at the marginal distribution relative to the scale parameter by integrating over time. By simple manipulations we obtain that

$$\int_{-\infty}^{+\infty} \Omega_x(t, a; f) \, dt =$$

$$= a \int_{-\infty}^{+\infty} \psi(0, a\nu) \left(\int_{-\infty}^{+\infty} \left[\int_{-\infty}^{+\infty} x\left(u + \frac{\tau}{2}\right) x^*\left(u - \frac{\tau}{2}\right) du \right] e^{-i2\pi\nu\tau} d\tau \right) d\nu$$

$$= \int_{-\infty}^{+\infty} \psi(0, \nu) \left| X\left(\frac{\nu}{a}\right) \right|^2 d\nu.$$

Provided that the parameter function satisfies

$$\psi(0,\nu) = \delta(\nu - \nu_0) \iff f(0,\tau) = e^{-i2\pi\nu_0\tau}, \qquad (2.111)$$

this implies that the marginal distribution has the form

$$\int_{-\infty}^{+\infty} \Omega_x(t,a;f)\, dt = \left|X\left(\frac{\nu_0}{a}\right)\right|^2. \qquad (2.112)$$

And thus it really corresponds to the spectral energy density of the signal subject to the identification "frequency = inverse of scale."

At this point it is appropriate to make some remarks concerning different types of distributions:

(i) When we consider the intersection of Cohen's class and the affine class, we end up with parameterizations having the form

$$f(\xi,\tau) = \varphi(\xi\tau)\, e^{-i2\pi\nu_0\tau}.$$

Hence, the forementioned identification is fully justified, and the correct marginal distribution is attained if

$$\varphi(0) = 1.$$

This coincides with the previous condition (cf. Eq. (2.67)), which was obtained in connection with the (time-frequency) analysis of Cohen's class.

(ii) For the class of affine Wigner distributions, as given by Eq. (2.63), the wanted property implies that

$$G(0) = 1 \quad \text{and} \quad H(0) = \nu_0.$$

(iii) Finally, the required condition can never be met by a scalogram because

$$\psi(0,\nu) = |H(\nu)|^2 \iff f(0,\tau) = \gamma_h(\tau).$$

Here the related marginal distribution is biased according to the relation

$$\int_{-\infty}^{+\infty} \Omega_x(t,a;f)\, dt = \int_{-\infty}^{+\infty} \left|X\left(\frac{\nu}{a}\right)\right|^2 |H(\nu)|^2\, d\nu. \qquad (2.113)$$

Chapter 2 Classes of Solutions

Let us next turn to the marginal distribution relative to the time. By integrating over all scales we obtain that

$$\int_{-\infty}^{+\infty} \Omega_x(t,a;f)\,\frac{da}{a^2} = \iint_{\mathbb{R}^2}\left[\int_{-\infty}^{+\infty} f\left(a\xi,\frac{\tau}{a}\right)\frac{da}{a^2}\right] A_x(\xi,\tau)\, e^{-i2\pi\xi t}\,d\xi\,d\tau$$

$$= \iint_{\mathbb{R}^2}\left(\int_{-\infty}^{+\infty}\left[\int_{-\infty}^{+\infty} f\left(a\xi,\frac{\tau}{a}\right)\frac{da}{a^2}\right] x\left(u+\frac{\tau}{2}\right) x^*\left(u-\frac{\tau}{2}\right) d\tau\right)$$

$$\times\, e^{i2\pi\xi(u-t)}\,d\xi\,du\ .$$

Here the marginal distribution will provide the correct value

$$\int_{-\infty}^{+\infty} \Omega_x(t,a;f)\,\frac{da}{a^2} = |x(t)|^2 \qquad (2.114)$$

if we have

$$\int_{-\infty}^{+\infty}\left[\int_{-\infty}^{+\infty} f\left(a\xi,\frac{\tau}{a}\right)\frac{da}{a^2}\right] x\left(u+\frac{\tau}{2}\right) x^*\left(u-\frac{\tau}{2}\right) d\tau = |x(u)|^2\ .$$

In other words, the condition

$$\int_{-\infty}^{+\infty} f\left(a\xi,\frac{\tau}{a}\right)\frac{da}{a^2} = \delta(\tau) \qquad (2.115)$$

must be satisfied for all frequencies ξ.

In the particular case of the affine Wigner distributions, considered in their time-frequency form, imposing this condition is equivalent to

$$\int_{-\infty}^{+\infty}\left[\frac{G(a\xi)}{a^2\frac{d}{da}\left(-\frac{H(a\xi)}{a}\right)}\right] \exp\left\{i2\pi\left(-\frac{H(a\xi)}{a}\right)\tau\right\} d\left(-\frac{H(a\xi)}{a}\right) = \frac{1}{\nu_0}\delta(\tau)\ .$$

This yields in consequence that

$$\nu_0\, G(a\xi) = a^2\,\frac{d}{da}\left(-\frac{H(a\xi)}{a}\right),$$

which by a proper simplification gives

$$\nu_0\, G(\xi) = H(\xi) - \xi\,\frac{dH}{d\xi}(\xi)\ . \qquad (2.116)$$

Reality. Just like the representations in Cohen's class, a time-scale distribution in the affine class is real-valued if and only if its parameter function is Hermitian. For the affine Wigner distributions (Eq. (2.63)) this is simply expressed by the parity relations

$$G(\xi) = G(-\xi) \quad \text{and} \quad H(\xi) = H(-\xi) \ . \tag{2.117}$$

Localization. The localization of a time-scale representation can have several meanings. The first is associated with the time only, and centers round the condition (introduced by P. and J. Bertrand)

$$X(\nu) = \nu^{-1/2} e^{-i2\pi\nu t_0} U(\nu) \implies \widetilde{C}_x(t,\nu;f) = \nu\,\delta(t-t_0)\,U(\nu) \ . \tag{2.118}$$

Let $x(t)$ be a signal matching the forementioned model. The frequential form (Eq. (2.62)) of the general definition of the affine class leads to the expression

$$\Omega_x(t,a;f) = a \int_{-\infty}^{+\infty} \left[\int_{a|\xi|/2}^{+\infty} \psi(a\xi,\nu) \left(\nu^2 - \left(\frac{a\xi}{2}\right)^2 \right)^{-1/2} d\nu \right] e^{-i2\pi\xi(t-t_0)} d\xi$$

which renders the desired result

$$\int_{|\xi|/2}^{+\infty} \psi(\xi,\nu) \left(\nu^2 - \left(\frac{\xi}{2}\right)^2 \right)^{-1/2} d\nu = \frac{1}{\nu_0} \ . \tag{2.119}$$

For the special class of affine Wigner distributions this constraint reduces to

$$\int_{|\xi|/2}^{+\infty} \nu_0 \, G(\xi) \, \delta(\nu - H(\xi)) \left(\nu^2 - \left(\frac{\xi}{2}\right)^2 \right)^{-1/2} d\nu = 1 \ ,$$

hence

$$\nu_0^2 \, G^2(\xi) = H^2(\xi) - \left(\frac{\xi}{2}\right)^2 \quad \text{and} \quad H(\xi) \geq \frac{|\xi|}{2} \ . \tag{2.120}$$

A second and more general constraint requires a time-scale representation of a signal to be perfectly localized to a certain curve in the plane, which describes the group delay of the signal. With respect to the identification $\nu = \nu_0/a$, this is equivalent to demanding that a signal of the form

$$X(\nu) = \nu^{-1/2} e^{i\Phi_x(\nu)} U(\nu)$$

Chapter 2 Classes of Solutions

have the time-frequency representation

$$\tilde{C}_x(t, \nu; f) = \nu\, \delta(t - t_x(\nu))\, U(\nu) = \nu\, \delta\left(t + \frac{1}{2\pi} \frac{d\Phi_x}{d\nu}(\nu)\right) U(\nu) \,. \quad (2.121)$$

In this form, the problem boils down to the ideal localization of a time-frequency representation to an arbitrary "chirp." [54] Without further restrictions, we can perform a direct computation and obtain

$$\Omega_x(t, a; f) = a \int_{-\infty}^{+\infty} \left(\int_{a|\xi|/2}^{+\infty} \psi(a\xi, \nu) \left(\nu^2 - \left(\frac{a\xi}{2}\right)^2 \right)^{-1/2} \right.$$

$$\left. \times\ \exp\left\{ i \left[\Phi_x\left(\frac{\nu}{a} - \frac{\xi}{2}\right) - \Phi_x\left(\frac{\nu}{a} + \frac{\xi}{2}\right) \right] \right\} d\nu \right) e^{-i2\pi\xi t}\, d\xi\,.$$

Hence, the desired condition is verified, if the inner integral with respect to ν has the value

$$\frac{1}{\nu_0} \exp\left\{ -i\xi \frac{d\Phi_x}{d\nu}\left(\frac{\nu_0}{a}\right) \right\}\,.$$

For the class of affine Wigner distributions this leads to the pair of constraints

$$\begin{cases} \Phi_x\left(\dfrac{H(a\xi)}{a} + \dfrac{\xi}{2}\right) - \Phi_x\left(\dfrac{H(a\xi)}{a} - \dfrac{\xi}{2}\right) = \xi \dfrac{d\Phi_x}{d\nu}\left(\dfrac{\nu_0}{a}\right) \\ \nu_0\, G(\xi) = \left(H^2(\xi) - \left(\dfrac{\xi}{2}\right)^2 \right)^{1/2} \,. \end{cases} \quad (2.122)$$

We can therefore deduce several consequences according to the type of the considered rule of the modulation:

(i) Constant group delay: $\Phi_x(\nu) = -2\pi\nu t_0$.

The first equation of Eq. (2.122) automatically holds in this case, and the only constraint refers to the second equation. We thus recover the previously established result for the purely temporal localization (cf. Eq. (2.120)).

(ii) Linear group delay: $\Phi_x(\nu) = -2\pi\left(\nu t_0 + \frac{\alpha}{2}\nu^2\right)$.

We can easily prove the necessity of $H(\xi) = \nu_0$, and consequently,

$$G(\xi) = \left(1 - \left(\frac{\xi}{2\nu_0}\right)^2 \right)^{1/2}\,.$$

The corresponding representation is nothing but a modified version of the Wigner-Ville distribution, as the (unique) specification of the parameterization leads to

$$\widetilde{C}_x(t,\nu;f) = \nu W_{x_{1/2}}(t,\nu) .$$

In the previous relation we employ the signal (of a pure phase)

$$X_{1/2}(\nu) = \nu^{1/2} X(\nu) .$$

We thus recover a characteristic feature of the Wigner-Ville distribution, namely its perfect localization on linear chirps. In the time-scale context, where it was just re-established, this property has a natural geometric interpretation. In fact, if we rewrite the constraint equation (2.122) regarding $H(\xi)$ as

$$\frac{1}{\xi}\left(\Phi_x\left(\frac{H(a\xi)}{a} + \frac{\xi}{2}\right) - \Phi_x\left(\frac{H(a\xi)}{a} - \frac{\xi}{2}\right)\right) = \frac{d\Phi_x}{d\nu}\left(\frac{\nu_0}{a}\right) , \quad (2.123)$$

it can be viewed as an identity of a finite difference and a derivative. In case the phase is a quadratic polynomial, both values are the same if the difference is centered around the point where we take the derivative. This meets exactly the condition from before with $H(\xi) = \nu_0$.

(iii) Hyperbolic group delay: $\Phi_x(\nu) = -2\pi \left[\nu t_0 + \alpha \log\left(\frac{\nu}{\nu_c}\right)\right]$.

In this case, the first constraint is verified if

$$H(\xi) = \frac{\xi}{2} \coth\left(\frac{\xi}{2\nu_0}\right) .$$

In conjunction with the second condition, this implies that

$$G(\xi) = \nu_0 \frac{(\xi/2\nu_0)}{\sinh(\xi/2\nu_0)} .$$

We thus recognize (see Table 2.3) the characterization of the Bertrand distribution, which turns out to be the only solution to the given problem. Hence, its importance relative to hyperbolic chirps is comparable to the rank of the Wigner-Ville distribution relative to linear chirps.

(iv) Group delay "by $1/\nu^2$": $\Phi_x(\nu) = -2\pi\left(\nu t_0 - \frac{\alpha}{\nu}\right)$.

An analogous calculation as before leads to the specifications

$$H(\xi) = \left(\nu_0^2 + \left(\frac{\xi}{2}\right)^2\right)^{1/2} \quad \text{and} \quad G(\xi) = 1 .$$

Here the definition of Unterberger's distribution is revealed in the so-called "active" form (cf. Table 2.3). Note that the related "passive" form of the distribution remains essentially localized to group delays of the type "$1/\nu^2$"; however, their localization cannot be perfect owing to $G(\xi) \neq 1$.

(v) Group delay "by $1/\sqrt{\nu}$": $\Phi_x(\nu) = -2\pi \left(\nu t_0 + 2\alpha \sqrt{\nu} \right)$.

Here we obtain the pair

$$H(\xi) = \nu_0 \left[1 + \left(\frac{\xi}{4\nu_0} \right)^2 \right] \quad \text{and} \quad G(\xi) = 1 - \left(\frac{\xi}{4\nu_0} \right)^2$$

as a solution, which, by definition, characterizes the D-distribution (cf. Table 2.3).

Unitarity. In respect of the natural measure of the affine group, the constraint of unitarity (Moyal's formula) for time-scale representations is expressed by the equation

$$\iint_{\mathbb{R}^2} \Omega_x(t, a; f) \, \Omega_y^*(t, a; f) \, \frac{dt \, da}{a^2} = \left| \int_{-\infty}^{+\infty} x(t) \, y^*(t) \, dt \right|^2. \qquad (2.124)$$

By reasoning similar to that used for the marginal distribution in time, and by some further calculations, we arrive at the condition

$$\int_{-\infty}^{+\infty} f\left(a\xi, \frac{\tau}{a} \right) f^*\left(a\xi, \frac{\tau'}{a} \right) \frac{da}{a^2} = \delta(\tau - \tau'), \qquad (2.125)$$

which must be fulfilled for every frequency ξ.

For the affine Wigner distributions, written in their time-frequency form, the same line of arguments leading to the marginal time distribution can be used here. It shows that Eq. (2.125) is equivalent to

$$\nu_0^2 \, G^2(\xi) = H(\xi) - \xi \, \frac{dH}{d\xi}(\xi). \qquad (2.126)$$

Group delay. Once more, we proceed by analogy with the time-frequency case, after associating the scale with the frequency in the usual way. The numerator in the expression of the local first-order moment can be computed as

$$\int_{-\infty}^{+\infty} t\,\Omega_x(t,a;f)\,dt$$

$$= a \iint_{\mathbb{R}^2} \psi(a\xi,a\nu)\,X\left(\nu - \frac{\xi}{2}\right) X^*\left(\nu + \frac{\xi}{2}\right) \left[\int_{-\infty}^{+\infty} t\,e^{-i2\pi\xi t}\,dt\right] d\xi\,d\nu$$

$$= \frac{a}{i2\pi} \int_{-\infty}^{+\infty} \frac{\partial}{\partial \xi}\left[\psi(a\xi,a\nu)\,X\left(\nu - \frac{\xi}{2}\right) X^*\left(\nu + \frac{\xi}{2}\right)\right]_{\xi=0} d\nu$$

$$= \int_{-\infty}^{+\infty} \left[\frac{a}{i2\pi}\,\frac{\partial \psi}{\partial \xi}(0,\nu)\left|X\left(\frac{\nu}{a}\right)\right|^2 - \frac{1}{2\pi}\,\psi(0,\nu)\,\mathrm{Im}\left\{\frac{dX}{d\nu}(\nu)\,X^*(\nu)\right\}\right] d\nu\;.$$

Let us restrict ourselves to the case where the condition of a correct marginal frequency distribution is satisfied (i.e., $\psi(0,\nu) = \delta(\nu - \nu_0)$; cf. Eq. (2.111)) and where

$$\frac{\partial \psi}{\partial \xi}(0,\nu) = 0\;. \tag{2.127}$$

Then the desired result, which identifies the group delay and the local first-order moment, follows in the form:

$$\frac{\displaystyle\int_{-\infty}^{+\infty} t\,\Omega_x(t,a;f)\,dt}{\displaystyle\int_{-\infty}^{+\infty} \Omega_x(t,a;f)\,dt} = t_x\left(\frac{\nu_0}{a}\right)\;. \tag{2.128}$$

For the subset of affine Wigner distributions, the additional constraint equation (2.127) simply reads as

$$\frac{dG}{d\xi}(0) = \frac{dH}{d\xi}(0)\;. \tag{2.129}$$

Narrowband limit. Finally, let us reconsider the foregoing constraints associating the group delay with the local center of gravity. Here we wish to investigate the behavior of time-scale distributions in the limiting case of narrowband signals. Provided that the preceding conditions are satisfied, the time-scale distributions reduce to the usual Wigner-Ville distribution of such signals.

Chapter 2 Classes of Solutions 141

In order to prove this assertion, let us make a simple substitution and rewrite the affine class (Eq. (2.62)) as

$$\Omega_x(t, a; f)$$
$$= \iint_{\mathbf{R}^2} \psi(a\xi, \nu) \, X\left(\frac{\nu}{a}\left(1 - \frac{a\xi}{2\nu}\right)\right) X^*\left(\frac{\nu}{a}\left(1 + \frac{a\xi}{2\nu}\right)\right) e^{-i2\pi\xi t} \, d\xi \, d\nu \; .$$

Assuming a narrowband signal signifies that the useful range of the integration with respect to the ξ-variable is confined to

$$1 \pm \frac{a\xi}{2\nu} \approx 1 \; .$$

This implies that the parameter function intervenes only by the behavior of $\psi(a\xi, \nu)$ in the neighborhood of $a\xi = 0$. When we limit ourselves to an expansion of first order, we can thus write

$$\psi(a\xi, \nu) \approx \psi(0, \nu) + a\xi \, \frac{\partial \psi}{\partial \xi}(0, \nu) \; .$$

Based on the hypotheses of Eqs. (2.111) and (2.127), namely

$$\psi(0, \nu) = \delta(\nu - \nu_0) \quad \text{and} \quad \frac{\partial \psi}{\partial \xi}(0, \nu) = 0 \; ,$$

we gain the approximation

$$\Omega_x(t, a; f) \approx \int_{-\infty}^{+\infty} X\left(\frac{\nu_0}{a} - \frac{\xi}{2}\right) X^*\left(\frac{\nu_0}{a} + \frac{\xi}{2}\right) e^{-i2\pi\xi t} \, d\xi = W_x\left(t, \frac{\nu_0}{a}\right) \; .$$

Recall that the group delay is a purely local characteristic in frequency. Hence, it is not surprising that it leads to the same constraints, which yield an interpretation of the Wigner-Ville distribution as the limit of representations of narrowband signals.

In Table 2.6 we give a summary of the discussed constraints and admissibility conditions. The marks in Table 2.7 point to the conditions that are verified by the affine Wigner distributions in Table 2.3.

Table 2.6

Affine Wigner distributions: typical constraints and the corresponding admissibility conditions for the parameter functions $H(\xi)$ and $G(\xi)$

Constraint	Admissibility Condition
Energy	$\nu_0 G(0) = |H(0)|$
Marginal in time	$\nu_0 G(\xi) = H(\xi) - \xi \dfrac{dH}{d\xi}(\xi)$
Marginal in frequency	$G(0) = 1 \; ; \quad H(0) = \nu_0$
Reality	$G(\xi) = G(-\xi) \; ; \quad H(\xi) = H(-\xi)$
Time localization	$\nu_0^2 G^2(\xi) = H^2(\xi) - \left(\dfrac{\xi}{2}\right)^2$
Unitarity	$\nu_0 G^2(\xi) = H(\xi) - \xi \dfrac{dH}{d\xi}(\xi)$
Group delay	$G(0) = 1 \; ; \quad H(0) = \nu_0 \; ; \quad \dfrac{dG}{d\xi}(0) = \dfrac{dH}{d\xi}(0)$
Narrowband limit	$G(0) = 1 \; ; \quad H(0) = \nu_0 \; ; \quad \dfrac{dG}{d\xi}(0) = \dfrac{dH}{d\xi}(0)$

2.3.3. Results of Exclusion and Conditional Uniqueness

Each constraint of a time-frequency representation can be expressed by an admissibility condition for the free parameter function. It is therefore easy to construct representations that are compatible with a preferred list of specifications. It is not always possible, however, to achieve the envisaged goal, as certain constraints can lead to mutually exclusive conditions. This can be observed by inspecting Tables 2.5 and 2.7, because no distribution verifies the full set of properties considered there. We will next explain some of these impossibilities, and we will further investigate the issue of uniqueness of a representation relative to a certain catalogue of constraints.

Table 2.7

Affine Wigner distributions: a list of constraints that are verified by several representations (B = Bertrand; Ua = Unterberger, "active" form; Up = Unterberger, "passive" form; D = D-distribution)

Constraint	B	Distributions Ua	Up	D
Energy	✓	✓	✓	✓
Marginal in time			✓	✓
Marginal in frequency	✓	✓	✓	✓
Reality	✓	✓	✓	✓
Time localization	✓	✓		✓
Unitarity	✓			
Group delay	✓	✓	✓	✓
Narrowband limit	✓	✓	✓	✓

Wigner's Theorem. [55] The prototype of impossibilities, that the time-frequency representations are always confronted, is a theorem from Wigner (which applies beyond Cohen's class). It is remarkable in that it puts an impossibility right to the core of what might be expected as minimal constraints of an energy distribution. Its formulation is very simple: *there exists no time-frequency representation that is bilinear, has correct marginal distributions, and is nonnegative everywhere.*

In order to prove this result, let us suppose that there exists a representation $\rho_x(t, \nu)$ that satisfies all three conditions. We must show that this leads to a contradiction. Let $x(t)$ be a signal, which vanishes identically outside a certain time interval. For every instant t outside this interval, our assumptions imply

$$0 = |x(t)|^2 = \int_{-\infty}^{+\infty} \rho_x(t, \nu) \, d\nu \ .$$

As the representation is supposed to be nonnegative, the expression on the right-hand side of this equation can only vanish, if the representation is zero on the whole cross section. We can thus deduce from our hypotheses that the representation vanishes wherever the signal vanishes.

Let us next consider two signals $x(t)$ and $y(t)$, which are zero outside two disjoint intervals T_1 and T_2; hence

$$x_1(t)\, x_2(t) = 0\ .$$

If we form the linear combination

$$x(t) = a x_1(t) + b x_2(t)\ ,$$

the first hypothesis (bilinearity) ensures that the resulting representation has the form

$$\rho_x(t,\nu) = |a|^2 \rho_1(t,\nu) + a^* b\, \rho_{12}(t,\nu) + a b^* \rho_{21}(t,\nu) + |b|^2 \rho_2(t,\nu)$$

(with obvious notations used here). Because $\rho_1(t,\nu)$ is zero at every instant outside T_1, the nonnegativity of $\rho_x(t,\nu)$ can only be guaranteed for all a and b, if the mixed distributions $\rho_{12}(t,\nu)$ and $\rho_{21}(t,\nu)$ both vanish outside T_1. The same argument, when applied to T_2, leads to the necessary nullity of $\rho_{12}(t,\nu)$ and $\rho_{21}(t,\nu)$ outside T_2. It thus follows that $\rho_{12}(t,\nu)$ and $\rho_{21}(t,\nu)$ must vanish everywhere, and we obtain that

$$\rho_x(t,\nu) = |a|^2 \rho_1(t,\nu) + |b|^2 \rho_2(t,\nu)\ .$$

If we next use the second hypotheses (correct marginal distributions), we end up with

$$|X(\nu)|^2 = |a|^2\, |X_1(\nu)|^2 + |b|^2\, |X_2(\nu)|^2\ ,$$

and this demands that

$$X_1(\nu)\, X_2^*(\nu) = 0$$

by the construction of $x(t)$. As this evidently contradicts the assumption of a finite duration of $x_1(t)$ and $x_2(t)$, the proof is completed.

There are several consequences of Wigner's theorem regarding the three exclusive properties:

(*i*) If a bilinear time-frequency representation has correct marginal distributions, it must attain negative (or complex) values. This happens for the distributions of Wigner-Ville, Rihaczek, Born-Jordan, Choï-Williams, etc.

(*ii*) If a bilinear time-frequency representation is positive, its marginal distributions cannot be correct. This is the case for the spectrograms and all their positive–linear combinations.

(*iii*) If a time-frequency representation has correct marginal distributions, while being nonnegative everywhere, it cannot be a bilinear transformation of the signal.

Let us note, however, that there is no obstruction to the existence of solutions of the last type, if one accepts leaving Cohen's class (in the strict sense of our present definition), or if one allows the parameter function of Cohen's class to depend on the analyzed signal. We will turn to this issue in Subsection 3.3.3.

Some results of exclusion. Inspecting the different properties of the various time-frequency representations of Cohen's class, the positivity seems to play a completely singular role, as it excludes most of the others.[56] This point can be formalized by considering several examples.

Positivity and marginal distributions. As already seen, the positivity of representations in Cohen's class is incompatible with the simultaneous attainment of correct marginal distributions in time and frequency. If we restrict ourselves to square integrable parameterizations, we can prove this result without appealing to Wigner's theorem. For this purpose let

$$f(\xi, \tau) = \sum_{k=1}^{+\infty} c_k A^*_{h_k}(\xi, \tau) \quad \text{with} \quad c_k > 0 \quad \text{and} \quad \sum_{k=1}^{+\infty} c_k = 1$$

be the parameter function of a positive representation. Taking the constraints $f(\xi, 0) = 1$ and $f(0, \tau) = 1$ relative to the marginal distributions into account, we obtain

$$\sum_{k=1}^{+\infty} c_k A^*_{h_k}(\xi, 0) = \sum_{k=1}^{+\infty} c_k A^*_{h_k}(0, \tau) = 1 .$$

Hence, the simultaneous identities

$$|h_k(t)|^2 = \delta(t) \quad \text{and} \quad |H_k(\nu)|^2 = \delta(\nu)$$

follow. This system of equations obviously has no solution.

Positivity and dilation, filtering, or modulation. We find as a corollary of the previous result that the positivity is incompatible with the covariance relative to dilations (or change of scales), the filtering, or the modulations. This comes from the fact that each of these constraints has the attainment of the marginal distributions as a by-product.

Positivity and supports. A positive distribution cannot preserve the support of a signal, not even in the wide sense. Indeed, if we consider a signal that is restricted to the time interval $[-T, +T]$, then the conservation of its support by a positive distribution lends itself to

$$\int_{-\infty}^{+\infty} C_x(t, \nu; f) \, d\nu = 0 , \quad |t| > T .$$

This can be combined with the general relation of representations in Cohen's class, which holds that

$$\int_{-\infty}^{+\infty} C_x(t,\nu;f)\,d\nu = \int_{-\infty}^{+\infty} F(s-t,0)\,|x(s)|^2\,ds \ .$$

Hence, the parameter function must be such that $f(\xi,0)$ is constant. This implies that a correct marginal time distribution is attained. Hence it is incompatible with the positivity. An analogous conclusion can be drawn from the conservation of the frequency support.

Remark. This impossibility is rather intuitive, in fact. In all cases where a positive distribution is a positive–linear combination of spectrograms, it is based on a set of "window" analyses that have nonzero supports (in time and frequency) by construction. Consequently, the supports of the distribution are enlarged by the size of the largest "window." In the limiting case, when the analyzed signal is a Dirac impulse, the positive distribution has the form

$$C_x(t,\nu;f) = \sum_{k=1}^{+\infty} c_k |h_k(t)|^2 \ .$$

Hence, its support cannot be a null set in the time domain.

Positivity and unitarity. It is impossible for a nonzero time-frequency energy distribution to be positive and unitary (i.e., verify Moyal's formula). Indeed, if the second condition is fulfilled, then

$$\iint_{\mathbb{R}^2} C_x(t,\nu;f)\,C_y^*(t,\nu;f)\,dt\,d\nu = \left| \int_{-\infty}^{+\infty} x(t)\,y^*(t)\,dt \right|^2$$

holds for every pair of signals $x(t)$ and $y(t)$. This must remain true when $x(t)$ and $y(t)$ become orthogonal on the real line, which brings about the nullity of the right-hand side of the previous equation. It thus implies that

$$\iint_{\mathbb{R}^2} C_x(t,\nu;f)\,C_y^*(t,\nu;f)\,dt\,d\nu = 0 \ ,$$

and this can happen only for a nonnegative distribution, if it vanishes identically. However, this was excluded by our assumption.

Positivity and instantaneous frequency or group delay. The instantaneous frequency and the group delay cannot be obtained as the local first-order moments of a positive distribution. This is an immediate consequence of the fact that these constraints imply the attainment of the correct marginal distributions beforehand, and this is impossible in the case of a positive distribution.

Chapter 2 Classes of Solutions 147

Some results on conditional uniqueness. If we accept giving up the positivity, some of the remaining constraints can be satisfied simultaneously. A larger number of such constraints will naturally cut back on the set of solutions. One can even identify certain combinations of constraints that imply the uniqueness of the representation. We will explain several such situations as an illustration.

s-Wigner. The s-Wigner distributions are unique solutions of different collections of constraints:

(i) The s-Wigner distributions are the only time-frequency distributions that are unitary and compatible with filterings and modulations. Indeed, it has already been mentioned (cf. Eq. (2.88)) that the last two constraints require the special form

$$f(\xi, \tau) = e^{\alpha \xi \tau}$$

of the parameter function. Combined with the condition of being unitary, which requires $|f(\xi, \tau)| = 1$, this shows that the free parameter must be purely imaginary, so $\alpha = i2\pi s$.

(ii) The s-Wigner distributions are the only time-scale distributions that are unitary and have a correct marginal distribution in time. Indeed, according to Eqs. (2.126) and (2.116), imposing these two properties simultaneously amounts to

$$\nu_0 G^2(\xi) = H(\xi) - \xi \frac{dH}{d\xi}(\xi)$$

and

$$\nu_0 G(\xi) = H(\xi) - \xi \frac{dH}{d\xi}(\xi) \ .$$

Therefore, we must have $G(\xi) = 1$, thus leading to the differential equation

$$\xi \frac{dH}{d\xi}(\xi) = H(\xi) - \nu_0 \ .$$

Its solution has the form

$$H(\xi) = -s\xi + \nu_0 \ , \quad s \in \mathbb{R} \ ,$$

and the related parameter function can be written as

$$f(\xi, \tau) = e^{i2\pi(s\xi - \nu_0)\tau} \ .$$

This defines an s-Wigner distribution in its time-scale form, where the scale a is associated with the frequency $\nu = \nu_0/a$.

Wigner-Ville. Several combinations of conditions ensure the uniqueness of the Wigner-Ville distribution. Here are some of them:

(i) *The Wigner-Ville distribution is the only s-Wigner distribution that is real-valued.* As we have seen in Eq. (2.68), the reality of a distribution is a result of the Hermitian symmetry of its parameter function $f(\xi, \tau)$. In case of an s-Wigner distribution it follows that

$$e^{i2\pi s\xi\tau} = e^{-i2\pi s\xi\tau},$$

and this can only be true if $s = 0$. This is precisely the Wigner-Ville case.

(ii) *The Wigner-Ville distribution is the only s-Wigner distribution whose local first-order moments furnish the instantaneous frequency and the group delay.* Indeed, as an s-Wigner distribution has correct marginal distributions, it suffices to verify the conditions concerning the derivatives of the parameter function in Eqs. (2.99) and (2.101). We simply obtain

$$\frac{\partial f}{\partial \tau}(\xi, 0) = i2\pi s\xi \quad \text{and} \quad \frac{\partial f}{\partial \xi}(0, \tau) = i2\pi s\tau.$$

Each of these expressions can only be zero if $s = 0$, and this corresponds to the Wigner-Ville distribution.

(iii) *The Wigner-Ville distribution is the only distribution in Cohen's class that is compatible with filterings and yields the instantaneous frequency as its local first-order moment.* Indeed, according to Eq. (2.83), the first condition imposes

$$f(\xi, \tau) = e^{G(\xi)\tau}.$$

Hence, the second condition in Eq. (2.99) takes the form $G(\xi) = 0$. This gives the announced result.

(iv) *The Wigner-Ville distribution is the only distribution in Cohen's class that is compatible with modulations and yields the group delay as its local first-order moment.* This result is dual to the previous one and established analogously.

(v) *The Wigner-Ville distribution is the only distribution in Cohen's class that is ideally concentrated on linear chirps.* This was shown in the preceding subsection (cf. Eq. (2.103)).

Page. The Page distribution is the only distribution in Cohen's class that preserves the temporal support (in the wide sense) and is causal, unitary, and compatible with modulations. Indeed, according to Eqs. (2.97) and (2.87), the last two constraints demand from the parameter function to be of the form

$$f(\xi, \tau) = e^{i\pi\xi g(\tau)} \implies F(s, \tau) = \delta\left(s + \frac{g(\tau)}{2}\right).$$

In these expressions $g(\tau)$ is a real function. As a consequence, the constraint equation (2.90) of conservation of the temporal support gives $|g(\tau)| \leq |\tau|$, and the constraint equation (2.73) of causality implies that $|g(\tau)| \geq |\tau|$. Hence the result follows.

Bertrand. The Bertrand distribution is the only affine Wigner distribution that is unitary and localized in time. In fact, according to Eq. (2.126), the first condition requires that

$$\nu_0 \, G^2(\xi) = H(\xi) - \xi \frac{dH}{d\xi}(\xi) \,,$$

and with Eq. (2.120) the second condition by itself implies that

$$\nu_0^2 \, G^2(\xi) = H^2(\xi) - \left(\frac{\xi}{2}\right)^2 .$$

If we introduce the auxiliary functions

$$U(\xi) = H(\xi) - \frac{\xi}{2} \,, \quad V(\xi) = H(\xi) + \frac{\xi}{2} \,,$$

these two constraints lead to the differential equation

$$U(\xi) \frac{dV}{d\xi}(\xi) - V(\xi) \frac{dU}{d\xi}(\xi) = \frac{1}{\nu_0} U(\xi) V(\xi) \,.$$

By putting $W(\xi) = U(\xi)/V(\xi)$ this gives

$$\frac{dW}{d\xi}(\xi) = -\frac{1}{\nu_0} W(\xi) \,.$$

Hence, we derive that

$$W(\xi) = \frac{H(\xi) - (\xi/2)}{H(\xi) + (\xi/2)} = C e^{-\xi/\nu_0} \,,$$

which implies $W(0) = 1$. This also gives $C = 1$. It follows that

$$H(\xi) = (\xi/2) \coth(\xi/2\nu_0) \quad \text{and} \quad G(\xi) = \frac{(\xi/2\nu_0)}{\sinh(\xi/2\nu_0)} \,.$$

This defines the Bertrand distribution (cf. Table 2.3).

D-distribution. The D-distribution is the only real distribution in the affine class that has a correct marginal distribution in time and is localized in time. Indeed, according to Eqs. (2.116) and (2.120), meeting both constraints (and the assumption that $H(\xi) \geq |\xi|/2$) leads to the identity

$$\nu_0 \, G(\xi) = H(\xi) - \xi \frac{dH}{d\xi}(\xi) = \left[H^2(\xi) - \left(\frac{\xi}{2}\right)^2 \right]^{1/2}.$$

Hence, we infer the differential equation

$$\frac{dH}{d\xi}(\xi) = \frac{H(\xi)}{\xi} + \left[\left(\frac{H(\xi)}{\xi}\right)^2 - \frac{1}{4} \right]^{1/2}.$$

The constraint of reality requires that $H(-\xi) = H(\xi)$, and it suffices to find a solution of the previous equation for $\xi \geq 0$. We thus obtain that

$$H(\xi) = \nu_0 \left[1 + \left(\frac{\xi}{4\nu_0}\right)^2 \right] \implies G(\xi) = 1 - \left(\frac{\xi}{4\nu_0}\right)^2,$$

and this gives the result.

§2.4. The Power Distributions

The time-frequency (or time-scale) representations considered so far were intended mainly to deal with *deterministic finite energy* signals. Formally, they can still be used in a wider setting, for example, including generalized functions. When we now wish to consider *random* signals, the first thing to be done is to give a more general notion of the *power* spectrum by making it time-dependent. As in the deterministic case, this generalization is not unique *a priori*, and a multitude of possible solutions exists, each of which comes with specific advantages and disadvantages.

2.4.1. From Deterministic to Random Signals

Decompositions and fluctuations. A first approach, taking the stochastic character of the signal into account, is based on our knowledge of the deterministic case. When we consider linear decompositions, the stochastic character introduces fluctuations of the coefficients of the decomposition. Recall that the decomposition acts like a linear filtering of the signal. Hence, it is easy to relate the fluctuations of the coefficients to those of the

Chapter 2 Classes of Solutions

analyzed signal. Let us look more closely at the continuous decomposition equation (2.11), for instance. It is given by

$$L_x(t,\lambda) = \int_{-\infty}^{+\infty} x(s)\, h_{t\lambda}^*(s)\, ds\;,$$

where $\lambda \in \mathbb{R}$ is an auxiliary variable denoting frequency or scale. Provided that the fourth–order moments of $x(t)$ exist (and $x(t)$ has zero mean), we can write

$$\mathbf{E}\left\{|L_x(t,\lambda)|^2\right\} = \iint_{\mathbb{R}^2} r_x(s,s')\, h_{t\lambda}^*(s)\, h_{t\lambda}(s')\, ds\, ds'\;. \tag{2.130}$$

Here we make use of the autocovariance function $r_x(s,s') = \mathbf{E}\left\{x(s)x^*(s')\right\}$.

The foregoing quantity defines a veritable power distribution, because the property equation (2.13) of the continuous bases implies that

$$\iint_{\mathbb{R}^2} \mathbf{E}\left\{|L_x(t,\lambda)|^2\right\} d\mu_G(t,\lambda)$$

$$= \iint_{\mathbb{R}^2} r_x(s,s') \left[\iint_{\mathbb{R}^2} h_{t\lambda}^*(s)\, h_{t\lambda}(s')\, d\mu_G(t,\lambda)\right] ds\, ds'$$

$$= \iint_{\mathbb{R}^2} r_x(s,s')\, \delta(s-s')\, ds\, ds'$$

$$= \int_{-\infty}^{+\infty} \mathbf{E}\left\{|x(s)|^2\right\} ds\;. \tag{2.131}$$

Distributions and expectation values. Let us recall that the squared modulus of a linear decomposition is just a special case of a bilinear representation. Hence, another possible way of defining a time-dependent (power) spectrum is by taking an ensemble average of an energy distribution with respect to all possible realizations

$$\rho_x(t,\nu) \quad\longrightarrow\quad \mathbf{E}\left\{\rho_x(t,\nu)\right\}\;. \tag{2.132}$$

Accordingly, we can generalize Cohen's class (Eq. (2.53)) to the stochastic setting. Assuming, as before, that the fourth–order moments of the analyzed signal exist, we obtain a general class via the definition

$$C_x(t,\nu;f) = \iint_{\mathbb{R}^2} F(s-t,\tau)\, r_x\!\left(s+\frac{\tau}{2}, s-\frac{\tau}{2}\right) e^{-i2\pi\nu\tau}\, ds\, d\tau\;. \tag{2.133}$$

The choice of the parameter function can be guided, *mutatis mutandis*, by the admissibility conditions that were established for deterministic signals. Special constraints for the stochastic situations may be added as well. So it seems natural to demand that in case of a stationary signal (whose stochastic properties are time-invariant) the spectrum reduces to the power spectrum in each instant. Because the autocorrelation has the form $r_x(t,t') = \gamma_x(t-t')$ in this case, the preceding condition is equivalent to the identity

$$\iint_{\mathbb{R}^2} F(s-t,\tau)\,\gamma_x(\tau)\,e^{-i2\pi\nu\tau}\,ds\,d\tau = \Gamma_x(\nu)\ ,$$

which yields
$$f(0,\tau) = 1\ . \qquad (2.134)$$

We thus recover the condition of Eq. (2.67), which ensures the correct marginal frequency distribution of a deterministic signal.

Cramér and beyond. Finally, a third and less *ad hoc* possibility is to go back to the fundamental characterization of stationary signals and to find ways of deviating from it. We already mentioned in Subsection 1.2.2, that a random signal $x(t)$ is stationary in the wide sense, if and only if it admits a spectral representation (called Cramér decomposition)

$$x(t) = \int_{-\infty}^{+\infty} e^{i2\pi\nu t}\,dX(\nu)\ . \qquad (2.135)$$

It underlies a double orthogonality in the sense that

$$\begin{cases} \displaystyle\int_{-\infty}^{+\infty} e^{i2\pi\nu t}\left(e^{i2\pi\xi t}\right)^*\,dt = \delta(\xi-\nu) \\ \mathbf{E}\left\{dX(\nu)\,dX^*(\xi)\right\} = \delta(\nu-\xi)\,\Gamma_x(\nu)\,d\xi\,d\nu\ . \end{cases} \qquad (2.136)$$

When we next wish to consider nonstationary signals, the offered alternative is simple. Because we cannot maintain both the spectral representation and the double orthogonality simultaneously, we must give up at least one of them. This defines the two large classes of approaches leading to an explicit construction of time-dependent representations: those which prioritize the orthogonality, and those giving preference to the frequential interpretation. [57]

2.4.2. The Orthogonal (or Almost Orthogonal) Solutions

Karhunen decompositions. We first describe the solution that maintains the double orthogonality of the decomposition of a nonstationary signal. It is obtained by replacing the complex exponentials with other basis functions $\psi(t,\nu)$, so that

$$x(t) = \int_{-\infty}^{+\infty} \psi(t,\nu)\, dX(\nu)\;, \tag{2.137}$$

and the orthogonality relation

$$\int_{-\infty}^{+\infty} \psi(t,\nu)\, \psi^*(t,\xi)\, dt = \delta(\xi-\nu) \tag{2.138}$$

holds. Such solutions exist and are called Karhunen decompositions: They result from employing the *eigenfunctions* of the autocovariance kernel as basis functions.

In fact, we can immediately derive from Eq. (2.137) that the autocovariance has the form

$$\begin{aligned} r_x(t,s) &= \iint_{\mathbb{R}^2} \psi(t,\nu)\, \psi^*(s,\xi)\, \mathbf{E}\left\{dX(\nu)\, dX^*(\xi)\right\} \\ &= \int_{-\infty}^{+\infty} \psi(t,\xi)\, \psi^*(s,\xi)\, \Gamma_x(\xi)\, d\xi\;. \end{aligned} \tag{2.139}$$

The multiplication of both sides of this equation by $\psi(s,\nu)$, integration over the time variable s, and an application of the orthogonality assumptions of Eq. (2.138) furnish

$$\begin{aligned} \int_{-\infty}^{+\infty} r_x(t,s)\, \psi(s,\nu)\, ds &= \int_{-\infty}^{+\infty} \psi(t,\xi) \left[\int_{-\infty}^{+\infty} \psi(s,\nu)\, \psi^*(s,\xi)\, ds\right] \Gamma_x(\xi)\, d\xi \\ &= \int_{-\infty}^{+\infty} \psi(t,\xi)\, \delta(\xi-\nu)\, \Gamma_x(\xi)\, d\xi \\ &= \Gamma_x(\nu)\, \psi(t,\nu)\;. \end{aligned}$$

The advantage of such Karhunen representations lies in their double orthogonality. Their major drawback, however, stems from replacing the complex exponentials with the eigenfunctions $\psi(t,\nu)$, as the variable ν of the decomposition can no longer be interpreted as a frequency. Keeping this reservation in mind, the decomposition of the autocovariance leads to the identity

$$\mathbf{E}\{|x(t)|^2\} = \int_{-\infty}^{+\infty} |\psi(t,\nu)|^2 \, \Gamma_x(\nu) \, d\nu \, .$$

Each "spectral" contribution $\Gamma_x(\nu)$ is thus weighted by a time-dependent function, and this leads to the formal definition of a time-dependent "power spectrum" by

$$K_x(t,\nu) = |\psi(t,\nu)|^2 \, \Gamma_x(\nu) \, . \tag{2.140}$$

We thus dispose of a truly time-dependent quantity, which is non-negative everywhere. More importantly, it reduces to the ordinary power spectrum in the stationary case. However, its spectral interpretation is questionable. Nevertheless, the "Karhunen spectra" provide a prototype of representations for nonstationary signals, on which variants of more tangible interpretations can be built.

Priestley spectrum. [58] Such a viewpoint was introduced by Priestley. He started from a formal (but not necessarily orthogonal) setting of a Karhunen decomposition, supposing that the basis functions submit to the generic form

$$\psi(t,\nu) = A(t,\nu) \, e^{i2\pi\nu t} \, . \tag{2.141}$$

The stochastic processes, which give rise to such (not necessarily unique) representations, are called *oscillatory*. Physically, they are directed at taking the temporal evolution of the different spectral contributions of a signal into account, by providing a model in a frequency-by-frequency manner. Here the function $A(t,\nu)$ operates like an amplitude modulation of each complex exponential.

If $A(t,\nu)$ is slowly varying in time for each frequency, the introduction of the notion of oscillatory processes stands for a compromise between orthogonality and frequential interpretation because the basis functions $\psi(t,\nu)$ are almost orthogonal under this assumption. Indeed, we infer that

$$\int_{-\infty}^{+\infty} \psi(t,\nu) \, \psi^*(t,\xi) \, dt = \int_{-\infty}^{+\infty} A(t,\nu) \, A^*(t,\xi) \, e^{i2\pi(\nu-\xi)t} \, dt \, ,$$

and this tends to $\delta(\xi - \nu)$ when $A(t,\nu) A^*(t,\xi)$ tends to 1. This last condition can be rephrased by referring to the notion of quasi-stationary signals, which signifies the fact that $A(t,\nu)$ slowly varies in time as compared to the oscillations of the frequency ν.

Chapter 2 Classes of Solutions

The corresponding time-dependent spectrum

$$\Xi_x(t,\nu) = |A(t,\nu)|^2 \, \Gamma_x(\nu) \qquad (2.142)$$

is derived from the special form of $\psi(t,\nu)$. It is called the *evolutionary spectrum* (in the sense of Priestley).

Just like the Karhunen spectrum, it has several interesting properties: It is nonnegative everywhere, reduces to the ordinary power spectrum in the stationary case ($A(t,\nu) = 1$), and has a correct marginal distribution in time because

$$\int_{-\infty}^{+\infty} \Xi_x(t,\nu)\, d\nu = \int_{-\infty}^{+\infty} |A(t,\nu)|^2 \, \Gamma_x(\nu)\, d\nu = \mathbf{E}\left\{|x(t)|^2\right\} \, .$$

Furthermore, we can observe a satisfactory behavior for simple non-stationary signals. For this purpose let us consider a *uniformly modulated* signal of Eq. (1.45) type, hence,

$$y(t) = c(t)\, x(t)$$

where $c(t)$ denotes the modulation and $x(t)$ is supposed to be weakly stationary. Then we obtain the representation

$$y(t) = \int_{-\infty}^{+\infty} c(t)\, e^{i2\pi\nu t}\, dX(\nu) \, .$$

This defines an oscillatory signal characterized by $A(t,\nu) = c(t)$. Consequently, the associated Priestley spectrum has the value

$$\Xi_y(t,\nu) = c^2(t)\, \Gamma_x(\nu) \, , \qquad (2.143)$$

and this readily describes the anticipated behavior of the spectrum of a stationary signal, whose amplitude is uniformly modulated in time.

These advantages of the Priestley spectrum, however, should not let us forget about its drawbacks, which have restricted, and will always restrict, its use. One of them is the fact that the class of oscillatory signals on which the definition of the Priestley spectrum relies is not well defined. In fact, there are no simple criteria by which we may verify if a given signal is oscillatory or not. Even worse is the fact that the class is not stable under addition (i.e., the sum of two oscillatory signals need not be an oscillatory signal).

The introduction of the evolutionary spectrum was proposed in the Fourier domain. We can also pursue a complementary approach in the time domain. This offers a new interpretation in a different light. In fact, let us consider the Fourier decomposition of the modulating function

$$A(t,\nu) = \int_{-\infty}^{+\infty} h(t,s)\, e^{-i2\pi\nu s}\, ds \ .$$

The *stationary* signal $x_{\rm s}(t)$, which can be constructed from the spectral increments $dX(\nu)$, is

$$x_{\rm s}(t) = \int_{-\infty}^{+\infty} e^{i2\pi\nu t}\, dX(\nu) \ .$$

Hence, we obtain the identities

$$\begin{aligned}
x(t) &= \int_{-\infty}^{+\infty} A(t,\nu)\, e^{i2\pi\nu t}\, dX(\nu) \\
&= \int_{-\infty}^{+\infty} \left[\int_{-\infty}^{+\infty} h(t,s)\, e^{-i2\pi\nu s}\, ds\right] e^{i2\pi\nu t}\, dX(\nu) \\
&= \int_{-\infty}^{+\infty} h(t,s)\, x_{\rm s}(t-s)\, ds \ .
\end{aligned}$$

This signifies that an oscillatory nonstationary signal can be regarded as the output of a linear time-varying filter whose input is a stationary signal. The impulse response of the generating filter is nothing but the Fourier transform of the modulating function defining the oscillatory signal.

This interpretation is interesting from a physical point of view as well. It suggests following a slightly different approach than the one by Priestley, as it directly refers to a temporal decomposition of the signal.

Tjøstheim, Mélard, Grenier approach. The general idea behind this modified approach is to start from a Cramér-type decomposition of a non-stationary random signal, which is an extension of the Wold decomposition for the stationary case. For the sake of simplicity, we will be contented with a heuristic description using discrete signals only, in order to avoid the problems of *multiplicity* related to the continuous time.[59]

It is known that every discrete stationary random signal $\{x[n],\ n \in \mathbb{Z}\}$ admits a decomposition

$$x[n] = \sum_{m=-\infty}^{n} h[n-m]\, e[m]$$

(called Wold decomposition), where $e[n]$ is a discrete-time normalized and white noise, such that $\mathbf{E}\{e[n]\,e[m]\} = \delta_{nm}$.

This signifies that every discrete stationary signal can be represented as the output of a causal moving average (MA) filter, with an eventually infinite memory, whose input is white noise. An extension to the nonstationary case yields the correct form

$$x[n] = \sum_{m=-\infty}^{n} h[n,m]\,e[m] \qquad (2.144)$$

for every discrete nonstationary signal.

The generating filter varies in time, as could be foreseen. Because the input noise is stationary and white, it can be written as

$$e[n] = \int_{-1/2}^{+1/2} e^{i2\pi\nu n}\,dE(\nu)$$

where

$$\mathbf{E}\{dE(\nu)\,dE^*(\xi)\} = \delta(\nu-\xi)\,d\nu\,d\xi\;.$$

We conclude that the signal $x[n]$ itself has the form

$$x[n] = \int_{-1/2}^{+1/2} \psi[n,\nu]\,dE(\nu)$$

by defining

$$\psi[n,\nu] = \sum_{m=-\infty}^{n} h[n,m]\,e^{i2\pi\nu m}\;.$$

Such a decomposition is the formal analogue of the one by Karhunen (Eq. (2.137)) for the discrete case. Proceeding in the same way as for the Priestley spectrum, we can further define an *evolutionary spectrum* (now in the sense of Tjøstheim and Mélard [60]) by

$$\Theta_x[n,\nu] = \left|\sum_{m=-\infty}^{n} h[n,m]\,e^{i2\pi\nu m}\right|^2\;. \qquad (2.145)$$

Once more, this defines a real quantity, which is nonnegative everywhere and reduces to the ordinary power spectrum when the signal is stationary. Moreover, we can prove that

$$\int_{-1/2}^{+1/2} \Theta_x[n,\nu]\,d\nu = \sum_{m=-\infty}^{n} |h[n,m]|^2 = \mathbf{E}\{|x[n]|^2\}\;,$$

and this guarantees a correct marginal distribution in time.

Let us first look at uniformly modulated signals $y[n] = c[n]\,x[n]$ (with $x[n]$ being stationary). The (time-varying) impulse response $h_y[n, m]$ of the generating filter admits a representation of the type

$$h_y[n, m] = c[n]\, h_x[n - m]\,,$$

where $h_x[n]$ is the (time-invariant) impulse response associated with the stationary signal $x[n]$. Hence, the definition of the evolutionary spectrum of Tjøstheim and Mélard leads directly to the anticipated result

$$\Theta_x[n, \nu] = c^2[n]\,\Gamma_x(\nu)\,.$$

In contrast to this example, cases exist for which the obtained results cannot be considered satisfactory. For instance, if we look at a piecewise–stationary signal

$$x[n] = \begin{cases} x_1[n] & \text{for } n < 0, \\ x_2[n] & \text{for } n \geq 0, \end{cases}$$

where $x_1[n]$ and $x_2[n]$ are two different stationary signals with respective power spectra $\Gamma_1(\nu)$ and $\Gamma_2(\nu)$, we have that

$$\Theta_x[n, \nu] = \Gamma_1(\nu)\,, \quad n < 0\,,$$

but only

$$\lim_{n \to +\infty} \Theta_x[n, \nu] = \Gamma_2(\nu)\,.$$

Failure to reproduce the individual power spectra clearly stems from the basically *causal* character of the MA representation, on which the definition of the evolutionary spectrum relies. A second drawback of the MA structure underlying the Tjøstheim and Mélard spectrum is the difficulty of its estimation due to the order of the model, which might be infinite *a priori*.

In order to gain access to a more flexible tool, Grenier [61] introduced a modified version of the Tjøstheim and Mélard spectrum. So as to eradicate the possibly infinite order of the MA model, it is replaced with (or *realized by*) an autoregressive moving average (ARMA) model of finite order

$$\sum_{j=0}^{p} a_j[n-j]\,x[n-j] = \sum_{j=0}^{q} b_j[n-j]\,e[n-j]\,. \qquad (2.146)$$

This new form is amenable to different assumptions. More precisely, a (causal) MA system with impulse response $h[n, m]$ admits an ARMA

Chapter 2 Classes of Solutions

representation of the foregoing type, if and only if there exist two integers p and q, and p functions $a_j[n]$, such that

$$\sum_{j=0}^{p} a_j[n-j]\, h[n-j, n-m] = 0 , \quad m > q ,$$

and $a_0[n] = 1$. The obtained ARMA equation is called "in synchronous form." It gives rise to a representation of the signal in the form of an observable state by means of the relations

$$\mathbf{y}[n] = \begin{pmatrix} -a_1[n-1] & 1 & 0 & \cdots & 0 \\ -a_2[n-1] & 0 & 1 & \ddots & \vdots \\ \vdots & \vdots & \ddots & \ddots & 0 \\ & & & & 1 \\ -a_m[n-1] & 0 & \cdots & & 0 \end{pmatrix} \mathbf{y}[n-1] + \begin{pmatrix} b_0[n] \\ b_1[n] \\ \vdots \\ b_{m-1}[n] \end{pmatrix} e[n]$$

and
$$x[n] = \begin{pmatrix} 1 & 0 & \cdots & 0 \end{pmatrix} \mathbf{y}[n] ,$$

where $\mathbf{y}[n] = (y_1[n]\ y_2[n]\ \cdots\ y_m[n])^T$.

The way the time indices (which were arbitrary *a priori*) are written, this representation depicts a situation which is frozen at the moment n. Hence, the nonstationary model seems to define a *tangential* stationary model at each moment n, which is characterized by the coefficients $a_j[n-1]$ and $b_j[n]$. The spectrum of this tangential model can be used as a natural definition of the rational spectrum, or *relief*, by letting

$$G_x[n, \nu] = \left| \frac{B_n(z) B_n(z^{-1})}{A_n(z) A_n(z^{-1})} \right|_{z=e^{i 2\pi \nu}} \qquad (2.147)$$

with

$$A_n(z) = \sum_{j=0}^{p} a_j[n-1]\, z^{-j} , \qquad B_n(z) = \sum_{j=0}^{q} b_j[n]\, z^{-j} .$$

This spectrum preserves the majority of properties, which are known to be satisfied by the Tjøstheim and Mélard spectrum. Moreover, it also meets the condition of sectional locality.

Another important point is related to the possibilities that it offers regarding its estimation. In particular, let us consider the case of a time-dependent AR model, that is, such that $b_0[n] = 1$ and $b_j[n] = 0$, $j \neq 0$.

Let us further assume that the coefficients themselves can be decomposed relative to a basis of functions $\{f_k[n];\ k = 0, \ldots, K\}$,

$$a_j[n] = \sum_{k=0}^{K} a_{jk}\, f_k[n] \ .$$

Then the defining equations of the evolutionary model can be rewritten in the equivalent form

$$x[n] + \left(X_{n-1}^T \ \cdots \ X_{n-p}^T\right) \cdot \theta = e[n] \tag{2.148}$$

where

$$X_n = \left(f_0[n]\, x[n]\, , \ \ldots\, , \ f_K[n]\, x[n]\right)^T ,$$

$$\theta = \left(a_{10}, \ldots, a_{1K}, a_{20}, \ldots, a_{pK}\right)^T .$$

This shows that the projection of the time-varying coefficients onto a basis of functions causes the *scalar nonstationary* model equation (2.146) to be mapped onto a *vectorial stationary* model equation (2.148). The time-invariance of the coefficients of this new model (the a_{jk}'s) allows us to employ estimation techniques (based on *one* observed sample) as in the classical stationary case. The only arbitrariness lies in the choice of the basis functions, and this depends *a priori* on the supposed type of the nonstationary behavior.

Because the parametric methods are beyond the scope of this book, we will not dwell on such evolutionary models. Instead, we refer the reader to other books on the subject matter for a more complete study of this problem. [62]

2.4.3. The Frequency Solutions

Harmonizable signals. [63] The second feasible strategy regarding a relaxation of the stationary assumptions is described here. It involves retaining the idea of a frequential decomposition, while introducing a correlation between the spectral increments. On the assumption that the spectral distribution function Φ_x is (absolutely) integrable, this corresponds to an investigation of the *harmonizable* signals (defined in Subsection 1.2.2). Their autocovariance function admits the representation

$$r_x(t_1, t_2) = \iint_{\mathbf{R}^2} \Phi_x(\nu_1, \nu_2)\, e^{i2\pi(\nu_1 t_1 - \nu_2 t_2)}\, d\nu_1\, d\nu_2 \ . \tag{2.149}$$

This function and the spectral distribution function $\Phi_x(\nu_1, \nu_2)$ constitute a pair of Fourier transforms. This can be regarded as an extension to

Chapter 2 Classes of Solutions

the nonstationary case of the pair "autocorrelation function — power spectrum," which is described in the Wiener-Khinchin Theorem. It certainly reduces to the latter in the stationary case. Indeed, for this case we recall from Eqs. (1.33) and (1.41) that

$$r_x(t_1, t_2) = \gamma_x(t_1 - t_2) \quad \text{and} \quad \Phi_x(\nu_1, \nu_2) = \delta(\nu_1 - \nu_2)\Gamma_x(\nu_1) .$$

Hence, the identity equation (2.149) turns into

$$\gamma_x(\tau) = \int_{-\infty}^{+\infty} \Gamma_x(\nu) e^{i2\pi\nu\tau} d\nu .$$

The harmonizable signals admit two equivalent descriptions of second–order. One is given in the time domain, the other in the frequency domain. Therefore, it is natural to look upon the time-dependent spectrum $\rho_x(t, \nu)$ that we try to construct here as a third description, lying in between the two and being a mixed function of time and frequency. For a better explanation, let us consider the Fourier relation that connects $r_x(t_1, t_2)$ and $\Phi_x(\nu_1, \nu_2)$. Then we require the searched spectrum $\rho_x(t, \nu)$ to support a linear relation with the autocovariance (or the spectral distribution), so that the new description remains of second–order. The problem thus reduces to finding a distribution, which ensures the commutativity of the diagram

$$
\begin{array}{ccc}
r_x(t_1, t_2) & \longleftrightarrow & \Phi_x(\nu_1, \nu_2) \\
& \searrow \quad \swarrow & \\
& \rho_x(t, \nu) &
\end{array}
$$

where the transformations are defined by total or partial Fourier transforms. Hence, we formulate as an appropriate requirement that the relations

$$r_x(t_1, t_2) = \int_{-\infty}^{+\infty} \rho_x(t, \nu) e^{i2\pi\nu\tau} d\nu \qquad (2.150)$$

and

$$\rho_x(t, \nu) = \int_{-\infty}^{+\infty} \Phi_x(a\nu + b\xi, c\nu + d\xi) e^{i2\pi\xi t} d\xi \qquad (2.151)$$

be verified for all linear transformations of the coordinates

$$\begin{pmatrix} t \\ \tau \end{pmatrix} = M_t \begin{pmatrix} t_1 \\ t_2 \end{pmatrix} \quad \text{with} \quad M_t = \begin{pmatrix} \alpha & \beta \\ \gamma & \delta \end{pmatrix}$$

and

$$\begin{pmatrix} \nu_1 \\ \nu_2 \end{pmatrix} = M_\nu \begin{pmatrix} \nu \\ \xi \end{pmatrix} \quad \text{with} \quad M_\nu = \begin{pmatrix} a & b \\ c & d \end{pmatrix} .$$

Furthermore, the invariance of the area elements defined by the normalized time-time or frequency-frequency basis (which is physically needed for the energy conservation of the transformations) imposes the additional constraint of isometry

$$|\det M_t| = |\det M_\nu| = 1 . \tag{2.152}$$

The two partial Fourier transforms thus brought into play are compatible with the two-dimensional transform connecting r_x and Φ_x, if the identity $\xi t + \nu\tau = \nu_1 t_1 - \nu_2 t_2$ holds. This implies that the matrices M_t and M_ν of the transformation are linked by the relation

$$\begin{pmatrix} 0 & 1 \\ 1 & 0 \end{pmatrix} M_\nu^T \begin{pmatrix} 1 & 0 \\ 0 & -1 \end{pmatrix} = M_t ,$$

leading to

$$\begin{pmatrix} a & b \\ c & d \end{pmatrix} = \begin{pmatrix} \gamma & \alpha \\ -\delta & -\beta \end{pmatrix} .$$

The frequency transformations can thus be derived from the time transformations and vice versa.

Using Eqs. (2.150) and (2.151) and the forementioned parameterization, the autocovariance equation (2.149) can be rewritten as

$$r_x(t_1, t_2) = \iint_{\mathbb{R}^2} \Phi_x \left(\nu - \frac{\xi}{2}, \nu + \frac{\xi}{2} \right) \tag{2.153}$$

$$\times \exp\left\{ i2\pi \left[(\alpha + \beta)\tau + \frac{\alpha - \beta}{2}\xi\tau - (\gamma + \delta)\nu t + \frac{\delta - \gamma}{2}\xi t \right] \right\} d\nu\, d\xi .$$

By assumption, the searched distribution supports a linear association with the spectral distribution function (and with the autocovariance). By analogy with the deterministic case, imposing the compatibility with the time-frequency shifts leads to a class of admissible definitions

$$\rho_x(t,\nu) = \iiint_{\mathbb{R}^3} e^{i2\pi\tau(\zeta - \nu)} f(\xi,\tau)\, \Phi_x\left(\zeta - \frac{\xi}{2}, \zeta + \frac{\xi}{2}\right) e^{-i2\pi\xi t}\, d\xi\, d\zeta\, d\tau ,$$

$$\tag{2.154}$$

Chapter 2 Classes of Solutions

where $f(\xi, \tau)$ is an arbitrary parameter function. Under the given hypothesis and for the considered transformations, we can deduce that the autocovariance function has the form

$$r_x(t_1, t_2) = \iint_{\mathbb{R}^2} e^{i2\pi\tau\nu} f(\xi, \tau) \, \Phi_x \left(\nu - \frac{\xi}{2}, \nu + \frac{\xi}{2}\right) e^{-i2\pi\xi t} \, d\xi \, d\nu \; .$$

When we compare this expression to that of Eq. (2.153), we find the necessary conditions

$$\alpha + \beta = 1 \; , \quad \frac{\gamma - \delta}{2} = 1 \; ,$$

or equivalently

$$\frac{\alpha - \beta}{2} = \alpha - \frac{1}{2} \; , \quad \gamma + \delta = 2(\gamma - 1) \; .$$

Thus the admissible distributions fall in the general class of representations, which are covariant with respect to translations and have a parameter function of the form

$$f(\xi, \tau) = \exp\left\{i2\pi\left[\left(\alpha - \frac{1}{2}\right)\xi\tau - 2(\gamma - 1)\nu t\right]\right\} \; .$$

As this function must be independent of t and ν, we must further insist on $\gamma = 1$. This is compatible with the condition of isometry, because it implies $\det M_t = -1$. We end up with the set of parameter values

$$\{ \, \alpha, \; \beta = 1 - \alpha, \; \gamma = 1, \; \delta = -1 \, \} \; .$$

Subject to the given constraints, the class of admissible representations is characterized by the parameter function

$$f(\xi, \tau) = e^{i2\pi s \xi \tau} \; , \quad s = \alpha - \frac{1}{2} \; . \tag{2.155}$$

Here we recognize the precise form of a distribution of s-Wigner type. Its explicit representation can be given in two equivalent ways, namely by

$$\mathbf{W}_x^{(s)}(t, \nu) = \int_{-\infty}^{+\infty} \Phi_x\left(\nu - \left(s + \tfrac{1}{2}\right)\xi, \, \nu - \left(s - \tfrac{1}{2}\right)\xi\right) e^{-i2\pi\xi t} \, d\xi \tag{2.156}$$

or

$$\mathbf{W}_x^{(s)}(t, \nu) = \int_{-\infty}^{+\infty} r_x\left(t - \left(s - \tfrac{1}{2}\right)\tau, \, t - \left(s + \tfrac{1}{2}\right)\tau\right) e^{-i2\pi\nu\tau} \, d\tau \; . \tag{2.157}$$

These two equations define a class of representations that depend on one scalar parameter s. As in the deterministic case, the specification of this parameter leads to different definitions and can be guided by some additional constraints. In particular, the cases $s = 0$ and $s = 1/2$, respectively, define the *Wigner-Ville spectrum*

$$\mathbf{W}_x(t, \nu) = \int_{-\infty}^{+\infty} \Phi_x\left(\nu - \frac{\xi}{2}, \nu + \frac{\xi}{2}\right) e^{-i2\pi \xi t} d\xi$$
$$= \int_{-\infty}^{+\infty} r_x\left(t + \frac{\tau}{2}, t - \frac{\tau}{2}\right) e^{-i2\pi \nu \tau} d\tau$$
(2.158)

and, by analogy with Eq. (2.5), the *Rihaczek spectrum*

$$\mathbf{R}_x(t, \nu) = \int_{-\infty}^{+\infty} \Phi_x(\nu - \xi, \nu) e^{-i2\pi \xi t} d\xi$$
$$= \int_{-\infty}^{+\infty} r_x(t, t - \tau) e^{-i2\pi \nu \tau} d\tau \ .$$
(2.159)

In general, there is a one-to-one correspondence between the s-Wigner spectra on the one side and the autocovariance and spectral distribution functions on the other side because

$$r_x(t_1, t_2) = \int_{-\infty}^{+\infty} \mathbf{W}_x^{(s)}\left(\left(\tfrac{1}{2} + s\right) t_1 + \left(\tfrac{1}{2} - s\right) t_2, \nu\right) e^{i2\pi \nu(t_1 - t_2)} d\nu \ ,$$

$$\Phi_x(\nu_1, \nu_2) = \int_{-\infty}^{+\infty} \mathbf{W}_x^{(s)}\left(t, \left(\tfrac{1}{2} - s\right) \nu_1 + \left(\tfrac{1}{2} + s\right) \nu_2\right) e^{-i2\pi(\nu_1 - \nu_2)t} dt \ .$$

In particular, the relations of marginal distributions

$$\begin{cases} \mathbf{E}\left\{|x(t)|^2\right\} = \int_{-\infty}^{+\infty} \mathbf{W}_x^{(s)}(t, \nu) \, d\nu \\ \Phi_x(\nu, \nu) = \int_{-\infty}^{+\infty} \mathbf{W}_x^{(s)}(t, \nu) \, dt \end{cases}$$
(2.160)

hold for all values of s.

Remark. These two properties could have been extrapolated directly from the deterministic case by means of the form of the parameter function: It suffices to notice that $f(\xi, 0) = f(0, \tau) = 1$. Similarly, we can assert that the s-Wigner spectra are compatible with the changes of scale, linear filterings, and multiplicative modulations. They finally reduce to the usual power spectrum in the case of stationary signals, and have the property of conservation of supports (in the wide sense) if $|s| \leq 1/2$.

We can thus see that the s-Wigner spectra feature a large number of attractive properties. The main missing properties (in the general case) are the reality and the positivity. However, we will see that the imposition of certain of these constraints (or of new ones) is enough to restrict the free parameter s to be 0.

Wigner-Ville spectrum. Within the class of s-Wigner spectra, the task of finding conditions for the uniqueness of the Wigner-Ville spectrum is trivial, as it reduces to the search for constraints that require s to be zero. Several such constraints can be imagined, and they lead to the following assertions.

(i) *The Wigner-Ville spectrum is the only s-Wigner spectrum that is real-valued.* Indeed, if we look at the definition equation (2.157) based on the autocovariance, the requirement of reality of the spectrum reduces to the identity

$$r_x\left(t - \left(s - \tfrac{1}{2}\right)\tau, t - \left(s + \tfrac{1}{2}\right)\tau\right) = r_x\left(t + \left(s + \tfrac{1}{2}\right)\tau, t + \left(s - \tfrac{1}{2}\right)\tau\right),$$

which must be attained for every signal. This leaves as the only solution $s = 0$.

(ii) *The Wigner-Ville spectrum is the only s-Wigner spectrum whose local first-order moment (center of gravity) in the frequency domain is the mean instantaneous frequency.* In order to obtain this result, it is suitable to adapt the definitions of instantaneous frequency and group delay to random signals. Let us reconsider the instantaneous frequency of a real signal. As we have seen in Subsection 1.2.1, the usual definition for a deterministic signal proceeds by computing the derivative of the phase of the analytic signal, hence

$$\nu_x(t) = \frac{1}{2\pi} \frac{d \arg z_x}{dt}(t)$$

with $z_x(t) = x(t) + i\,\mathrm{H}\{x(t)\}$, and H denoting the Hilbert transform. Equivalently, we find that

$$\nu_x(t) = \frac{1}{2\pi} \frac{d}{dt} \arctan\left\{\frac{\mathrm{H}\{x(t)\}}{x(t)}\right\},$$

hence,

$$\nu_x(t) = \frac{1}{2\pi|z_x(t)|^2} \operatorname{Im}\left\{\frac{dz_x}{dt}(t)\, z_x^*(t)\right\}.$$

An appropriate definition of the instantaneous frequency of a random signal can thus be given by the stochastic quantity

$$\nu_x(t) = \frac{1}{2\pi \mathbf{E}\{|z_x(t)|^2\}} \operatorname{Im}\left\{\frac{dz_x}{dt}(t)\, z_x^*(t)\right\},$$

if the random signal is differentiable in the sense of a quadratic mean.[64] Its expectation value equals

$$\mathbf{E}\{\nu_x(t)\} = \frac{1}{2\pi \mathbf{E}\{|z_x(t)|^2\}} \operatorname{Im}\left\{\left.\frac{\partial r_{z_x}}{\partial s}(t,s)\right|_{s=t}\right\}. \tag{2.161}$$

Consequently, a direct computation shows that

$$2\pi \int_{-\infty}^{+\infty} \nu W_x^{(s)}(t,\nu)\, d\nu = i\left.\frac{\partial r_{z_x}}{\partial t}\left(t - (s-\tfrac{1}{2})\tau,\, t - (s+\tfrac{1}{2})\tau\right)\right|_{\tau=0}$$

$$= \frac{1}{2i}\left[(1-2s)\frac{\partial r_{z_x}}{\partial s}(t,s) - (1+2s)\frac{\partial r_{z_x}^*}{\partial s}(t,s)\right]_{s=t}.$$

This yields the announced result

$$\frac{1}{|z_x(t)|^2}\int_{-\infty}^{+\infty} \nu \mathbf{W}_x^{(s)}(t,\nu)\, d\nu = \mathbf{E}\{\nu_x(t)\}, \tag{2.162}$$

provided that the condition $s = 0$ is satisfied.

Remark. Again one could have drawn one's inspiration directly from the deterministic case by noticing that

$$f(\xi,\tau) = e^{i2\pi s\xi\tau} \implies \frac{\partial f}{\partial \tau}(\xi,0) = i2\pi s\xi.$$

This quantity is zero if and only if $s = 0$.

As in the deterministic case, it is worthwhile to remember that the uniqueness of the Wigner-Ville spectrum makes sense only relative to the imposed constraints. Some different viewpoints could lead to other solutions with their own advantages and disadvantages. As an example, the additional constraint of causality in the Fourier relation between the spectrum and the autocovariance function would lead to the definition

$$p_x(t,\nu) = \int_0^{+\infty} r_x(t, t-\tau)\, e^{-i2\pi\nu\tau}\, d\tau.$$

This is the counterpart of the deterministic Page distribution for the case of random signals. Apart from the causality, however, the Page spectrum

Chapter 2 Classes of Solutions   167

does not possess as many good features for applications as the Wigner-Ville spectrum.

In order to emphasize these good properties once more, it might be interesting to illustrate the behavior of the Wigner-Ville spectrum and its possible features in signal analysis by looking at some simple and typical examples.

Locally stationary signals. The corresponding form of Eq. (1.44) for the autocovariance is

$$\mathbf{E}\left\{x\left(t+\frac{\tau}{2}\right)x^*\left(t-\frac{\tau}{2}\right)\right\} = m_x(t)\,\gamma_x(\tau)\;.$$

It leads immediately to the result

$$\mathbf{W}_x(t,\nu) = m_x(t)\,\Gamma_x(\nu)\;,$$

and this agrees nicely with the physical interpretation of a stationary power spectrum, which is modulated in time.

Uniformly modulated signals. In this case we have

$$\mathbf{W}_x(t,\nu) = \int_{-\infty}^{+\infty} c\left(t+\frac{\tau}{2}\right)c\left(t-\frac{\tau}{2}\right)\gamma_x(\tau)\,e^{-i2\pi\nu\tau}\,d\tau\;.$$

Consequently, if the evolution of $c(t)$ is slow in comparison with the radius of correlation of $x(t)$, we obtain the approximate form

$$\mathbf{W}_x(t,\nu) \approx c^2(t)\,\Gamma_x(\nu)\;.$$

This can almost be identified with a Priestley spectrum (cf. Eq. (2.143)).

Filtered white noise. Let us consider the quantity

$$x(t) = \int_{-\infty}^{t} h(t,s)\,dB(s)\;,\quad \mathbf{E}\left\{dB(t)\,dB(s)\right\} = \sigma^2\,\delta(t-s)\,dt\,ds\;.$$

Physically it corresponds to the filtering of a white noise ("derivative of Brownian motion $B(t)$") by a filter with an (eventually time-varying) impulse response $h(t,s)$. In general, it also furnishes a Cramér representation of nonstationary random signals in continuous time with multiplicity one. For such signals the Wigner-Ville spectrum can be written as

$$\mathbf{W}_x(t,\nu) = \sigma^2 \int_{-\infty}^{+\infty} \left[\int_{-\infty}^{t-|\tau|/2} h\left(t+\frac{\tau}{2},s\right)h^*\left(t-\frac{\tau}{2},s\right)ds\right] e^{-i2\pi\nu\tau}\,d\tau\;.$$

Its explicit form depends on the nature of the considered filter.

Let us first look at the special case where

$$h(t,s) = h_\mathrm{s}(t-s)\, U(t)\, U(s)\ ,$$

which corresponds to the activation of a time-invariant linear filter with impulse response $h_\mathrm{s}(t)$ at the initial moment $t=0$. Then we obtain the simple form of the spectrum

$$\mathbf{W}_x(t,\nu) = \sigma^2 \int_0^t W_{h_\mathrm{s}}(s,\nu)\, ds\ U(t)\ . \qquad (2.163)$$

The preceding relation shows, in particular, that the evolutionary character of the response to a white noise is restricted to a duration T, provided that the filter has a finite impulse response with the same duration T. Beyond this transient period, the steady state of the filter is effectively attained: It is manifested in the Wigner-Ville spectrum by the time-invariance of a certain frequential property, which is closely related to the power density of the frequency response of the filter. Indeed, a straightforward argument shows that

$$t > T \quad \Longrightarrow \quad \mathbf{W}_x(t,\nu) = \sigma^2 |H(\nu)|^2\ .$$

On the other hand, if we consider the perfect integrator

$$h_\mathrm{s}(t) = U(t)\ ,$$

the transient period lasts *forever*, and it actually defines the Brownian motion or *Wiener-Lévy process*

$$B(t) = \int_0^t h(t,s)\, dB(s)\ U(t)\ .$$

In this case, the Wigner-Ville spectrum admits the simple (positive) form

$$\mathbf{W}_B(t,\nu) = 2\sigma^2 t^2 \left(\frac{\sin 2\pi|\nu|t}{2\pi|\nu|t}\right)^2 U(t)\ . \qquad (2.164)$$

Here the time-dependence is evident and expresses the nonstationary character of the Brownian motion.

Let us also observe that the Wigner-Ville spectrum puts us into a position to define the notion of a *mean spectrum* of a nonstationary signal. We simply let

$$\overline{\Gamma}_x(\nu) = \lim_{T\to+\infty} \frac{1}{T} \int_0^T \mathbf{W}_x(s,\nu)\, ds\ . \qquad (2.165)$$

Chapter 2 Classes of Solutions

In the case of the Brownian motion, a direct computation shows that

$$\overline{\Gamma}_B(\nu) = \frac{\sigma^2}{4\pi^2\nu^2}, \qquad (2.166)$$

and this establishes a nice correspondence with the behavior "like $1/f^2$" of the "empirical" power spectrum of a Brownian motion.

Impulsive noise. A last and simple (but physically important) example of a nonstationary signal shall be considered next: It is called *impulsive noise* and can be regarded as a filtering of a Poisson point process. Its representation in time has the general form

$$x(t) = \sum_{k=-\infty}^{+\infty} h(t - t_k),$$

where $h(t)$ is the impulse response of a linear filter, and the instants t_k are distributed according to a Poisson rule of variable density $\lambda(t)$. It follows that the associated Wigner-Ville spectrum acquires the form

$$\mathbf{W}_x(t,\nu) = \int_{-\infty}^{+\infty} W_h(s,\nu)\,\lambda(t-s)\,ds.$$

Hence, it is the convolution of the Wigner-Ville distribution of the impulse response of the filter and the density of the Poisson impulses. In case of a slow variation of the density in comparison with the temporal support of the filter, the spectrum has the approximate value

$$\mathbf{W}_x(t,\nu) \approx \lambda(t)\,|H(\nu)|^2.$$

It thus exposes a locally stationary situation, which of course reduces to the stationary case when the density is constant. When we consider the intermediate case

$$\lambda(t) = \lambda\,U(t),$$

where a constant density is used from the activation of the filter (initially at rest) at time $t = 0$, we obtain that

$$\mathbf{W}_x(t,\nu) = \lambda \int_0^t W_h(s,\nu)\,ds\,U(t).$$

Hence, one can recognize a situation which is formally equivalent to the filtered white noise (cf. Eq. (2.163)), except that a replacement of the power spectrum of the white noise with the density of the Poisson impulses has been performed.

2.4.4. Some Links Between the Different Spectra

Although the different spectra discussed before were obtained by arguments of a different nature (orthogonality vs harmonizability), we can still try to establish certain links between them, at least in special cases.

Continuous time. [65] Let us first dwell on the doubly orthogonal decomposition of equation (2.137). We already noticed (cf. Eq. (2.139)) that the associated autocovariance function can be written as

$$r_x(t,s) = \int_{-\infty}^{+\infty} \psi(t,\xi)\, \psi^*(s,\xi)\, \Gamma_x(\xi)\, d\xi \ .$$

By inserting this expression into the definition of the Wigner-Ville spectrum, we derive the result

$$\mathbf{W}_x(t,\nu) = \iint_{\mathbb{R}^2} \psi\left(t+\frac{\tau}{2},\xi\right) \psi^*\left(t-\frac{\tau}{2},\xi\right) e^{-i2\pi\nu\tau}\, \Gamma_x(\xi)\, d\xi\, d\tau \ .$$

Written in a different form, this is the same as

$$\mathbf{W}_x(t,\nu) = \int_{-\infty}^{+\infty} V(t,\nu;\xi)\, K_x(t,\xi)\, d\xi \ , \qquad (2.167)$$

where the notation for the Karhunen spectrum $K_x(t,\xi) = |\psi(t,\xi)|^2\, \Gamma_x(\xi)$ is used as before, and where we let

$$V(t,\nu;\xi) = \frac{1}{|\psi(t,\xi)|^2} \int_{-\infty}^{+\infty} \psi\left(t+\frac{\tau}{2},\xi\right) \psi^*\left(t-\frac{\tau}{2},\xi\right) e^{-i2\pi\nu\tau}\, d\tau \ . \qquad (2.168)$$

We have thus established a formal relation that enables us to pass from the Karhunen spectrum, which is based on the double orthogonality, to the Wigner-Ville spectrum, which prioritizes the frequential interpretation. Let us next extrapolate this relation to oscillatory signals. They are defined, as we recall from Eq. (2.141), by letting

$$\psi(t,\nu) = A(t,\nu) e^{i2\pi\nu t} \ .$$

Then we can find the analogous identity

$$\mathbf{W}_x(t,\nu) = \int_{-\infty}^{+\infty} V(t,\nu;\xi)\, \Xi_x(t,\xi)\, d\xi \ ; \qquad (2.169)$$

here $\Xi_x(t,\xi)$ is the Priestley spectrum and

$$V(t,\nu;\xi) = \frac{1}{|A(t,\xi)|^2} \int_{-\infty}^{+\infty} A\left(t+\frac{\tau}{2},\xi\right) A^*\left(t-\frac{\tau}{2},\xi\right) e^{-i2\pi(\nu-\xi)\tau} d\tau \ . \tag{2.170}$$

Let us restrict our attention to oscillatory signals with sufficiently slow variation, so that we can use the approximation

$$A\left(t+\frac{\tau}{2},\xi\right) A^*\left(t-\frac{\tau}{2},\xi\right) \approx |A(t,\xi)|^2 \ .$$

Then the identity equation (2.169) can be simplified to

$$\mathbf{W}_x(t,\nu) \approx \Xi_x(t,\nu) \ . \tag{2.171}$$

This simplification is in tandem with previous statements regarding the equivalence of both spectra, the Priestley and the Wigner-Ville, in quasi-stationary situations.

Discrete time. If we start from the Cramér decomposition (which underlies the Tjøstheim and Mélard spectrum) of a real and discrete signal

$$x[n] = \sum_{m=-\infty}^{n} h[n,m]\, e[m] \ ,$$

we obtain the representation

$$r_x[j,k] = \sum_{m=-\infty}^{\min(j,k)} h[j,m]\, h[k,m]$$

for the autocovariance. Having such a model at our disposal, we can incorporate it into the definitions of spectra, which are based on the autocovariance. In particular, if we consider the discrete-time version of the Wigner-Ville spectrum, which reads as

$$\mathbf{W}_x[n,\nu] = 2 \sum_{k=-\infty}^{+\infty} r_x[n+k, n-k]\, e^{-i4\pi\nu k} \ , \tag{2.172}$$

we derive that

$$\mathbf{W}_x[n,\nu] = 2 \sum_{k=-\infty}^{+\infty} \sum_{m=-\infty}^{n-|k|} h[n+k,m]\, h[n-k,m]\, e^{-i4\pi\nu k} \ . \tag{2.173}$$

This expression should be compared with the Tjøstheim and Mélard spectrum (Eq. (2.145)), which can be rewritten as

$$\Theta_x[n,\nu] = 2 \sum_{k=-\infty}^{+\infty} \sum_{m=-\infty}^{n-|k|} h[n, m-k]\, h[n, m+k]\, e^{-i4\pi\nu k} \ . \qquad (2.174)$$

These two representations reflect the same structure. They are only distinguishable by the blend of their indices for the summation: this exhibits the fundamentally causal (noncausal, respectively) character of the Tjøstheim and Mélard spectrum (the Wigner-Ville spectrum, respectively). We can extend this comparison by referring to another causal representation, which belongs to the same general class as the Wigner-Ville spectrum: the Page spectrum. By analogy with the definition in continuous time, let us use the relation

$$P_x[n,\nu] = 2\,\mathrm{Re}\left\{ \sum_{m=-\infty}^{n} r_x[n,m]\, e^{-i2\pi\nu(n-m)} \right\} - r_x[n,n]$$

as its definition in the discrete case. This leads to an expansion of the form

$$P_x[n,\nu] = \sum_{k=-\infty}^{+\infty} \sum_{m=-\infty}^{n-|k|} h[n-|k|, m]\, h[n, m]\, e^{-i2\pi\nu k} \ . \qquad (2.175)$$

Again, there appears a structural similarity and a distinction in the blend of indices for the summation.

Let us regard the Cramér decomposition of a nonstationary signal as being characterized by an (infinite) lower triangular matrix

$$H = \begin{pmatrix} \ddots & & & & \\ \cdots & h[n-1, n-1] & & 0 & \\ \cdots & h[n, n-1] & h[n, n] & & \\ \cdots & h[n+1, n-1] & h[n+1, n] & h[n+1, n+1] & \\ \cdots & \vdots & \vdots & \vdots & \ddots \end{pmatrix} \qquad (2.176)$$

The three spectra (Tjøstheim-Mélard, Page, and Wigner-Ville) are distinguishable by the information content of H that they use. More precisely,

 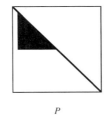 

TM   P   WV

Figure 2.12. Different spectra.

Symbolic comparison of the information used by the definitions of Tjøstheim-Mélard (TM), Page (P) and Wigner-Ville (WV) for the construction of a time-dependent spectrum, which is based on the Cramér decomposition of a nonstationary discrete signal. In each case, the square represents the matrix H, and the portion drawn in black depicts the part of that matrix used by the analysis.

for calculating the spectrum at the moment n, the approach by Tjøstheim and Mélard uses only the elements with a row index n, the one by Page also brings the elements of the upper triangle into play, and the Wigner-Ville spectrum is the only one that, by its noncausal character, uses the information of the lower triangle as well (cf. Fig. 2.12).

This highlights the differences that can result from causal or noncausal approaches. The causality can intervene on two different levels:

(i) The causality of the system generating the signal: Only the past innovation can affect the present state; and

(ii) The causality of the transformation, which determines the spectrum of the signal at each moment: Only the signal's past can affect the present value of the spectrum.

Chapter 2 Notes

2.1.1.

[1] It is difficult to locate precisely the first appearance of the methods of "instantaneous spectra" in the literature. The allusion to Sommerfeld is borrowed from the book by Pimonow (1962), where one can find an extremely important bibliography on the subject, which includes in particular measuring apparatus. [It is surprising, however, that there is no reference to the fundamental article (Ville, 1948).] The first articles on the "sound spectrograph" or sonagraph were Koenig, Dunn, and Lacy (1946) and Potter, Kopp, and Green (1947). We mentioned the FTAN method proposed in Levshin, Pisarenko, and Pogrebinsky (1972). But we should mention that this variant is very similar, in principle, to other approaches such as Gambardella (1968).

[2] There are many examples of applications of spectrograms/sonagrams. The most significant and best known ones, for which these methods were invented, concern human speech, or in a wider sense, any audio-acoustic signal. There are books such as the one by Potter, Kopp, and Green (1947) devoted to the "visible speech," or Pimonow (1962). A more general synopsis of the application of time-frequency methods to speech signals can be found in d'Alessandro and Demars (1992).

[3] The reading from a spectrogram was particularly well developed by Zue and the group at MIT (see, for example, Zue and Cole, 1979; or Lamel, 1988).

[4] As far as the auditory models in terms of filter banks are concerned, we refer to Flanagan (1972) and d'Alessandro (1989) for a first study.

2.1.2.

[5] The first continuous generalizations, which followed from the initial work by Gabor (1946), appeared in Helström (1966) and Montgomery and Reed (1967). The decomposition into the prolate spheroidal wave functions is due to Lacoume and Kofman (1975). The original articles are the ones by Bastiaans (1980) and Balian (1981).

[6] Although "wavelet theory" emerged only recently as an established scientific field, it has already undergone a rich history. An important impact comes from the fact that it offers a forum to workers with rather different backgrounds (physics, mathematics, signal, etc.). One of the inaugurating articles was presented by Grossmann and Morlet (1984). An excellent introduction to the theory, coupled with a fascinating historical perspective, can be found in the small book by Meyer (1993a).

2.1.3.

[7] The Ville distribution was defined for the first time in the article by Ville (1948) quoted here, and the Wigner distribution in Wigner (1932). Bass was one of the first who devoted much of his work to the Wigner distribution (during the same period as Ville, but obviously without any connection to him). He introduced the notion of the local momentum in Bass (1945).

[8] The original articles are by Claasen and Mecklenbräuker (1980a,b,c). They constitute the first complete synopsis of the usefulness of the Wigner-Ville distribution to signal theory. They were first published in the *Philips Journal of Research*, and were later more widely disseminated in international conferences (Claasen and Mecklenbräuker, 1981). Indirectly, they invigorated the more confidential work carried out at the same time by some French groups (cf. Escudié and Gréa, 1976; Bouachache, Escudié, and Komatitsch, 1979; and Flandrin and Escudié, 1980). One can assess the development of the studies in this area that followed until 1985 by consulting the cumulative bibliography in the survey article by Mecklenbräuker (1987). It can also be noticed that this date marked a slowdown of the exponentially growing activity experienced until then. As a matter of fact, this slowdown was intensified by the growing interest in the new wavelet theory. On the other hand, but only later on, the excursion to the land of the wavelets gave a fresh impetus to the methods "à la Wigner," which is to be regarded as a natural balancing effect. One can thus experience a renewed interest in these methods since the beginning of the 1990s, as is now testified by the compilation of references in the survey article by Hlawatsch and Boudreaux-Bartels (1992) and in the volume by Mecklenbräuker and Hlawatsch (1998).

[9] The original article is by Page (1952), but the same ideas were reintroduced several times, see, for example, Grace (1981). The related "anti-causal" version was initially introduced in Blanc-Lapierre and Picinbono (1955), then independently in Levin (1967).

[10] The Rihaczek distribution was proposed in Rihaczek (1968) based on purely theoretical arguments. As far as we know, the original publication was left totally isolated and has not led to any follow-up studies, not even by the author. An attempt was made at the beginning of the 1980s to use it for an application called a "phonochrome" (Johannesma *et al.*, 1981). But it only found mediocre success due to the bad visual quality of the representation. Extensions or modifications were proposed more recently in Hippenstiel and de Oliveira (1990) and Auger and Doncarli (1992).

[11] There is a very interesting discussion of the sheer notion of a "spectrum" of a nonstationary process in Loynes (1968). [One should also read the

discussion that follows this article.] Among other general and available references we also might mention those by Priestley (1988), Grenier (1987), or Flandrin (1989a).

[12] The original article is by Cohen (1966). It was presented in the setting of quantum mechanics. The same author gave a "signal" formulation later on (which was introduced in Escudié and Gréa (1976) in parallel). Cohen (1984) has further proceeded with the development of its properties. Since then he has made numerous contributions. A good source is Cohen's review article (1989) and his book (1995). There also exist other general articles and/or books dealing with this subject, for example, Longo and Picinbono (1989), Hlawatsch and Boudreaux-Bartels (1992), Boashash (1992c), or Mecklenbräuker and Hlawatsch (1998). The approach by J. Bertrand and P. Bertrand was started in P. Bertrand (1983), and the tomographic construction was especially laid out by J. Bertrand and P. Bertrand (1987); the same authors gave a recent survey of their work (1992a,b).

2.1.4.

[13] Besides the already quoted and more recent work by L. Cohen (1992), there is a comparative approach of "quantum mechanics/signal theory" in Mourgues (1987). The referenced work by Born and Jordan is from their original article (1925).

[14] The quantum theoretical distribution, which anticipates the Rihaczek distribution, was proposed in Margenau and Hill (1961). On the other hand, the reason that there was never any analogue of the Page distribution comes from the difficulty of giving a suitable interpretation of the notion of causality *in space*. (The same difficulty occurs in image processing.) A similar difficulty arises, though different by nature, when one tries to transfer the notion of the spectrogram, as it was done in Kuryshkin (1972; 1973). We refer to Ivanovic (1983) for a possible interpretation in terms of *propensity*.

2.2.1.

[15] For a more rigorous treatment of the approach presented here, and to know more about the theory of representations in connection with the transformation groups, the reader is referred to Grossmann, Morlet, and Paul (1985; 1986), for example. These references also provide an introduction to the notions of invariant measure and reproducing kernel.

[16] A more detailed description of the theory of frames ("structures obliques" in French) and their applications are given in Daubechies, Grossmann, and Meyer (1985) and Daubechies (1992).

2.2.2.

[17] Among the abundant literature on the short-time Fourier transform we quote the surveys in Allen and Rabiner (1977), Portnoff (1981), and Nawab and Quatieri (1988).

[18] It is difficult to identify the roots of the condition for the mixed reconstruction equation (2.20) precisely. It "seems to have been part of the common knowledge for a very long time, just like the tales of the brothers Grimm" (A. Grossmann, private communication).

[19] The conditions on the density of the sampling lattice for the short-time Fourier transform are discussed, for instance, in Bastiaans (1981).

[20] The obstruction by Balian and Low was originally described by Balian (1981). A complete proof is given in Daubechies (1992).

[21] We can refer to Jensen, Hoholdt, and Justesen (1988) as one of the examples for the construction of orthonormal bases "à la Gabor." Nevertheless, it must be noted that the Balian-Low obstruction can be circumvented by using *Wilson bases*. They employ suitably adjusted sines and cosines instead of the complex exponentials entering Eq. (2.22) (Daubechies, Jaffard, and Journé, 1991).

[22] The solution can be looked up in Bastiaans (1980).

[23] A summary of the Zak transform, viewed from a signal perspective, can be found in Janssen (1988b).

[24] Gabor's original article (1946) proposes an iterative computation of the coefficients of the decomposition. The issue of how the dual basis depends on the density of the time-frequency lattice is explained in Daubechies (1992). See also the collection in Feichtinger and Strohmer (1998).

2.2.3.

[25] The continuous wavelet transform was originally proposed by Morlet who called it "cycle-octave analysis" (1982). The first theoretical paper, which explores its capabilities and the admissibility conditions, is by Grossmann and Morlet (1984). A general presentation of the subject matter is Torrésani (1995).

[26] One can consult Grossmann, Morlet, and Paul (1986), for example, for a description of the affine group and its invariant measures.

[27] The conditions for the discretization of a wavelet transform are extensively investigated in Heil and Walnut (1989) and Daubechies (1992).

[28] The quoted numerical examples are taken from Daubechies (1992).

[29] The principle of an almost perfect reconstruction by oversampling and the use of several "voices per octave" is used, for instance, in Kronland-Martinet, Morlet, and Grossmann (1987).

[30] Our presentation faithfully follows Daubechies' book (1992). One can equally gain profit from reading the books by Chui (1992), A. Cohen (1992), Abry (1997), or Mallat (1998), not to forget those by Meyer (1990; 1993a,b), of course, which are just as exhaustive as introductory to the subject. These different books also deal with important extensions and/or variants of the theory (such as *wavelet packets*, *local trigonometric bases* or *Malvar wavelets*). We will not touch on any of those in this book.

[31] The concept of multiresolution analysis was introduced and perfected by Mallat and Meyer (see Mallat, 1989a,b).

[32] The "quadrature mirror filters" were originally introduced by Esteban and Galand (1977). They were the subject of many developments, which can be associated with the names of Smith and Barnwell (1986), Vetterli (1986), and Vaidyanathan (1993). The connection with wavelets was first observed by Mallat and explored more deeply by A. Cohen (1992). These aspects receive close attention in the books by Vetterli and Kovacevic (1995) and Strang and Nguyen (1996).

[33] The notion of a pyramidal algorithm, such as the Laplacian pyramid in Burt and Adelson (1983), existed in image processing before the advent of wavelets. Daubechies (1988a) was the first who brought these works together with those by Mallat.

[34] It should be remarked that a first basis of wavelets was discovered by Meyer (1990) shortly before the introduction of the Battle-Lemarié wavelets. At first sight it looked unique, as far as its construction relied on some "miraculous" properties. This is very well explained in Lemarié (1989).

[35] See, for example, Daubechies (1992) or Chui (1992).

[36] The first construction of orthonormal wavelet bases with compact support is due to Daubechies (1988a).

[37] See Daubechies (1992), where other consequences of the compact support are also investigated in more detail (such as the necessary growth of support as a function of regularity). As far as the location of the poles for a best possible approximation of a linear phase is concerned, one can also look at Dorize and Villemoes (1991). Let us finally remark that we only considered the case of orthonormal bases here. More flexible solutions can actually be obtained in a *biorthogonal* setting. This is described in Daubechies (1992) as well.

2.2.4.

[38] We quote Whalen (1971) and Van Trees (1968) as examples of general books that deal with the optimal decision theory (detection/estimation) and related concepts (maximum likelihood, matched filtering, etc.).

[39] There are two references in French that specifically deal with the processing of radar and/or sonar signals: Le Chevalier (1989) and Bouvet (1992). A more mathematical approach is given in Blahut Miller, and Wilcox (1991). Here one also finds a thorough treatment of the modeling of echoes in radar or sonar from the equations of physics.

[40] The definition was introduced by Woodward (1953). The notion of the ambiguity function initiated an abundant literature, particularly during the 1960s. As far as we know, however, there exists no book that is specially devoted to this subject. For learning more about it, one can still profit from the books by Vakman (1968) or Rihaczek (1969), which contain many illustrations.

[41] The original concept of wideband ambiguity functions goes back to Kelly and Wishner (1965) and Speiser (1967). A more profound investigation of the properties was performed by Altes (1971; 1973).

[42] An iterative selection algorithm for choosing the most pertinent atoms in a large dictionary was proposed by Mallat and Zhang (1993). The paper by Altes (1985) deals with the same issue and includes an easily accessible description.

2.3.

[43] No doubt the principle of these constructive approaches appears for the first time in the literature on quantum mechanics, as seen in Krüger and Poffyn (1976) and Ruggeri (1971). Later it is presented in different contexts of signal theory, such as in P. Bertrand (1983), Flandrin (1982; 1987), Hlawatsch (1988), or Rioul and Flandrin (1992).

2.3.1.

[44] See for example Duvaut (1991).

[45] Our presentation here follows Rioul and Flandrin (1992) for establishing the general form of a class of bilinear time-scale distributions. We have chosen it for at least two reasons. First, it accentuates the parallel with Cohen's class. Secondly, it simplifies its interpretation while explicitly depending on the scale parameter a, which also plays a role in the wavelet transform. However, when we endow the time-scale class with a time-frequency interpretation (an issue to which we will come back later),

we must observe that a suitable change of the parameterization can, in certain cases, lead to an equivalent formulation as Bertrand's class, which is considered in J. Bertrand and P. Bertrand (1988; 1992a) (see also Footnote 47).

2.3.2.

[46] More complete lists of definitions can be found in the surveys by Hlawatsch and Boudreaux-Bartels (1992), Auger (1991), and Boudreaux-Bartels (1996). Historically, the Ackroyd distribution was proposed in Ackroyd (1970), the s-Wigner distribution in Janssen (1982), the distribution with a separable kernel in Martin and Flandrin (1983) and Flandrin (1984), the one by Choï and Williams in 1989, by Zhao, Atlas, and Marks in 1990, and the Butterworth representation in Papandreou and Boudreaux-Bartels (1992). In a sense the Born-Jordan distribution was never really defined in its explicit form, as we mentioned in Subsection 2.1.4, prior to its actual use in Flandrin (1984).

[47] The concept of "affine" or "wideband time-frequency" distributions is due to J. Bertrand and P. Bertrand. They were the first to construct a family of time-frequency solutions based on a covariance principle relative to the affine group (see for example P. Bertrand, 1983; J. Bertrand and P. Bertrand, 1988; or Boashash, 1992a,b). The class of "affine Wigner distributions" defined here, with a different parameterization equation (2.63) than the one used by J. Bertrand and P. Bertrand, fits into the same family of representations. We should note that the general class of these distributions includes several definitions, such as the one by Unterberger (1984), which were proposed in a totally different context (symbol calculus).

[48] The systematic exploration of the properties of a distribution based on its parameter function was first started in Cohen (1966), then pursued and completed in Claasen and Mecklenbräuker (1980c) and subsequent work. Nowadays the (more or less complete) list of these results is contained in almost all surveys (see for example Flandrin and Martin, 1983a; Flandrin, 1987, 1991; Auger, 1991; Auger and Doncarli, 1992; Rioul and Flandrin, 1992; Hlawatsch and Boudreaux-Bartels, 1992; Cohen, 1995; Boudreaux-Bartels, 1996). Except for some special cases we do not try to retrace their genealogy systematically.

[49] The properties of positivity of the time-frequency distributions were especially studied by Janssen (1984b; 1985; 1988a; 1998).

[50] See, for example, Picinbono (1977).

[51] This result and the argument for its proof are borrowed from Auger (1991).

Chapter 2 Classes of Solutions

[52] Moyal's formula first appeared in Moyal (1949).

[53] The chosen approach was proposed in Flandrin and Escudié (1982). But one can as well consider other ways in which a good localization in the plane may be guaranteed. Some of them will be explained in the context of affine Wigner distributions (Eq. (2.122)). We wish to mention that there exist further solutions that are based on a *multilinear* extension of the Wigner-Ville distribution, for example, from Boashash and Ristic (1992) and Amblard and Lacoume (1992).

[54] There is a more systematic and formal treatment of this problem in J. Bertrand and P. Bertrand (1992b). The present objective is to simplify the approach in order to render its geometric interpretation more easily accessible. This will be further developed in Subsection 3.2.2.

2.3.3.

[55] The original proof of Wigner's Theorem was given in Wigner (1971). Its presentation and a complete discussion are contained in Mugur-Schächter (1977).

[56] A rather systematic discussion of the results of exclusion owing to the positivity of the distribution is given in Janssen (1988a).

2.4.1.

[57] Among the general references that provide a synthetical description of these two large classes of methods we refer to Grenier (1987), Flandrin (1989a), or Flandrin and Martin (1998). The paper by Matz, Hlawatsch, and Kozek (1997) offers a complementary point of view.

2.4.2.

[58] The notion of the evolutionary spectrum (in the sense of Priestley) was introduced in Priestley (1965). It is also contained in Priestley (1981; 1988). The discussion of the problems connected with the definition of oscillatory signals can be found in Battaglia (1979).

[59] For a description of the general notions of decomposition and multiplicity we refer to Cramér (1971).

[60] See Tjøstheim (1976), Mélard (1978).

[61] For an overview over these works one can consult Grenier (1983; 1986; 1987).

[62] Among the other approaches, which are not mentioned here, the *adaptive* methods come first. When they are used in a nonstationary setting,

they allow one to follow the spectral parameters in time. There is an abundant literature that is concerned with adaptive signal processing. We content ourselves with the quotation of Michaut (1992) from a general point of view. Furthermore, a synthesized description of the adaptive algorithms, used in an explicitly time-frequency context, is given in Basseville, Flandrin, and Martin (1992).

2.4.3.

[63] The approach based on the harmonizability was introduced by Martin (1982) at the beginning of the 1980s, and then further developed during subsequent years (see, for example, Martin and Flandrin, 1985b). We should mention, however, that the resulting definitions lead to the earlier solutions of Bastiaans (1978) and Mark (1970), at least in some cases such as the stochastic extension of the Wigner distribution.

[64] This definition is due to Broman (1981).

[65] See Hammond and Harrison (1985).

Chapter 3
Issues of Interpretation

The objective of this third chapter is a more profound study of the potential, the characteristics, and/or the limitations of several time-frequency tools that have already been constructed. At the same time we attempt to give some guidance concerning their application (which representation should be chosen?) and facilitate their interpretation (which information can be extracted?). As not all approaches can be covered in detail, we focus on the *bilinear* representations (Cohen's class and affine class) in this chapter.

More precisely, Section 3.1 describes different viewpoints concerning the bilinear classes of time-frequency and time-scale representations. Subsection 3.1.1 is devoted to a physical interpretation of their various parameterizations, depending on whether the parameter function is expressed in the time-frequency, time-time, frequency-frequency, or frequency-time plane. This emphasizes the complementarity of the interpretations in terms of energy distributions, on the one hand, and correlations, on the other hand. It also leads to generalizations of the well-understood case of stationary signals.

Then we take a different look at the parameter functions of the bilinear classes in Subsection 3.1.2 (which can be omitted during a first study). Here we explain how the arbitrariness of a time-frequency representation, as conveyed by the multiplicity of its parameterizations, is closely associated with the noncommutativity of the elementary operators related to the time and frequency variables. This arbitrariness in the choice of the parameterization is compared with the ambiguity in associating operators with more general time-frequency functions.

Subsection 3.1.3 partially uses this approach in order to reconsider the link between the two concepts of frequency and scale. This problem will be (briefly) discussed by other approaches like the Mellin transform as well.

In regard to its numerous theoretical properties, the Wigner-Ville transform plays a central role in the time-frequency theory. A better identification of its properties of "visualization" is therefore indispensable, if one wants to use it for the practical purpose of analyzing signals with an unknown structure. This issue is addressed in Section 3.2. Here we attack the main problems associated with the special structure of the Wigner-Ville distribution from a geometric point of view.

First (in Subsection 3.2.1) we compare the Wigner-Ville distribution with the spectrogram. We show by elaborating on simple examples, that it has certain advantages over the latter (supports, localization). However, we also underline certain improvements of the spectrogram (reassignment).

Then (in Subsection 3.2.2) we show that the same reason (nonlocal bilinearity) that guarantees the good theoretical properties of the Wigner-Ville distribution also affects it with interference structures (or cross-terms). This can reduce its readability and usefulness. This point is studied in detail by analyzing two simple models. One is related to separated time-frequency "atoms," while another deals with frequency-modulated signals. We succeed in establishing the construction principles of the Wigner-Ville transform explicitly (either exactly or by approximations by the method of stationary phase). We also discuss possible generalizations to other representations and further implications of this geometric approach concerning the localization in the plane.

The analysis carried out in Subsection 3.2.2 provides a rationale for the reduction of the interferences. Further considerations of this issue from several angles are the contents of Subsection 3.2.3 (use of the analytic signal, atomic decompositions, fixed or adapted smoothing in the bilinear classes).

It is often desirable to reduce the importance of the cross-terms in a time-frequency representation for improving its readability. On the other hand, Subsection 3.2.4 addresses the complementary issue, that these terms also carry some useful information that should be preserved for certain reasons. They allow us to find some phase information, and their pure existence leads to a better understanding of what a "component" of a signal might be.

We can think of the reduction of interferences as a problem of *geometric* estimation of the time-frequency structure of a signal. Another problem of *statistical* estimation is attacked in Subsection 3.2.5: Here we seek for estimates of the Wigner-Ville *spectrum* (which was defined in Chapter 2) from *one* observed realization. Under reasonable hypotheses such as the slow evolution of the nonstationary properties, a remarkable observation can be made. It tells that the general bilinear classes provide some natural families of estimators for the Wigner-Ville spectrum. This is connected

Chapter 3 Issues of Interpretation

with the geometric approach from before, but relies on a different perspective. We investigate the stochastic properties of first and second order of these estimators. Thus the arbitrariness of the parameterization is no longer expressed in terms of localization or reduction of interferences, but in terms of bias and variance of the estimator.

The eventual existence of negative values in a bilinear time-frequency representation is a severe obstacle. It sometimes renders a thorough interpretation of the cross-terms impossible, and it also prohibits a strict analogy with the notion of a joint probability density function. Several problems related to this question of positivity of the representation are discussed in Section 3.3.

Subsection 3.3.1 begins with an inventory of certain difficulties that arise from the nonpositivity. They occur, in particular, when we try to measure the (local or global) dispersion by means of the second-order moments.

In Subsection 3.3.2 we investigate the issue of positivity based on the analyzed signal. We show that, in general, the positivity is the exceptional case for deterministic signals (while justifying that the situation is less critical for the spectrum of random signals).

Finally, Subsection 3.3.3 addresses the problem of positivity by considering the distribution itself. Two different solutions are developed. The first consists in leaving the bilinear setting, which permits (without violating Wigner's Theorem) reconciling positivity and marginal properties. The second solution can be found inside Cohen's class. It is expressed by a smoothing of the Wigner-Ville distribution. This point is related to the forementioned stochastic approach, showing that a sufficient degree of disorder can assure the positivity of the spectrum.

§3.1. About the Bilinear Classes

The bilinear classes lend themselves to many interpretations, each rooted in some special interest. We will describe the most important ones in the following Subsections. [1]

3.1.1. The Different Parameterizations

The bilinear classes (Cohen's and the affine class) are, by definition, bilinear forms of the signal that depend on an arbitrary function of two variables. This function can evidently take different (though equivalent) forms according to the type of the attributed variables. Using the same notations as in Subsection 2.3.1, we obtain the following diagram in which each arrow represents a partial Fourier transform:

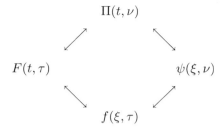

Here t and τ (ξ and ν, respectively) are the time (and frequency) variables.

In order to guarantee the consistency of this diagram when passing from one function to another, we must fix the following conventions for the signs of the involved partial or total Fourier transforms:

$$\Pi(t,\nu) = \iint_{\mathbf{R}^2} f(\xi,\tau) e^{i2\pi(\nu\tau+\xi t)} \, d\xi \, d\tau \tag{3.1}$$

$$F(t,\tau) = \iint_{\mathbf{R}^2} \psi(\xi,\nu) e^{-i2\pi(\nu\tau-\xi t)} \, d\xi \, d\nu \tag{3.2}$$

$$\psi(\xi,\nu) = \iint_{\mathbf{R}^2} F(t,\tau) e^{i2\pi(\nu\tau-\xi t)} \, dt \, d\tau \tag{3.3}$$

$$f(\xi,\tau) = \iint_{\mathbf{R}^2} \Pi(t,\nu) e^{-i2\pi(\nu\tau+\xi t)} \, dt \, d\nu \tag{3.4}$$

$$\Pi(t,\nu) = \int_{-\infty}^{+\infty} \psi(\xi,\nu) e^{i2\pi\xi t} \, d\xi = \int_{-\infty}^{+\infty} F(t,\tau) e^{i2\pi\nu\tau} \, d\tau \tag{3.5}$$

$$F(t,\tau) = \int_{-\infty}^{+\infty} f(\xi,\tau) e^{i2\pi\xi t} \, d\xi = \int_{-\infty}^{+\infty} \Pi(t,\nu) e^{-i2\pi\nu\tau} \, d\nu \tag{3.6}$$

$$\psi(\xi,\nu) = \int_{-\infty}^{+\infty} \Pi(t,\nu) e^{-i2\pi\xi t} \, dt = \int_{-\infty}^{+\infty} f(\xi,\tau) e^{i2\pi\nu\tau} \, d\tau \tag{3.7}$$

$$f(\xi,\tau) = \int_{-\infty}^{+\infty} F(t,\tau) e^{-i2\pi\xi t} \, dt = \int_{-\infty}^{+\infty} \psi(\xi,\nu) e^{-i2\pi\nu\tau} \, d\nu \; . \tag{3.8}$$

We thus have four parameterizations available (time-frequency, time-time, frequency-frequency, and frequency-time), which result in as many different ways of looking upon the bilinear classes.

Chapter 3 Issues of Interpretation

Time-frequency. We can write the bilinear classes employing the *time-frequency* parameterization $\Pi(t,\nu)$, which yields (cf. Subsection 2.3.1)

$$C_x(t,\nu;f) = \iint_{\mathbb{R}^2} \Pi(s-t,\xi-\nu)\, W_x(s,\xi)\, ds\, d\xi \,, \tag{3.9}$$

$$\Omega_x(t,a;f) = \iint_{\mathbb{R}^2} \Pi\left(\frac{s-t}{a}, a\xi\right) W_x(s,\xi)\, ds\, d\xi \,. \tag{3.10}$$

Likewise they can be defined as the *correlation* of the parameter function with the Wigner-Ville distribution of the signal. This interpretation results from the fact that both bilinear classes can be written in the form of an inner product in the time-frequency plane,

$$\rho_x(t,\lambda) = \iint_{\mathbb{R}^2} \Pi_{t\lambda}(s,\xi)\, W_x(s,\xi)\, ds\, d\xi \,,$$

where a reference object (the Wigner-Ville distribution of the signal) is compared with a family of analyzing objects. These analyzing objects are constructed by the action of a natural transformation group according to

$$\begin{cases} \lambda = \nu & \Longrightarrow \quad \Pi(s,\xi) \rightarrow \Pi_{t\nu}(s,\xi) = \Pi(s-t,\xi-\nu) \,, \\ \lambda = a & \Longrightarrow \quad \Pi(s,\xi) \rightarrow \Pi_{ta}(s,\xi) = \Pi\left(\dfrac{s-t}{a}, a\xi\right) . \end{cases}$$

In this sense, and by analogy with the discussion of Subsection 2.2.4, one could call the representation ρ_x (narrow- or wideband) *quadratic cross-ambiguity function* relative to the *time-frequency object* that is the Wigner-Ville distribution of the signal, not the signal itself.

Remark 1. The chosen *correlative* form was quite arbitrary. It could as well be replaced with a *convolutive* form, at least for Cohen's class. An interesting feature of the correlative form is its homogeneity: it allows a description of both classes by precisely those transformations that constitute the underlying group of the representations.

Remark 2. No matter if we regard this operation as correlation or convolution, we are not allowed, in general, to interpret it as a *smoothing* of the Wigner-Ville distribution; that is, it might not correspond to a bivariate filtering of *lowpass* type in the time-frequency plane. (As an example consider the Rihaczek distribution (cf. Table 2.1) for which $\Pi(t,\nu) = 2\exp(-i4\pi\nu t)$.) We must therefore refrain from drawing any conclusion that says that the Wigner-Ville distribution has better localization properties among all representations in the bilinear classes by virtue of the previous relations, because the other distributions look like smoothed versions of it.

Remark 3. Let us finally note that the central role, which the Wigner-Ville distribution seems to play within the bilinear classes, is quite arbitrary as well. Both classes can be constructed in a similar way by starting from any other invertible representation. We can use, for instance, any s-Wigner distribution (cf. Table 2.1), denoted $W_x^{(s)}(t,\nu)$ here, and obtain

$$\rho_x(t,\lambda) = \iint_{\mathbb{R}^2} \Pi_{t\lambda}^{(s)}(u,\xi)\, W_x^{(s)}(u,\xi)\, du\, d\xi \ .$$

We only need to modify the time-frequency parameter function according to

$$\Pi^{(s)}(t,\nu) = \int_{-\infty}^{+\infty} F(t-s\tau,\tau)\, e^{i2\pi\nu\tau}\, d\tau \ .$$

In particular, while the previous relations yield the expression

$$R_x(t,\nu) = 2 \iint_{\mathbb{R}^2} e^{-i4\pi(s-t)(\xi-\nu)}\, W_x(s,\xi)\, ds\, d\xi$$

for the Rihaczek distribution $R_x(t,\nu)$, we can also write, vice versa,

$$W_x(t,\nu) = 2 \iint_{\mathbb{R}^2} e^{i4\pi(s-t)(\xi-\nu)}\, R_x(s,\xi)\, ds\, d\xi \ .$$

This simple example demonstrates that the notion of a *generating* distribution of the bilinear classes must be further qualified. Moreover, it illustrates the fact that correlating an oscillatory kernel with a function, which itself can have negative values, does not permit any kind of interpretation in terms of a hierarchy of resolutions.

Time-time. If we start from the *time-time* parameterization $F(t,\tau)$, we can immediately rewrite the bilinear classes according to

$$C_x(t,\nu;f) = \iint_{\mathbb{R}^2} F(s-t,\tau)\, x\!\left(s+\frac{\tau}{2}\right) x^*\!\left(s-\frac{\tau}{2}\right) e^{-i2\pi\nu\tau}\, ds\, d\tau \ , \quad (3.11)$$

$$\Omega_x(t,a;f) = |a|^{-1} \iint_{\mathbb{R}^2} F\!\left(\frac{s-t}{a},\frac{\tau}{a}\right) x\!\left(s+\frac{\tau}{2}\right) x^*\!\left(s-\frac{\tau}{2}\right) ds\, d\tau \ . \quad (3.12)$$

A first interpretation of these relations, as in the previous case of a time-frequency representation, is in terms of a "correlation" or "inner product" in the parameter plane. For this purpose let us introduce the product

$$q_x(s,\tau) = x\!\left(s+\frac{\tau}{2}\right) x^*\!\left(s-\frac{\tau}{2}\right) \ . \quad (3.13)$$

Chapter 3 Issues of Interpretation

Then we can express every representation in the bilinear classes by the action of a linear operator on this function, namely

$$\rho_x(t,\lambda) = \iint_{\mathbb{R}^2} F_{t\lambda}(s,\tau)\, q_x(s,\tau)\, ds\, d\tau$$

where

$$\begin{cases} \lambda = \nu & \implies \quad F(s,\tau) \to F_{t\nu}(s,\tau) = F(s-t,\tau)\, e^{-i2\pi\nu\tau}, \\ \lambda = a & \implies \quad F(s,\tau) \to F_{ta}(s,\tau) = |a|^{-1} F\left(\dfrac{s-t}{a}, \dfrac{\tau}{a}\right). \end{cases}$$

The preceding form gives rise to another interpretation as a "Fourier transform of a local autocorrelation function." Indeed, we can rewrite $C_x(t, \nu; f)$ as

$$C_x(t,\nu;f) = \int_{-\infty}^{+\infty} \left[\int_{-\infty}^{+\infty} F(s-t,\tau)\, x\left(s+\frac{\tau}{2}\right) x^*\left(s-\frac{\tau}{2}\right) ds \right] e^{-i2\pi\nu\tau}\, d\tau.$$

If $F(s-t,\tau)$ were the constant 1, the term in brackets would simply reduce to the deterministic autocorrelation function

$$\gamma_x(\tau) = \int_{-\infty}^{+\infty} x\left(s+\frac{\tau}{2}\right) x^*\left(s-\frac{\tau}{2}\right) ds.$$

Taking its Fourier transform would thus yield the power spectrum of the signal; that is, $C_x(t,\nu;f) = \Gamma_x(\nu) = |X(\nu)|^2$. The presence of the additional term $F(s-t,\tau) \neq 1$ leads to two modifications of this scheme. The first is related to the first variable and indicates that the evaluation of the autocorrelation function is "local"; this means that it must be associated with the current date t. The other modification accounts for the existence of the second variable τ, whose role is an eventual weighting prior to the Fourier transform. For rendering these ideas more transparent, let us consider the special case of a separable parameterization

$$\Pi(t,\nu) = g(t) H(-\nu),$$

in which $g(t)$ (and $H(-\nu)$, respectively) ascertain a temporal (or frequential) smoothing. Then the associated representation takes the form

$$C_x(t,\nu;f) = \int_{-\infty}^{+\infty} h(\tau) \left[\int_{-\infty}^{+\infty} g(s-t)\, x\left(s+\frac{\tau}{2}\right) x^*\left(s-\frac{\tau}{2}\right) ds \right] e^{-i2\pi\nu\tau}\, d\tau.$$

Hence we recover the two essential and well-known ingredients that arise in nonparametric spectral estimation by a "weighted correlogram."[2] The representations of Cohen's class thus appear as natural extensions of the power spectrum to nonstationary signals, making it time-dependent by virtue of a local estimation.

In order that this point of view can be carried over to the affine class, we have to use the formal identification $\nu = \nu_0/a$. This can be achieved by putting

$$\Pi(t,\nu) = \Pi_0(t, \nu - \nu_0)$$

where $\Pi_0(t,\nu)$ is a time-frequency function of lowpass type with respect to the variable ν. Calling its partial (time-time) Fourier transform $F_0(t,\tau)$, we obtain

$$\Omega_x(t,a;f)$$
$$= \int_{-\infty}^{+\infty} \left[\int_{-\infty}^{+\infty} |a|^{-1} F_0\left(\frac{s-t}{a}, \frac{\tau}{a}\right) x\left(s+\frac{\tau}{2}\right) x^*\left(s-\frac{\tau}{2}\right) ds \right] e^{-i2\pi(\nu_0/a)\tau} d\tau.$$

The term in brackets once again measures the local and weighted autocorrelation function. The only difference is the scale-dependence of the local evaluation of the autocorrelation function and the weighting, which precedes the Fourier transform. When we consider the forementioned separable case, we can write

$$\Omega_x(t,a;f) = |a|^{-1/2} \int_{-\infty}^{+\infty} h_0\left(\frac{\tau}{a}\right) q_x^{(a)}(t,\tau) e^{-i2\pi(\nu_0/a)\tau} d\tau$$

with

$$q_x^{(a)}(t,\tau) = |a|^{-1/2} \int_{-\infty}^{+\infty} g\left(\frac{s-t}{a}\right) x\left(s+\frac{\tau}{2}\right) x^*\left(s-\frac{\tau}{2}\right) ds .$$

We thus see that the time-horizon of the smoothing brought into effect by the function g is not fixed, but depends on the scale (being shorter on finer scales). Likewise the useful range of delays τ, on which the Fourier transform operates (and which is fixed by the window h) is a function of the scale within the same proportions. From a spectral point of view, we capture the observation that the representation performs a broader frequential smoothing at higher frequencies.

Chapter 3 Issues of Interpretation

Frequency-frequency. The preceding description in the time-time setting has a natural counterpart of *frequency-frequency* type. Here we use the parameterization $\psi(\xi, \nu)$, leading to the expressions

$$C_x(t, \nu; f) = \iint_{\mathbf{R}^2} \psi(\xi, \zeta - \nu) \, X\left(\zeta - \frac{\xi}{2}\right) X^*\left(\zeta + \frac{\xi}{2}\right) e^{-i2\pi\xi t} \, d\xi \, d\zeta \;, \tag{3.14}$$

$$\Omega_x(t, a; f) = |a| \iint_{\mathbf{R}^2} \psi(a\xi, a\zeta) \, X\left(\zeta - \frac{\xi}{2}\right) X^*\left(\zeta + \frac{\xi}{2}\right) e^{-i2\pi\xi t} \, d\xi \, d\zeta \;. \tag{3.15}$$

By the definition of

$$Q_X(\xi, \zeta) = X\left(\zeta - \frac{\xi}{2}\right) X^*\left(\zeta + \frac{\xi}{2}\right)$$

we can write

$$\rho_x(t, \lambda) = \iint_{\mathbf{R}^2} \psi_{t\lambda}(\xi, \zeta) \, Q_X(\xi, \zeta) \, d\xi \, d\zeta$$

in the general case. Here the transformation rules are given by

$$\begin{cases} \lambda = \nu & \Longrightarrow \quad \psi(\xi, \zeta) \;\rightarrow\; \psi_{t\nu}(\xi, \zeta) = \psi(\xi, \zeta - \nu) \, e^{-i2\pi\xi t} \;, \\ \lambda = a & \Longrightarrow \quad \psi(\xi, \zeta) \;\rightarrow\; \psi_{ta}(\xi, \zeta) = |a| \, \psi(a\xi, a\zeta) \, e^{-i2\pi\xi t} \;. \end{cases}$$

Although a *frequential* autocorrelation function is less frequently used than its temporal counterpart, an analogous interpretation like the one in the time-time case can be given here, *mutatis mutandis*. We content ourselves with the remark that a frequential autocorrelation function emphasizes the spectral periodicity of a signal, while a usual autocorrelation function underlines the temporal periodicity. This point becomes important in the study of so-called *cyclo-stationary* signals, which are characterized by periodic stochastic properties without being stationary.[3]

Frequency-time. The fourth interpretation concerns the *frequency-time* parameterization as given by the function $f(\xi, \tau)$. The symmetrical ambiguity function

$$A_x(\xi, \tau) = \int_{-\infty}^{+\infty} x\left(s + \frac{\tau}{2}\right) x^*\left(s - \frac{\tau}{2}\right) e^{i2\pi\xi s} \, ds \tag{3.16}$$

plays a prominent role in the corresponding relations. It coincides with the two-dimensional Fourier transform of the Wigner-Ville distribution, that is,

$$A_x(\xi, \tau) = \iint_{\mathbf{R}^2} W_x(t, \nu) \, e^{i2\pi(\nu\tau + \xi t)} \, dt \, d\nu \;. \tag{3.17}$$

Owing to the fact that the Fourier transform maps a convolution into a product, we obtain (observing the conventions of the signs) that

$$C_x(t, \nu; f) = \iint_{\mathbb{R}^2} f(\xi, \tau) A_x(\xi, \tau) e^{-i2\pi(\nu\tau + \xi t)} d\xi\, d\tau\ , \qquad (3.18)$$

$$\Omega_x(t, a; f) = \iint_{\mathbb{R}^2} f\left(a\xi, \frac{\tau}{a}\right) A_x(\xi, \tau) e^{-i2\pi\xi t} d\xi\, d\tau\ . \qquad (3.19)$$

All representations in Cohen's class therefore admit a particularly simple interpretation: They are just the two-dimensional Fourier transform of a unique object (the ambiguity function) weighted by an arbitrary parameter function.

This point of view completes the previously mentioned approaches, insofar as the ambiguity function naturally extends the notion of an autocorrelation function to a nonstationary context, in which temporal and spectral properties must be taken into account simultaneously. Let us dwell on this point by reconsidering the usual (deterministic and symmetrical) autocorrelation functions. When operating in the time-domain, we can obviously write

$$\int_{-\infty}^{+\infty} x\left(s + \frac{\tau}{2}\right) x^*\left(s - \frac{\tau}{2}\right) ds = \int_{-\infty}^{+\infty} x(s)\, x^*_{\tau,0}(s)\, ds\ .$$

This amounts to measuring the level of resemblance (in terms of an inner product) of the signal with its different copies that are shifted in time. In a dual fashion, if we operate in frequency, we obtain

$$\int_{-\infty}^{+\infty} X\left(\zeta - \frac{\xi}{2}\right) X^*\left(\zeta + \frac{\xi}{2}\right) d\zeta = \int_{-\infty}^{+\infty} x(s)\, x^*_{0,-\xi}(s)\, ds\ .$$

Consequently, a combination of both shifts, in time and frequency, renders a "mixed autocorrelation function" accessible. It is identical to the symmetrical ambiguity function, apart from a phase, as we have

$$\int_{-\infty}^{+\infty} x(s)\, x^*_{\tau,-\xi}(s)\, ds = \int_{-\infty}^{+\infty} x(s)\, x^*(s - \tau)\, e^{i2\pi\xi s}\, ds = e^{i\pi\xi\tau} A_x(\xi, \tau)\ .$$

The preceding discussion justifies anew the structure of the representations in Cohen's class. In a "stationary" situation (where stationary means that the spectral properties of the signal do not change in time) the power spectrum provides a sufficient energetic representation. This is a univariate function with a frequency variable, which is the Fourier transform of

Chapter 3 Issues of Interpretation 193

the temporal autocorrelation function. For practical purposes we can enlarge this class of solutions by introducing a weight function $w(\tau)$ to the autocorrelation $\gamma_x(\tau)$ prior to taking its Fourier transform. This leads to estimators of the form

$$\widehat{\Gamma}_x(\nu) = \int_{-\infty}^{+\infty} w(\tau)\,\gamma_x(\tau)\,e^{-i2\pi\nu\tau}\,d\tau \ .$$

When we now turn to "nonstationary" situations, the time should appear as an explicit parameter in the description. The construction of a joint representation in time and frequency thus reduces to making the previous quantity time-dependent. One possible way to achieve this is by sticking to the idea of "autocorrelation + weighting + Fourier transform," but rendering it in a *two-dimensional* time-frequency form. Hence, it is the question of passing from the temporal autocorrelation function to an analogous quantity, which combines time and frequency. The ambiguity function is a natural candidate in this regard, and this makes Cohen's class turn up immediately.

There is a comparable situation for the affine class, considered in its time-frequency interpretation. The forementioned change of the time-frequency parameterization, which brings a reference frequency ν_0 into play, has a frequency-time analogue of the form

$$\Pi(t,\nu) = \Pi_0(t,\nu-\nu_0) \quad \Longleftrightarrow \quad f(\xi,\tau) = f_0(\xi,\tau)\,e^{-i2\pi\nu_0\tau} \ .$$

Hence, we obtain

$$\Omega_x(t,a;f) = \iint_{\mathbb{R}^2} f_0\left(a\xi,\frac{\tau}{a}\right) A_x(\xi,\tau)\,e^{-i2\pi[(\nu_0/a)\tau+\xi t]}\,d\xi\,d\tau \ .$$

Remark 1. Our preceding heuristic arguments should not make us believe that the ambiguity function is a perfect time-frequency generalization of a univariate autocorrelation function. Even the fact that both univariate functions in time and in frequency can be obtained as special cross sections, namely

$$\int_{-\infty}^{+\infty} x\left(s+\frac{\tau}{2}\right) x^*\left(s-\frac{\tau}{2}\right) ds = A_x(0,\tau)$$

and

$$\int_{-\infty}^{+\infty} X\left(\zeta-\frac{\xi}{2}\right) X^*\left(\zeta+\frac{\xi}{2}\right) d\zeta = A_x(\xi,0) \ ,$$

does not justify such an idea. A major defect of the ambiguity function is its lack of positive definiteness, which is one of the essential characteristics of any true autocorrelation function. Indeed, the assumption of this property would imply that its two-dimensional Fourier transform would be nonnegative everywhere. As this directly concerns the Wigner-Ville distribution, the hypothesis cannot be true.

Remark 2. Giving preference to the affine class is sometimes justified for signals with a *wideband* structure. However, we should observe that the affine class is based on the same *narrowband* ambiguity function as Cohen's class. Hence, it does not invoke a *wideband* generalization of this function, although this would be deemed more appropriate *a priori*. Let us thus keep in mind that the affine character of this class is only introduced by the operation of the weight function, which multiplies the ambiguity function by a scale-dependent measure.

3.1.2. Parameterizations, Operators and Correspondence Rules

We already mentioned the formal link between the time-frequency representations of signals and the position-momentum representations in quantum mechanics in Subsections 1.1.2 and 2.1.4. Drawing our inspiration from the approaches in the latter field, we can in fact introduce an operational calculus to signal theory. It is based on the two elementary operators of time $\hat{t}$ and frequency $\hat{\nu}$. As we readily recall, they act by

$$(\hat{t}x)(t) \equiv tx(t), \qquad (\hat{\nu}x)(t) \equiv \frac{1}{i2\pi}\frac{dx}{dt}(t). \qquad (3.20)$$

These two (self-adjoint) operators do not commute. On the contrary, they fulfill the relation

$$[\hat{t},\hat{\nu}] \equiv \hat{t}\hat{\nu} - \hat{\nu}\hat{t} = \frac{i}{2\pi}\widehat{I} \qquad (3.21)$$

where $\widehat{I}$ denotes the identity. Without going into the details of this operational approach, we will briefly explain some of the issues originating from this perspective on the following pages. We will also dwell on the intimate connection between the noncommutativity of the elementary operators and the arbitrariness of the choice of a joint representation via its parameter function. [4]

Why operators? Before we start dealing with these questions, it might be helpful to explain to which extent an operational formalism can be beneficial to signal theory. Among the numerous reasons we emphasize only three. First of all, the transformations that are *imposed* on a signal for changing the representational space play an important role. This can be, for instance, the projection of a signal onto a basis of the representational space, or the construction of a family of basic signals from one reference element, and many other situations encountered in Chapter 2 that are liable to be described in terms of a linear operator acting on a signal.

The second reason of equally great importance concerns the transformation that a signal *undergoes* and which can be expressed by the action of an operator as well: filtering, dispersion, delay, Doppler effect, etc. In a nonstationary context the description of such situations gains a very particular significance (and difficulty) if both the temporal and frequential aspects are considered. As an example we already mentioned the ambiguity functions. The same happens with linear time-varying "filters." Their response to a signal $x(t)$ can be written as

$$(\widehat{H}x)(t) = \int_{-\infty}^{+\infty} h(t,s)\, x(s)\, ds \;.$$

The different forms in time and/or frequency of the "bivariate impulse response" $h(t,s)$ carry some complementary information (time-dependent transfer function, diffusion function, etc.).

Finally, the third reason for an operational description is supported by the fact that the expectation value of an operator, defined by

$$\langle \widehat{A} \rangle_x = \langle \widehat{A}x, x \rangle = \int_{-\infty}^{+\infty} (\widehat{A}x)(s)\, x^*(s)\, ds \;, \qquad (3.22)$$

provides some information about the inherent properties of the signal. (Thus it is immediately clear that $\langle \hat{t} \rangle_x$ and $\langle \hat{\nu} \rangle_x$ coincide with the time and frequency centers of the signal, respectively. Likewise, $\langle \hat{t}^2 \rangle_x$ and $\langle \hat{\nu}^2 \rangle_x$ yield the time and frequency spread in Gabor's sense, etc.)

The operator of time-frequency shifts. As a first example let us consider the linear operator, which is the exponential of $\hat{\nu}$. Then we obtain

$$\begin{aligned}\left(e^{i2\pi\tau\hat{\nu}}x\right)(t) &= \left[\left(\widehat{I} + i2\pi\tau\hat{\nu} + \frac{1}{2!}(i2\pi\tau\hat{\nu})^2 + \cdots\right)x\right](t) \\ &= x(t) + \tau\frac{dx}{dt}(t) + \frac{\tau^2}{2}\frac{d^2x}{dt^2}(t) + \cdots \\ &= x(t+\tau) \;,\end{aligned} \qquad (3.23)$$

and this defines a temporal shift of length τ. We realize as a by-product that the expectation value of this operator is just the autocorrelation of the signal, hence

$$\langle e^{i2\pi\tau\hat{\nu}} \rangle_x = \int_{-\infty}^{+\infty} x(t)\, x^*(t-\tau)\, dt \;.$$

Analogously, the exponential of the operator $\hat{t}$ is given by

$$(e^{i2\pi\xi\hat{t}}x)(t) = e^{i2\pi\xi t}\, x(t)\;, \qquad (3.24)$$

which corresponds to a frequential shift.

By the noncommutativity of $\hat{t}$ and $\hat{\nu}$ the situation becomes more complicated, when we bring *both* elementary operators into play. As our next example let us consider the operator

$$e^{i2\pi(\xi\hat{t}+\tau\hat{\nu})}\;.$$

Unlike the ordinary calculus of functions (or commuting operators), this operator cannot be factored in the most obvious way. However, a classical result (*Lemma of Glauber and Jordan*[5]) states that the following is true: If the two noncommuting operators $\widehat{A}$ and $\widehat{B}$ are such that

$$[\widehat{A},[\widehat{A},\widehat{B}]] = [\widehat{B},[\widehat{A},\widehat{B}]] = 0\;,$$

that is, if they commute with their commutator, then we may write

$$e^{\widehat{A}+\widehat{B}} = e^{\widehat{A}}\, e^{\widehat{B}}\, e^{-(1/2)[\widehat{A},\widehat{B}]}\;.$$

By applying this lemma twice and using the relation of (non)commutativity of $\hat{t}$ and $\hat{\nu}$, we obtain as one possible factorization

$$e^{i2\pi(\xi\hat{t}+\tau\hat{\nu})} = e^{i\pi\tau\hat{\nu}}\, e^{i2\pi\xi\hat{t}}\, e^{i\pi\tau\hat{\nu}}\;. \qquad (3.25)$$

Consequently, the explicit computation of the *expectation* value of this operator by virtue of this factorization leads to

$$\langle e^{i2\pi(\xi\hat{t}+\tau\hat{\nu})}\rangle_x = \int_{-\infty}^{+\infty} \left[e^{i\pi\tau\hat{\nu}}\, e^{i2\pi\xi\hat{t}}\, x\left(s+\frac{\tau}{2}\right)\right] x^*(s)\, ds$$

$$= \int_{-\infty}^{+\infty} \left[e^{i\pi\tau\hat{\nu}}\, e^{i2\pi\xi s}\, x\left(s+\frac{\tau}{2}\right)\right] x^*(s)\, ds$$

$$= \int_{-\infty}^{+\infty} e^{i2\pi\xi(s+\tau/2)}\, x(s+\tau)\, x^*(s)\, ds \qquad (3.26)$$

$$= \int_{-\infty}^{+\infty} x\left(s+\frac{\tau}{2}\right) x^*\left(s-\frac{\tau}{2}\right) e^{i2\pi\xi s}\, ds$$

$$= A_x(\xi,\tau)\;.$$

Chapter 3 Issues of Interpretation 197

Hence, we have laid open the form of the operator of a combined time-frequency shift by showing that its expectation value is identical to the symmetrical ambiguity function of Eq. (3.16).

Correspondence rules. In a rather general sense, it is conceivable to identify the following two quantities: On the one side, we consider the expectation value of an *operator* $\widehat{G}$ composed of $\hat{t}$ and $\hat{\nu}$ relative to a *signal* $x(t)$; on the other side, we use a mean value of a *function* G of the variables t and ν weighted by the analogue of a *probability density* of the same signal in the time-frequency plane. This amounts to the introduction of a joint time-frequency representation ρ_x of the signal via the functional equation

$$\langle \widehat{G} \rangle_x = \iint_{\mathbb{R}^2} G(t,\nu)\, \rho_x(t,\nu)\, dt\, d\nu \;. \tag{3.27}$$

The association of $\widehat{G}$ with the function $G(t,\nu)$ is known as a *correspondence rule* that maps functions to operators. When the considered operator algebra is not commutative, in contrast to the algebra of functions, generally there cannot be an unambiguous correspondence rule of this type. This property is closely related to the arbitrariness in picking a time-frequency representation of the signal. We can achieve a precise formulation by restricting ourselves to representations in Cohen's class. Then we can write

$$C_x(t,\nu;f) = \iint_{\mathbb{R}^2} f(\xi,\tau)\, A_x(\xi,\tau)\, e^{-i2\pi(\nu\tau+\xi t)}\, d\xi\, d\tau$$

$$= \iint_{\mathbb{R}^2} f(\xi,\tau)\, \langle e^{i2\pi(\xi\hat{t}+\tau\hat{\nu})}\rangle_x\, e^{-i2\pi(\nu\tau+\xi t)}\, d\xi\, d\tau$$

$$= \left\langle \iint_{\mathbb{R}^2} f(\xi,\tau)\, e^{i2\pi[\xi(\hat{t}-t\widehat{I})+\tau(\hat{\nu}-\nu\widehat{I})]}\, d\xi\, d\tau \right\rangle_x \;.$$

Hence, we infer that

$$\langle \widehat{G} \rangle_x = \iint_{\mathbb{R}^2} G(t,\nu)\, C_x(t,\nu;f)\, dt\, d\nu$$

$$= \left\langle \iint_{\mathbb{R}^2} f(\xi,\tau)\, g(\xi,\tau)\, e^{i2\pi(\xi\hat{t}+\tau\hat{\nu})}\, d\xi\, d\tau \right\rangle_x \;.$$

by employing the definition

$$g(\xi,\tau) = \iint_{\mathbb{R}^2} G(t,\nu)\, e^{-i2\pi(\nu\tau+\xi t)}\, dt\, d\nu \ .$$

When we identify proper terms and include the dependence on the parameter function in our notation, this furnishes the anticipated result

$$\widehat{G}_f = \iint_{\mathbb{R}^2} f(\xi,\tau)\, g(\xi,\tau)\, e^{i2\pi(\xi\hat{t}+\tau\hat{\nu})}\, d\xi\, d\tau \ . \tag{3.28}$$

We have thus developed an explicit rule for associating an operator with a function. This rule depends strongly on the choice of a joint representation. This brings the parameter function f into play for the representation as well as for the correspondence rule.

In Chapter 2 we found that certain desirable properties of the time-frequency representation are guaranteed by imposing suitable conditions on the parameter function. Analogously, we can assure a desired type of a correspondence rule between functions and operators by adequate specifications of the same parameter function.

Example 1. The association of self-adjoint operators (thus having real expectation values) with real-valued functions imposes the condition

$$f(\xi,\tau) = f^*(-\xi,-\tau) \ .$$

Recall that the same condition implies that the representation in Cohen's class is real-valued itself.

Example 2. Let us next consider the special case where $f(\xi,\tau) = 1$ and $G(t,\nu) = t^k$. Then we can derive

$$g(\xi,\tau) = \delta(\tau)\left(\frac{i}{2\pi}\right)^k \delta^{(k)}(\xi) \ ,$$

and this leads to the correspondence rule

$$t^k \;\mapsto\; \hat{t}^k \ .$$

(One can show in the same way that an analogous rule applies to the frequency.)

Contemplating this simple and satisfactory situation, we might believe that every correspondence rule at least ensures

$$t \mapsto \hat{t} \quad \text{and} \quad \nu \mapsto \hat{\nu} \ .$$

Chapter 3 Issues of Interpretation   199

However, one can only conclude, in general, that

$$t \mapsto f(0,0)\,\hat{t} + \frac{1}{i2\pi}\frac{\partial f}{\partial \xi}(0,0)\,\widehat{I}\;,$$

and a dual relation for $\hat{\nu}$ holds. In fact, the parameter function needs to meet certain criteria concerning its function value and its derivatives at the origin in order for the desired relations to hold. Such conditions are verified, for example, by the correspondence rules stemming from the Wigner-Ville, Ackroyd, or Born-Jordan parameterizations. However, the precise rule derived for a spectrogram involves the mean value of its window $h(t)$ according to

$$t \mapsto \hat{t} - \langle \hat{t}\rangle_h\,\widehat{I}\;.$$

Hence, it yields only the desired correspondence, if the window has zero mean.

Kernels. Let us pause for a moment, before we give more elaborate examples for the construction of operators associated with particular time-frequency functions. We can use the preceding result in order to characterize the *action* of the operator associated with a given function $G(t,\nu)$ by a *kernel*. We again denote the dependence on the parameter function f by an index in our notations. A simple computation shows

$$(\widehat{G}_f x)(t) = \int_{-\infty}^{+\infty} \gamma_f(t,s)\,x(s)\,ds\;,$$

where the kernel $\gamma_f(t,s)$ of the operator $\widehat{G}_f$ is defined by

$$\gamma_f(s,t) = \int_{-\infty}^{+\infty} F\left(\frac{t+s}{2} - \theta, t-s\right)\gamma(\theta, t-s)\,d\theta \qquad (3.29)$$

and

$$\gamma(t,\tau) = \int_{-\infty}^{+\infty} g(\xi,\tau)\,e^{i2\pi\xi t}\,d\xi = \int_{-\infty}^{+\infty} G(t,\nu)\,e^{-i2\pi\nu\tau}\,d\nu\;.$$

One can observe that this kernel admits an interpretation as a *mean value* of γ (in the sense of functions, not operators) whose definition depends on the parameter function f.[6] In fact, for specific choices of the parameterization and by putting $\gamma_*(\theta) = \gamma(\theta, s-t)$, we can easily establish the form of the kernel as listed in Table 3.1.

In particular, we conclude that the first three rules (Wigner-Ville, Ackroyd, Born-Jordan) yield identical results for all *linear* functions $\gamma_*(\theta)$; in other words, it gives the same operators $\widehat{G}_f$ for all functions $G(t,\nu)$ of the form

$$G(t,\nu) = A(\nu) + tB(\nu)\;.$$

Table 3.1
Kernels of operators for different correspondence rules

Wigner-Ville	$f(\xi,\tau) = 1$	$\gamma_W(t,s) = \gamma_*\left(\dfrac{t+s}{2}\right)$
Ackroyd	$f(\xi,\tau) = \cos \pi\xi\tau$	$\gamma_A(t,s) = \dfrac{1}{2}[\gamma_*(t) + \gamma_*(s)]$
Born-Jordan	$f(\xi,\tau) = \dfrac{\sin \pi\xi\tau}{\pi\xi\tau}$	$\gamma_{BJ}(t,s) = \dfrac{1}{\|t-s\|} \displaystyle\int_{\min(t,s)}^{\max(t,s)} \gamma_*(\theta)\,d\theta$
Page	$f(\xi,\tau) = e^{i\pi\xi\|\tau\|}$	$\gamma_P(t,s) = \gamma_*(\max(t,s))$
Spectrogram	$f(\xi,\tau) = A_h^*(\xi,\tau)$	$\gamma_S(t,s) = \displaystyle\int_{-\infty}^{+\infty} h^*(t-\theta)h(s-\theta)\gamma_*(\theta)d\theta$

More generally, these three correspondence rules provide almost the same results when the evolution of $\gamma_*(\theta)$ is *quasi-linear* as compared to the temporal extent of $\gamma(t,\tau)$ with respect to the variable τ.

Weyl calculus. Let us now investigate the 'Wigner-Ville' case more deeply. Here we have $f(\xi,\tau) = 1$, and the kernel $\gamma_W(t,s)$ of the operator $\widehat{G}_W$, which is associated with a function $G(t,\nu)$ by means of the respective correspondence rule, can be written as

$$\gamma_W(t,s) = \gamma\left(\frac{t+s}{2}, s-t\right). \tag{3.30}$$

Conversely, if an operator $\widehat{G}_W$ with kernel $\gamma_W(t,s)$ is given, the associated function $G(t,\nu)$ is obtained via the transformation

$$G(t,\nu) = \int_{-\infty}^{+\infty} \gamma_W\left(t + \frac{\tau}{2}, t - \frac{\tau}{2}\right) e^{-i2\pi\nu\tau}\,d\tau. \tag{3.31}$$

Adopting the language of pseudo-differential calculus,[7] the function $G(t,\nu)$ is the *Weyl symbol* of the pseudo-differential operator $\widehat{G}_W$. Passing from an operator to its symbol is the inverse operation of the correspondence rule (also called *Weyl quantization*). It consists of associating a function with an operator. (We note in passing that the symbol of the *covariance* operator is exactly the *Wigner-Ville spectrum* as defined in Subsection 2.4.3.)

Chapter 3 Issues of Interpretation 201

Because the operator algebra is not commutative, the symbol of the product of two operators cannot be obtained directly by the ordinary product of the symbols. Its computation actually requires the introduction of a *twisted product*,[8] which is noncommutative and defined by the quantization

$$\widehat{A}_W \widehat{B}_W \mapsto (A \# B)(t, \nu) \ ;$$

here A and B are the symbols of $\widehat{A}_W$ and $\widehat{B}_W$, respectively. If we denote the kernels of $\widehat{A}_W$ and $\widehat{B}_W$ by α_W and β_W, the new kernel γ_W of the operator product $\widehat{G}_W = \widehat{A}_W \widehat{B}_W$ is given by

$$\gamma_W(t,s) = \int_{-\infty}^{+\infty} \alpha_W(t,r) \, \beta_W(r,s) \, dr$$

$$= \iiint_{\mathbb{R}^3} A\left(\frac{t+r}{2}, \xi\right) B\left(\frac{r+s}{2}, \zeta\right) e^{i2\pi[\xi(t-r)+\zeta(r-s)]} \, d\xi \, d\zeta \, dr \ .$$

Consequently, making use of the fact

$$(A \# B)(t, \nu) = \int_{-\infty}^{+\infty} \gamma_W\left(t + \frac{\tau}{2}, t - \frac{\tau}{2}\right) e^{-i2\pi\nu\tau} \, d\tau \ ,$$

we can derive the relation, after rearranging terms,

$$(A \# B)(t, \nu) = 4 \iiiint_{\mathbb{R}^4} A(t-\tau, \nu-\xi) B(t-\theta, \nu-\zeta) e^{i4\pi(\tau\zeta-\theta\xi)} \, d\tau \, d\theta \, d\xi \, d\zeta \ . \tag{3.32}$$

Moments. Let us now suppose that the function $G(t, \nu)$ under consideration has a power series expansion

$$G(t, \nu) = \sum_{k=0}^{+\infty} \sum_{l=0}^{+\infty} G_{kl} \, t^k \, \nu^l \ .$$

We wish to associate with it an operator $\widehat{G}_f$ in the sense of the correspondence rule defined by the parameter function $f(\xi, \tau)$. Owing to the linearity of the general rule established before, we can immediately find

$$\widehat{G}_f = \sum_{k=0}^{+\infty} \sum_{l=0}^{+\infty} G_{kl} \, \widehat{\mu}_f^{kl} \ ,$$

where each $\widehat{\mu}_f^{kl}$ denotes an operator that is associated with the ordinary product $t^k \nu^l$. Therefore, the analysis of the quantization can be restricted to these latter terms. This results in the study of the *moments* of the joint representation, as we have

$$\langle \widehat{\mu}_f^{kl} \rangle_x = \iint_{\mathbb{R}^2} t^k \nu^l \, C_x(t,\nu;f) \, dt \, d\nu \ . \tag{3.33}$$

By a well-known property of the Fourier transform, the moments of a function are connected with the behavior of the derivatives of its Fourier transform at the origin. By invoking the relation (3.18) between time-frequency representations and weighted ambiguity functions, it is easy to see that

$$\langle \widehat{\mu}_f^{kl} \rangle_x = \frac{1}{(i2\pi)^{k+l}} \left(\frac{\partial^{k+l}}{\partial \xi^k \partial \tau^l} f(\xi,\tau) A_x(\xi,\tau) \right)(0,0) \ .$$

Hence, we infer the formal result

$$\widehat{\mu}_f^{kl} = \frac{1}{(i2\pi)^{k+l}} \left(\frac{\partial^{k+l}}{\partial \xi^k \partial \tau^l} f(\xi,\tau) e^{i2\pi(\xi \hat{t} + \tau \hat{\nu})} \right)(0,0) \ . \tag{3.34}$$

An application of this relation to the general case can be quite tedious. The procedure can be slightly simplified by observing that

$$\left(\frac{\partial^{k+l}}{\partial \xi^k \partial \tau^l} f(\xi,\tau) e^{i2\pi(\xi \hat{t} + \tau \hat{\nu})} \right)(0,0)$$

$$= \sum_{p=0}^{k} \sum_{q=0}^{l} \binom{k}{p} \binom{l}{q} \frac{\partial^{p+q} f}{\partial \xi^p \partial \tau^q}(0,0) \left(\frac{\partial^{k-p+q-l}}{\partial \xi^{k-p} \partial \tau^{q-l}} e^{i2\pi(\xi \hat{t} + \tau \hat{\nu})} \right)(0,0)$$

$$= \sum_{p=0}^{k} \sum_{q=0}^{l} \binom{k}{p} \binom{l}{q} \frac{\partial^{p+q} f}{\partial \xi^p \partial \tau^q}(0,0) \frac{1}{(i2\pi)^{p+q}} \widehat{\mu}_W^{k-p,l-q} \ .$$

This enables us to deal with the (Weyl) correspondence governed by the parameter function $f(\xi,\tau) = 1$ and the Wigner-Ville distribution alone. For this matter we recall the identities

$$\langle e^{i2\pi(\xi \hat{t} + \tau \hat{\nu})} \rangle_x = A_x(\xi,\tau)$$

and

$$A_x(\xi,\tau) = \iint_{\mathbb{R}^2} W_x(t,\nu) \, e^{i2\pi(\xi t + \tau \nu)} \, dt \, d\nu \ .$$

Chapter 3 Issues of Interpretation 203

Hence, the Weyl correspondence imposes the association

$$e^{i2\pi(\xi t+\tau\nu)} \mapsto e^{i2\pi(\xi\hat{t}+\tau\hat{\nu})} . \tag{3.35}$$

By expanding both exponentials (of functions and operators) into power series and using the linearity of the quantization once again, we can find the searched result. In fact, we infer from

$$e^{i2\pi(\xi t+\tau\nu)} = \sum_{k=0}^{+\infty}\sum_{l=0}^{+\infty} \frac{(i2\pi)^{k+l}}{k!\,l!} (\xi t)^k (\tau\nu)^l$$

that the relation

$$e^{i2\pi(\xi\hat{t}+\tau\hat{\nu})} = \sum_{k=0}^{+\infty}\sum_{l=0}^{+\infty} \frac{(i2\pi)^{k+l}}{k!\,l!} \xi^k \tau^l \widehat{\mu}_W^{kl}$$

holds. Moreover, we can write

$$e^{i2\pi(\xi\hat{t}+\tau\hat{\nu})} = e^{i\pi\xi\hat{t}} e^{i2\pi\tau\hat{\nu}} e^{i\pi\xi\hat{t}}$$

$$= \sum_{k=0}^{+\infty}\sum_{l=0}^{+\infty}\sum_{m=0}^{+\infty} \frac{(i2\pi)^{k+l+m}}{k!\,l!\,m!\,2^{k+m}} \xi^{k+m}\tau^l \hat{t}^k \hat{\nu}^l \hat{t}^m .$$

A change of the summation indices and a termwise comparison with the previous expression leads to

$$\widehat{\mu}_W^{kl} = 2^{-k}\sum_{m=0}^{k}\binom{k}{m}\hat{t}^{k-m}\hat{\nu}^l\hat{t}^m . \tag{3.36}$$

Remark. If we had used the factorization

$$e^{i2\pi(\xi\hat{t}+\tau\hat{\nu})} = e^{i\pi\tau\hat{\nu}} e^{i2\pi\xi\hat{t}} e^{i\pi\tau\hat{\nu}}$$

instead, we would have obtained the equivalent form

$$\widehat{\mu}_W^{kl} = 2^{-l}\sum_{m=0}^{l}\binom{k}{l}\hat{\nu}^{l-m}\hat{t}^k \hat{\nu}^m .$$

By means of these expressions we can now determine the operators $\widehat{\mu}_f^{kl}$ for any correspondence rule whatsoever. However, it may be simpler in certain practical situations to follow a more direct approach. If we consider, for instance, the Ackroyd distribution, we find

$$f(\xi,\tau) A_x(\xi,\tau) = \cos\pi\xi\tau \left\langle e^{i2\pi(\xi\hat{t}+\tau\hat{\nu})} \right\rangle_x$$

$$= \left\langle \frac{1}{2} \left(e^{i\pi\xi\tau} e^{i2\pi(\xi\hat{t}+\tau\hat{\nu})} + e^{-i\pi\xi\tau} e^{i2\pi(\xi\hat{t}+\tau\hat{\nu})} \right) \right\rangle_x$$

$$= \left\langle \frac{1}{2} \left(e^{i2\pi\xi\hat{t}} e^{i2\pi\tau\hat{\nu}} + e^{i2\pi\tau\hat{\nu}} e^{i2\pi\xi\hat{t}} \right) \right\rangle_x .$$

Hence, the respective correspondence rule reads as

$$e^{i2\pi(\xi t + \tau \nu)} \mapsto \frac{1}{2} \left(e^{i2\pi\xi\hat{t}} e^{i2\pi\tau\hat{\nu}} + e^{i2\pi\tau\hat{\nu}} e^{i2\pi\xi\hat{t}} \right) . \tag{3.37}$$

This further implies

$$\widehat{\mu}_A^{kl} = \frac{\hat{t}^k \hat{\nu}^l + \hat{\nu}^l \hat{t}^k}{2} . \tag{3.38}$$

The Born-Jordan parameterization (for which $f(\xi,\tau) = \sin\pi\xi\tau/\pi\xi\tau$) on its part leads to the rule [9]

$$\widehat{\mu}_{BJ}^{kl} = \frac{1}{k+1} \sum_{m=0}^{k} \hat{t}^{k-m} \hat{\nu}^l \hat{t}^m = \frac{1}{l+1} \sum_{m=0}^{l} \hat{\nu}^{l-m} \hat{t}^k \hat{\nu}^m , \tag{3.39}$$

which can also be written in the more compact form

$$\widehat{\mu}_{BJ}^{kl} = \frac{2\pi}{i(k+1)(l+1)} \left[\hat{t}^{k+1}, \hat{\nu}^{l+1} \right] . \tag{3.40}$$

(The equivalence is proven by an inductive argument based on the relations for the commutators $[\hat{t}^k, \hat{\nu}] = (i/2\pi)k\,\hat{t}^{k-1}$ and $[\hat{t}, \hat{\nu}^l] = (i/2\pi)l\,\hat{\nu}^{l-1}$.)

The differences between the various correspondence rules originate from the properties of the underlying parameter functions as related to their derivatives at the origin. Hence, if we restrict ourselves to the moments of total degree $k + l < 4$, we can recognize without many complications that the three rules, Wigner-Ville (Weyl), Ackroyd (symmetry), and Born-Jordan (commutator), yield identical results. This comes from the fact that in each of these cases

$$f(0,0) = 1 \quad \text{and} \quad \frac{\partial^{k+l} f}{\partial \xi^k \partial \tau^l}(0,0) = 0 , \quad 1 \leq k + l < 4 .$$

Chapter 3 Issues of Interpretation

We can thus use the simplest rule (which is the symmetrical one in this case) in order to express the correspondences for all three cases by

$$t^k \nu^l \mapsto \frac{\hat{t}^k \hat{\nu}^l + \hat{\nu}^l \hat{t}^k}{2}, \qquad 0 \le k+l < 4 . \tag{3.41}$$

The first difference between the three rules turns up when $k = l = 2$; then we obtain

$$\frac{\partial^4 f}{\partial \xi^2 \, \partial \tau^2}(0,0) = \begin{cases} 0 & \text{(Wigner-Ville)} \\ -2\pi^2 & \text{(Ackroyd)} \\ -2\pi^2/3 & \text{(Born-Jordan)} , \end{cases} \tag{3.42}$$

and this results in

$$t^2 \nu^2 \mapsto \begin{cases} \dfrac{1}{2}(\hat{t}^2 \hat{\nu}^2 + \hat{\nu}^2 \hat{t}^2) + \dfrac{1}{8\pi^2} \hat{I} & \text{(Wigner-Ville)} \\ \dfrac{1}{2}(\hat{t}^2 \hat{\nu}^2 + \hat{\nu}^2 \hat{t}^2) & \text{(Ackroyd)} \\ \dfrac{1}{2}(\hat{t}^2 \hat{\nu}^2 + \hat{\nu}^2 \hat{t}^2) + \dfrac{1}{12\pi^2} \hat{I} & \text{(Born-Jordan)} . \end{cases} \tag{3.43}$$

Hence, the moments in all three cases have the generic form

$$\iint_{\mathbb{R}^2} t^2 \nu^2 \, C_x(t,\nu;f) \, dt \, d\nu = \frac{1}{4\pi^2} \left[\int_{-\infty}^{+\infty} t^2 \left| \frac{dx}{dt}(t) \right|^2 dt - \beta E_x \right] \tag{3.44}$$

where only the constant β differs according to

$$\beta = \frac{1}{2} - \frac{1}{4\pi^2} \frac{\partial^4 f}{\partial \xi^2 \, \partial \tau^2}(0,0) = \begin{cases} 1/2 & \text{(Wigner-Ville)} \\ 1 & \text{(Ackroyd)} \\ 2/3 & \text{(Born-Jordan)} . \end{cases} \tag{3.45}$$

Dilations and ambiguities. The simplest example of a combined function of time and frequency is the product $t\nu$. It allows of an interesting interpretation in terms of *dilations*.[10] In fact, as already stated (cf. Eqs. (3.23) and (3.24)), the exponential of the operator associated with t (or ν, respectively) corresponds to a shift in frequency (or in time). Let us now consider the exponential of the operator associated with $t\nu$ and determine its action. If we work in one of the previous settings where

$$t\nu \mapsto \frac{\hat{t}\hat{\nu} + \hat{\nu}\hat{t}}{2}, \tag{3.46}$$

we find the identity

$$e^{i2\pi(\log\eta)(\hat{t}\hat{\nu}+\hat{\nu}\hat{t})/2} = \sqrt{\eta}\,e^{i2\pi(\log\eta)\hat{t}\hat{\nu}}$$

$$= \sqrt{\eta}\sum_{m=0}^{+\infty}\frac{(\log\eta)^m}{m!}\,(\hat{t}\,\widehat{d_t})^m$$

for all $\eta > 0$; here we used the abbreviation

$$\widehat{d_t} = \frac{d}{dt}\;.$$

A straightforward argument gives

$$(\hat{t}\,\widehat{d_t})^m\,t^n = n\,(\hat{t}\,\widehat{d_t})^{m-1}\,t^n = \ldots = n^m\,t^n\;.$$

Hence, we derive

$$e^{i2\pi(\log\eta)\hat{t}\hat{\nu}}\,t^n = (\eta t)^n\;,$$

and this implies that for every signal $x(t)$, which has a power series expansion, the considered operator acts as

$$\left(e^{i2\pi(\log\eta)(\hat{t}\hat{\nu}+\hat{\nu}\hat{t})/2}\,x\right)(t) = \sqrt{\eta}\,x(\eta t)\;. \tag{3.47}$$

The so-formed operator, which is the exponential of the image of the time-frequency function $t\nu$, is just the *dilation* by a factor η. We came across this operator earlier in the framework of time-scale representations and the wideband ambiguity functions. In order to determine the time-frequency function $G(t,\nu)$, with which this operator is associated (in the sense of the correspondence rule of Wigner-Ville-Weyl, for instance), it is enough to consider the kernel associated with

$$G(t,\nu) = e^{i2\pi\alpha t\nu}\;.$$

A simple computation shows that in this case

$$\int_{-\infty}^{+\infty}\gamma_W(t,s)\,x(s)\,ds = \frac{1}{1-\alpha/2}\,x\!\left(\frac{1+\alpha/2}{1-\alpha/2}t\right),\quad |\alpha| < 2\;.$$

This leads to the searched quantization of 'operator-symbol' type, which reads as

$$e^{i2\pi(\log\eta)(\hat{t}\hat{\nu}+\hat{\nu}\hat{t})/2} \;\mapsto\; \sqrt{1-\alpha^2/4}\,e^{i2\pi\alpha t\nu}\;, \tag{3.48}$$

Chapter 3 Issues of Interpretation

where α solves the equation $\eta = (1 + \alpha/2)/(1 - \alpha/2)$.

We can go further and express the wideband ambiguity function as the expectation value of an operator combining the actions of a dilation and a translation. It thus takes the form

$$\sqrt{\eta} \int_{-\infty}^{+\infty} x(t)\, x^*(\eta(t-\tau))\, dt = \left\langle e^{i2\pi\tau\hat{\nu}}\, e^{-i2\pi(\log \eta)(\hat{t}\hat{\nu}+\hat{\nu}\hat{t})/2} \right\rangle_x . \quad (3.49)$$

By analogy with the situation of the narrowband ambiguity function, we can propose a slightly modified definition of the wideband ambiguity function by employing the substitution

$$e^{i\pi\tau\hat{\nu}}\, e^{i2\pi\xi\hat{t}}\, e^{i\pi\tau\hat{\nu}} \quad \to \quad e^{i\pi\tau\hat{\nu}}\, e^{i2\pi(\log \eta)(\hat{t}\hat{\nu}+\hat{\nu}\hat{t})/2}\, e^{i\pi\tau\hat{\nu}} ;$$

so we simply replace the frequency shift with a dilation. This results in a first *symmetrical* definition

$$A_x^C(\alpha, \tau) = \sqrt{1 - \alpha^2/4} \int_{-\infty}^{+\infty} x\left((1+\alpha/2)t + \frac{\tau}{2}\right) x^*\left((1-\alpha/2)t - \frac{\tau}{2}\right) dt \quad (3.50)$$

where we let $\eta = (1 + \alpha/2)/(1 - \alpha/2)$ as before. (From a physical point of view, if the parameter η is interpreted as the Doppler rate, α is identified as twice the ratio "relative velocity/velocity of the propagation.")

By using the definition of the twisted product one can show

$$e^{i\pi\tau\nu} \# e^{i2\pi\alpha t\nu} \# e^{i\pi\tau\nu} = e^{i2\pi(\tau+\alpha t)\nu} . \quad (3.51)$$

This means that the operator, whose expectation value is the wideband ambiguity function, has the symbol

$$G(t, \nu) = \sqrt{1 - \alpha^2/4}\; e^{i2\pi(\tau+\alpha t)\nu} .$$

Another simple relation can thus be established between the wideband ambiguity function in its symmetrical form and the Wigner-Ville distribution. It states that

$$A_x^C(\alpha, \tau) = \left\langle e^{i\pi\tau\hat{\nu}}\, e^{i2\pi(\log \eta)(\hat{t}\hat{\nu}+\hat{\nu}\hat{t})/2}\, e^{i\pi\tau\hat{\nu}} \right\rangle_x$$

$$= \iint_{\mathbb{R}^2} \sqrt{1 - \alpha^2/4}\; e^{i2\pi(\tau+\alpha t)\nu}\, W_x(t, \nu)\, dt\, d\nu .$$

Finally, we can apply the Fourier transform, which maps the Wigner-Ville distribution into the narrowband ambiguity function (here denoted by $A_x^T(\xi,\tau)$ for discriminating it from the wideband case). Then the transition relation [11]

$$A_x^C(\alpha,\tau) = \sqrt{1-\alpha^2/4} \iint_{\mathbf{R}^2} A_x^T(\xi, \tau+\alpha t)\, e^{-i2\pi\xi t}\, dt\, d\xi \qquad (3.52)$$

connecting the two ambiguity functions follows.

Remark. Another symmetrical form of the mixed operator of translation-dilation is conceivable. For instance, the operator product

$$e^{i2\pi(\log\sqrt{\eta})(\hat{t}\hat{\nu}+\hat{\nu}\hat{t})/2}\, e^{i2\pi\tau\hat{\nu}}\, e^{i2\pi(\log\sqrt{\eta})(\hat{t}\hat{\nu}+\hat{\nu}\hat{t})/2}$$

leads to a second symmetrical definition of the wideband ambiguity function by [12]

$$A_x^{C'}(\eta,\tau) = \int_{-\infty}^{+\infty} x\left(\frac{1}{\sqrt{\eta}}\left(t+\frac{\tau}{2}\right)\right) x^*\left(\sqrt{\eta}\left(t-\frac{\tau}{2}\right)\right) dt. \qquad (3.53)$$

Note that this new operator has a slightly more complicated symbol than the previous one. It has the form

$$\frac{1}{1+\alpha^2/4}\, \exp\left\{i\, \frac{2\pi}{1+\alpha^2/4}\, \left[2\alpha t + (1-\alpha^2/4)\tau\right]\nu\right\},$$

where we put $\alpha = 2(\sqrt{\eta}-1)/(\sqrt{\eta}+1)$.

3.1.3. Time-Frequency or Time-Scale?

The two bilinear classes (Cohen's and affine class) were introduced maintaining the distinction between frequency and scale, although the latter has a frequential interpretation as well (see Subsection 2.3.1). It is therefore important to identify the context in which the idea of the scale leads to appropriate tools. This also enables us to adopt several definitions while dealing with the pursued objectives.

Fourier scale. The formalism of the previous section was based on the time-frequency distributions of Cohen's class. Analogously, we can employ the time-scale distributions of the *affine* class for the computation of expectation values relative to time-scale functions. This could be performed according to the definition

$$\langle \widehat{G}_f \rangle_x = \iint_{\mathbf{R}^2} G(t,a)\, \Omega_x(t,a;f)\, \frac{dt\, da}{a^2}. \qquad (3.54)$$

Chapter 3 Issues of Interpretation

In this expression $\widehat{G}_f$ denotes the operator, which is associated with the (time-scale) function G in the sense of a correspondence rule, which on its part is based on the parameter function f of the distribution Ω_x. Although this operator is associated with a function of the time t and the scale a, it can be represented by means of the elementary operators $\hat{t}$ and $\hat{\nu}$. In fact, we recall from Eq. (3.19) that the affine class can be expressed in terms of the narrowband ambiguity function. We have also seen in Eq. (3.26) that this ambiguity function by itself is the expectation value of a correctly symmetrized shift-operator. As a consequence we can write

$$\Omega_x(t,a;f) = \iint_{\mathbb{R}^2} f\left(a\xi, \frac{\tau}{a}\right) A_x(\xi, \tau)\, e^{-i2\pi\xi t}\, d\xi\, d\tau$$

$$= \iint_{\mathbb{R}^2} f\left(a\xi, \frac{\tau}{a}\right) \left\langle e^{i2\pi(\xi\hat{t}+\tau\hat{\nu})} \right\rangle_x e^{-i2\pi\xi t}\, d\xi\, d\tau$$

$$= \left\langle \iint_{\mathbb{R}^2} f\left(a\xi, \frac{\tau}{a}\right) e^{i2\pi[\xi(\hat{t}-tI)+\tau\hat{\nu}]}\, d\xi\, d\tau \right\rangle_x.$$

Hence, we obtain

$$\langle \widehat{G}_f \rangle_x = \left\langle \iiint_{\mathbb{R}^3} f\left(a\xi, \frac{\tau}{a}\right) G_1(\xi, a)\, e^{i2\pi(\xi\hat{t}+\tau\hat{\nu})}\, d\xi\, d\tau\, \frac{da}{a^2} \right\rangle_x$$

by putting

$$G_1(\xi, a) = \int_{-\infty}^{+\infty} G(t, a)\, e^{-i2\pi\xi t}\, dt\ .$$

A comparison with Eq. (3.54) yields the result

$$\widehat{G}_f = \iiint_{\mathbb{R}^3} f\left(a\xi, \frac{\tau}{a}\right) G_1(\xi, a)\, e^{i2\pi(\xi\hat{t}+\tau\hat{\nu})}\, d\xi\, d\tau\, \frac{da}{a^2}\ . \tag{3.55}$$

If we now look at the special case in which the function $G(t,a)$ is just the scaling parameter a, a simple calculation brings out the correspondence rule

$$a \mapsto \frac{f(0,0)}{\hat{\nu}}\ . \tag{3.56}$$

This states that the operator associated with the scaling parameter a of the affine class is the same, apart from the constant $f(0,0)$, as the inverse of the elementary frequency-operator.

Conversely, if we start from the function $G(t,\nu) = \nu_0/\nu$, where ν_0 is an arbitrary (nonzero) reference frequency, we can show that

$$f(0,\tau) = 1 \quad \Longrightarrow \quad \frac{\nu_0}{\nu} \mapsto \frac{\nu_0}{\hat{\nu}} . \tag{3.57}$$

This double viewpoint leads to a justification of the interpretation "scale = inverse of frequency," and this motivates the use of the notion of "Fourier scale" for the parameter a.[13]

Although this time-frequency interpretation is of current use in wavelet analysis and for scalograms, it is important to observe that it can be restrictive and does not always support the most pertinent point of view. In particular, this happens when we loosen the bandpass character of the wavelet and allow spectra with several "humps." Then they cannot simply be attached to a *single* frequency. Rather, they must be associated with the *proportions* between the frequencies. Explaining this briefly in terms of music, the wavelet transform looks more like an analysis by *chords* rather than *notes*, and this renders the *scale* more meaningful than the *frequency*. In other words, better than employing the formal identification of "scale = inverse of frequency" in this case, we must consider the transformation rule

$$\frac{a}{a'} = \frac{\nu'}{\nu} . \tag{3.58}$$

Mellin scale. A different perspective concerning the notion of the scale consists of leaving the usual paradigm of the Fourier transform. Then we can adopt another viewpoint that is better suited to the operations of dilation/compression. It is based on a kind of *invariance* property, which is different from the covariance in the definition of the affine class. We shall explain this procedure very briefly. We know that the shift of a signal induces a modification of its Fourier transform involving its phase only: The modulus of the Fourier transform stays invariant under this class of operations. If we now wish to obtain a similar result relative to dilations/compressions, we must give up the Fourier transform as the appropriate tool. One can show that the *Mellin transform*[14] should be used instead. Its definition in terms of the frequential form of the signal, for instance, reads as

$$\underline{X}(s) = \int_0^{+\infty} X(\nu) \nu^{i2\pi s - 1} d\nu . \tag{3.59}$$

By means of this definition it is easy to show that

$$X_a(\nu) = X(a\nu) \quad \Longrightarrow \quad \underline{X}_a(s) = a^{-i2\pi s} \underline{X}(s) .$$

Chapter 3 Issues of Interpretation 211

In contrast to the Fourier transform, which decomposes a signal into pure complex exponentials, the Mellin transform uses the elementary waveforms

$$\nu^{-i2\pi s} = e^{-i2\pi s \log \nu} .$$

Their group delay is *hyperbolic* and not constant; that is, they verify

$$t_x(\nu) = \frac{s}{\nu} . \tag{3.60}$$

It follows that the Mellin parameter s (which cannot be associated with a dimension) can be regarded as a *hyperbolic modulation rate*. Thus it has the character of "time multiplied by frequency." Therefore, and by Eqs. (3.46), (3.47), it acquires the rank of a *scale* parameter denoted as the *Mellin scale*.

We can use this definition for the construction of new mixed representations. This was done, for example, by Marinovic.[15] He actually drew his inspiration from the structure of the Wigner-Ville distribution (construction of a bilinear kernel by multiplication of the delayed signal and the advanced signal, then taking the Fourier transform with respect to the displacement). He proposed, *mutatis mutandis*, the definition of a *scale-invariant Wigner distribution*, which performs a Mellin transform of a bilinear kernel obtained by multiplication of the compressed and the dilated signal. This amounts to the definition

$$\underline{W}_x(s,\nu) = \int_0^{+\infty} X(\lambda^{+1/2}\nu)\, X^*(\lambda^{-1/2}\nu)\, \lambda^{i2\pi s - 1}\, d\lambda . \tag{3.61}$$

Remark. A different way of introducing the same entity was proposed by Altes (called the *Q-distribution*).[16] It consists of a generalization of the duality relation equation (3.17), which exists between the ambiguity function and the Wigner-Ville distribution. One starts from the wideband definition of Eq. (3.53) and substitutes a Mellin transform for one of the Fourier transforms. It can easily be shown that Eq. (3.53) can be rewritten as

$$A_x^{C'}(\eta,\tau) = \int_{-\infty}^{+\infty} X(\eta^{+1/2}\xi)\, X^*(\eta^{-1/2}\xi)\, e^{i2\pi\xi\tau}\, d\xi . \tag{3.62}$$

Hence, we infer the relation

$$\underline{W}_x(s,\nu) = \int_{-\infty}^{+\infty} \int_0^{+\infty} A_x^{C'}(\eta,\tau)\, \eta^{i2\pi s - 1}\, e^{-i2\pi\nu\tau}\, d\tau\, d\eta . \tag{3.63}$$

As a consequence, the same procedure can generate other distributions by adopting any other suitable definition of a wideband ambiguity function.

The form of Eq. (3.50), for instance, admits the equivalent frequential expression

$$A_x^C(\alpha, \tau)$$
$$= \frac{1}{\sqrt{1-\alpha^2/4}} \int_{-\infty}^{+\infty} X\left(\frac{\xi}{1+\alpha/2}\right) X^*\left(\frac{\xi}{1-\alpha/2}\right) e^{i2\pi\xi\tau/(1-\alpha^2/4)} \, d\xi$$

(with $|\alpha| < 2$). Then, using the relation $\eta = (1+\alpha/2)/(1-\alpha/2)$ from the foregoing Subsection, we derive

$$\int_{-\infty}^{+\infty} \int_0^{+\infty} A_x^C\left(2\frac{\eta-1}{\eta+1}, \tau\right) \eta^{i2\pi s - 1} e^{-i2\pi\nu\tau} \, d\tau \, d\eta \qquad (3.64)$$

$$= \int_{-2}^{+2} X((1-\alpha/2)\nu) \, X^*((1+\alpha/2)\nu) \left(\frac{1+\alpha/2}{1-\alpha/2}\right)^{i2\pi s} \frac{d\alpha}{\sqrt{1-\alpha^2/4}} \, .$$

Among the infinity of conceivable definitions this provides one distribution, which is based on the frequency and the Mellin scale.

The distribution in Eq. (3.63) of Marinovic-Altes is called a *scale-invariant* Wigner distribution, as it has the property

$$X_a(\nu) = X(a\nu) \implies \underline{W}_{x_a}(s,\nu) = \underline{W}_x(s, a\nu) \, . \qquad (3.65)$$

Furthermore, in case of a hyperbolic chirp it gives

$$X(\nu) = \nu^{-i2\pi c} U(\nu) \implies \underline{W}_x(s,\nu) = \delta(s-c) U(\nu) \, , \qquad (3.66)$$

and this shows that it is perfectly localized to the modulation rate by means of its Mellin variable s. If we go back to the interpretation $s = t\nu$ motivated by Eq. (3.60), we can also derive a time-*frequency* representation $\widetilde{\underline{W}}_x(t,\nu) = \underline{W}_x(t\nu, \nu)$ for it. This admits a perfect localization to the corresponding rule of the group delay, which reads as

$$\widetilde{\underline{W}}_x(t,\nu) = \delta(t\nu - c) U(\nu) \, . \qquad (3.67)$$

It thus joins up with the characteristic property of the Bertrand distribution (cf. Subsection 2.3.2). This point, of course, is not fortuitous, and recent studies have shown that both distributions (Marinovic-Altes and Bertrand) can be subsumed under a common framework, the so-called *hyperbolic class*. Within this setting further generalizations and variations are conceivable, such as it was done for Cohen's and the affine class.[17]

Chapter 3 Issues of Interpretation

Analysis and decision statistics. A final and noteworthy point to be made here is the importance of the context in which we refer to the notions of frequency and scale. Schematically, we can differentiate between two typical situations: The first allows for the tasks of *analysis* and the second for the tasks of *decision*.

In the first case, the (frequency or scale) description results from a deliberate choice by the user, who is interested mainly in a better comprehension of the fine structure of the signal (finding the modulation patterns, "looking" at the details on finer scales, detecting the similarities on different resolution levels, etc.). In the second case, the situation is completely different. Here the user is *required* to introduce a description in terms of frequency or scale, if he wants to give account of a physical reality that he cannot control by himself (distinction between narrowband/wideband in presence of a Doppler effect, scale effects related to model simulations, etc.). Thus the specific context of the second situation can lead to a particular choice for the definition or interpretation.

§3.2. The Wigner-Ville Distribution and Its Geometry

The Wigner-Ville transformation plays a primordial role in theory and practice of time-frequency analysis. Although it is not the only element that can be used as a generator of Cohen's (or the affine) class *theoretically*, it yet occupies a central position *practically*, caused by the choice of its advantageous properties. The knowledge about this distribution, the analysis of its fine structure, and the comprehension about the reasons for its "good" properties were the driving forces for passing from a nice mathematical object to a family of operational tools. We are thus urged to take a closer look at this distribution.

3.2.1. Wigner-Ville versus Spectrogram

In order to emphasize some special properties of the Wigner-Ville distribution, it might be interesting to compare it with the spectrogram, which constitutes the first intuitive prototype of a time-frequency analysis.

Structure of the distributions. The spectrogram (with window $h(t)$) of a signal $x(t)$ is defined by

$$S_x(t,\nu) = \left| \int_{-\infty}^{+\infty} x(s)\, h^*(s-t)\, e^{-i2\pi\nu s}\, ds \right|^2. \tag{3.68}$$

Its evaluation combines a linear operation (Fourier transform of the weighted signal) with a quadratic operation (modulus squared). The opposite situation is in force for the Wigner-Ville distribution, which is defined by

$$W_x(t,\nu) = \int_{-\infty}^{+\infty} x\left(t+\frac{\tau}{2}\right) x^*\left(t-\frac{\tau}{2}\right) e^{-i2\pi\nu\tau} \, d\tau \;, \tag{3.69}$$

as we recall. It first uses a quadratic operation applied to the signal and then a linear transformation (Fourier transform). This constitutes an essential difference between the corresponding structures. This difference is even more pronounced by the fact that the Wigner-Ville transform, in its original form, does not require the introduction of a (more or less arbitrary) window function, which is external to the signal.

In spite of these differences, we can still bring both definitions closer together in several respects. First, of course, both belong to Cohen's class. Recall from Table 2.1 that the spectrogram and the Wigner-Ville distribution have the parameterizations $f(\xi,\tau) = A_h^*(\xi,\tau)$ and 1, respectively. As a consequence, we obtain

$$S_x(t,\nu) = \iint_{\mathbf{R}^2} W_x(s,\xi) \, W_h(s-t, \xi-\nu) \, ds \, d\xi \;. \tag{3.70}$$

A second level of comparing the two is rooted in the short-time Fourier transform. As a matter of fact, it is easy to rewrite both definitions as

$$S_x(t,\nu) = |F_x(t,\nu;h)|^2 \;, \tag{3.71}$$

$$W_x(t,\nu) = 2e^{i2\pi\nu t} \, F_x(2t, 2\nu; x_-) \;, \tag{3.72}$$

where we put $x_-(t) = x(-t)$. This shows that the spectrogram comes from a short-time Fourier transform with an external window function, while the Wigner-Ville distribution can be regarded as the same type of analysis with a "window," which is persistently matched with the signal. This second "window" is nothing but the mirror image of the signal itself. In other words, the Wigner-Ville transform amounts to the following two operations:

(i) multiplication of the signal, at each instant t, by the complex conjugate of its mirror image about this instant, in order to generate the quantity

$$q_x(t,\tau) = x\left(t+\frac{\tau}{2}\right) x^*\left(t-\frac{\tau}{2}\right) ;$$

(ii) Fourier transform of $q_x(t,\tau)$ with respect to the variable τ of the lag.

Chapter 3 Issues of Interpretation

Pseudo-Wigner-Ville. The computation of this Fourier transform may correspond to a possibly infinite time interval, and this clearly causes problems for practical applications. We can therefore try to modify the original definition of the Wigner-Ville distribution by imposing a restriction on the extension of $q_x(t, \tau)$ in the direction of τ. This can be achieved by multiplication by a window $p(\tau)$, which in turn amounts to a frequential smoothing owing to the identity

$$\int_{-\infty}^{+\infty} p(\tau)\, x\left(t + \frac{\tau}{2}\right) x^*\left(t - \frac{\tau}{2}\right) e^{-i2\pi\nu\tau}\, d\tau = \int_{-\infty}^{+\infty} P(\nu - \xi)\, W_x(t, \xi)\, d\xi \ .$$

In case this function $p(\tau)$ can be factored as

$$p(\tau) = h^*\left(\frac{\tau}{2}\right) h\left(-\frac{\tau}{2}\right) ,$$

it gives rise to the so-called *pseudo-Wigner-Ville distribution*.[18] While it stays in the spirit of the Wigner-Ville transform, it is a (moving) short-time analysis. Hence, it is also linked to the spectrogram in a certain sense.

In order to be more precise, let us introduce the shifted and weighted signal

$$x_t(s) = h^*(s)\, x(s + t) \ . \tag{3.73}$$

This enables us to define the pseudo-Wigner-Ville distribution by

$$PW_x(t, \nu) = \int_{-\infty}^{+\infty} h^*\left(\frac{\tau}{2}\right) h\left(-\frac{\tau}{2}\right) x\left(t + \frac{\tau}{2}\right) x^*\left(t - \frac{\tau}{2}\right) e^{-i2\pi\nu\tau}\, d\tau$$

$$= \int_{-\infty}^{+\infty} x_t\left(\frac{\tau}{2}\right) x_t^*\left(-\frac{\tau}{2}\right) e^{-i2\pi\nu\tau}\, d\tau \tag{3.74}$$

$$= W_{x_t}(0, \nu) \ .$$

The introduction of the auxiliary signal $x_t(s)$ (which can be associated with a moving reference mark) thus allows us to interpret the pseudo-Wigner-Ville distribution using the preceding notion of a "local mirror image," which is now restricted to a short-time neighborhood of the point of the evaluation. Indeed, we can easily see that

$$PW_x(t, \nu) = 2 \int_{-\infty}^{+\infty} x_t(\tau)\, x_t^*(-\tau)\, e^{-i4\pi\nu\tau}\, d\tau \ . \tag{3.75}$$

For every moment the pseudo-Wigner-Ville distribution is computed from *exactly the same information* as the corresponding spectrogram; but the difference remains that the latter has the form

$$S_x(t,\nu) = \left| \int_{-\infty}^{+\infty} x_t(\tau) \, e^{-i2\pi\nu\tau} \, d\tau \right|^2 . \tag{3.76}$$

Both distributions, spectrogram and pseudo-Wigner-Ville, thus use the same ingredient (the segment of the signal selected by means of a short-time window) and apply a Fourier transform together with a quadratic operation. The different order, however, in which these two operations are performed, leads to completely different properties of the distributions. This fact will be further explained by the two simple examples that follow.

Supports. The first example concerns the property of the temporal *support*. As we have seen in Chapter 2, the Wigner-Ville distribution preserves the supports of a signal in the wide sense; that is, a signal with finite duration (or bandwidth) has a Wigner-Ville transform restricted to the same duration (bandwidth, respectively). We have also observed that the same cannot be true for the spectrogram. We shall present a simple justification of this difference using the comparative approach developed here.

In fact, let a signal $x(t)$ be given, which is restricted to a time interval $[-T_x/2, +T_x/2]$, and let us apply a short-time analysis (spectrogram or pseudo-Wigner-Ville) based on a window function $h(t)$ with support $[-T_h/2, +T_h/2]$. Resuming the definition of the auxiliary signal $x_t(\tau)$, we can easily see that it has nonzero values for all instants t in the interval $[-T_x/2 - T_h/2, +T_x/2 + T_h/2]$, which is the support of the signal enlarged by the support of the window. The same must therefore be true for the spectrogram: by construction, the spectrogram begins to be nonzero, when the useful part of the signal "enters" the moving window. For a centered window this happens before the signal commences to exist by an amount of half the width of the window. The same argument applies to the end, where the spectrogram exceeds the termination of the signal by the same amount. This situation can be improved only by taking a shorter window, which in return causes a deterioration of the frequential resolution.

Let us compare this to the pseudo-Wigner-Ville distribution. Although the signal $x_t(\tau)$ is nonzero for all $-T_x/2 - T_h/2 \leq t < -T_x/2$, the "cross-product" $x_t(\tau)x_t^*(-\tau)$ vanishes identically, by construction. It only begins to be nonzero at the moment where the center of the "local mirror image" starts to "enter" the signal, and it becomes zero again when this center "leaves" the signal. Therefore, the pseudo-Wigner-Ville transform preserves the time support of a signal of finite duration. *This remains true*

Chapter 3 Issues of Interpretation

regardless of the size of the short-time window. We can thus use the freedom of enlarging the window in order to enhance the frequency resolution without affecting the temporal localization.

Localization to chirps. [19] A second example concerns the capability of a time-frequency distribution to be localized to the curve of the frequency evolution of a modulated single-component signal (chirp). Let us therefore consider the analysis of such a signal, which has an instantaneous frequency denoted by $\nu_x(t)$. The frequency spread in a fixed instant t of a spectrogram results from two sources: the maximal frequential deviation of $\nu_x(t)$ inside the window and the size of the window itself. The spectrogram is thus confronted with the well-known compromise concerning the choice for the observation window: The only way to improve the sharpness of the frequency analysis is by widening the time-window, which in return results in a greater incorporation of the instantaneous frequency. Conversely, if a better approximation of the "quasi-stationary" case is desired (small variations of $\nu_x(t)$), one needs to reduce the duration of the window, and this affects the sharpness of the frequency analysis. The best situation for the analysis is thus obtained, when the instantaneous frequency stays quasi-constant throughout the aperture of the time-window.

The situation is totally different for the pseudo-Wigner-Ville distribution. In fact, based on its definition

$$PW_x(t,\nu) = \int_{-\infty}^{+\infty} \left[h^*\left(\frac{\tau}{2}\right) h\left(-\frac{\tau}{2}\right)\right] \left[x\left(t+\frac{\tau}{2}\right) x^*\left(t-\frac{\tau}{2}\right)\right] e^{-i2\pi\nu\tau} d\tau$$

it can be viewed as a short-time Fourier transform of the modified signal

$$y_t(\tau) = x\left(t+\frac{\tau}{2}\right) x^*\left(t-\frac{\tau}{2}\right),$$

the analysis being based on the modified window

$$h'(\tau) = h^*\left(\frac{\tau}{2}\right) h\left(-\frac{\tau}{2}\right).$$

Let us suppose for the time being that the analyzed signal is given in its analytic form. We further assume that its instantaneous amplitude is quasi-constant relative to the essential support of the short-time window. Then the modified signal $y_t(\tau)$ has the remarkable property that its instantaneous

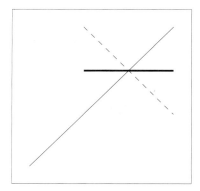 

Figure 3.1. Interpretation of the pseudo-Wigner-Ville distribution. The symbolic representation in the left-hand figure illustrates the fact that the pseudo-Wigner-Ville distribution operates like a short-time Fourier analysis at each instant, based on a *virtual* signal with an instantaneous frequency (thick line), which is the average of the instantaneous frequency of the analyzed signal (thin line) and its mirror image relative to the considered moment (dashed line). This results in a perfect localization to a linear frequency modulation. The right-hand figure depicts an actually computed distribution.

frequency is given by

$$\begin{aligned}
\nu_{y_t}(\tau) &= \frac{1}{2\pi} \frac{d}{d\tau} \arg\left\{ x\left(t + \frac{\tau}{2}\right) x^*\left(t - \frac{\tau}{2}\right) \right\} \\
&= \frac{1}{2\pi} \left\{ \frac{d}{d\tau} \arg x\left(t + \frac{\tau}{2}\right) - \frac{d}{d\tau} \arg x\left(t - \frac{\tau}{2}\right) \right\} \quad (3.77) \\
&= \frac{1}{2} \left[\nu_x\left(t + \frac{\tau}{2}\right) + \nu_x\left(t - \frac{\tau}{2}\right) \right] .
\end{aligned}$$

Hence, in the case of a *linear* frequency modulation, the modified short-time analysis encounters a *constant* frequency $\nu_x(t)$ during the whole period of the window centered around the moment t. More generally, the inherent trade-off between time- and frequency-resolution of the spectrogram is pushed by an order of magnitude, when we move to the pseudo-Wigner-Ville distribution. Here the best situation for the analysis is attained for a *quasi-linear* evolution of the instantaneous frequency during the aperture of the window (cf. Fig. 3.1). (Another interpretation of this result refers to the equation (3.72) in terms of the short-time Fourier transform with a window function, which is adapted to the signal at every instant by taking its local mirror image.)

Chapter 3 Issues of Interpretation 219

We can determine the limitations of the spectrogram more accurately when we compare its resolution for linear (and quasi-linear) frequency modulations with the pseudo-Wigner-Ville distribution that uses the same window function. Let us define a measure for the effective width $\delta t(h)$ of a short-time window $h(t)$ by the quantity

$$\delta t^2(h) = \frac{4\pi}{E_h} \int_{-\infty}^{+\infty} t^2 |h(t)|^2 \, dt \, .$$

Here $h(t)$ is supposed to have zero mean and finite energy E_h. Furthermore, we will measure the frequency spread (at a moment t) of a distribution $\rho_x(t, \nu)$ (spectrogram or pseudo-Wigner-Ville) by

$$\delta \nu^2(\rho_t) = 4\pi \left(\mu_2(t) - \mu_1^2(t) \right) \, ,$$

where we put

$$\mu_k(t) = \left(\int_{-\infty}^{+\infty} \nu^k \rho_x(t, \nu) \, d\nu \right) \Big/ \left(\int_{-\infty}^{+\infty} \rho_x(t, \nu) \, d\nu \right) \, .$$

For the sake of simplicity, let us consider the case of a Gaussian window; hence

$$h(t) = e^{-\pi \alpha t^2} \quad \Longrightarrow \quad \delta t(h) = \alpha^{-1/2} \, .$$

We further assume that the analyzed signal $x(\tau)$ can be locally approximated by a linear chirp in the vicinity of a fixed moment t; that is, its instantaneous frequency $\nu_x(\tau)$ and instantaneous amplitude $a_x(\tau)$ behave like

$$\nu_x(\tau) = \nu_0 + \beta(\tau - t) \, , \quad a_x(\tau) = 1 \quad \text{for} \quad |\tau - t| \leq \delta t(h)/2 \, .$$

A direct computation leads to

$$\delta \nu^2(S_t) = \frac{1}{\delta t^2(h)} + \beta^2 \, \delta t^2(h), \tag{3.78}$$

while at the same time

$$\delta \nu^2(PW_t) = \frac{1}{\delta t^2(h)} \, . \tag{3.79}$$

These formulas are valid for any slope of the modulation (cf. Fig. 3.2).

The slope β of the modulation plays a natural role in the spectrogram, being a control parameter of the "degree of the nonstationary behavior of the signal" relative to the observation window. During the period $\delta t(h)$ the chirp sweeps a frequency range of size $|\beta|\delta t(h)$. As the resolution of the Fourier analysis is of order $1/\delta t(h)$ for this same period, the condition of a quasi-stationary behavior takes the form

$$|\beta|\,\delta t^2(h) \ll 1 \;.$$

Hence, in the quasi-stationary case the frequency resolution of the spectrogram is inversely proportional to the time resolution. This relates to the discussion in Subsection 1.1.2 concerning the Heisenberg-Gabor uncertainty principle. However, for a nonzero slope of the modulation, the choice for $\delta t(h)$ underlies a certain compromise (see Fig. 3.2). In fact, for gaining better conformity to a quasi-stationary behavior we must reduce the size of the window; but this comes with a deterioration of the frequency resolution. Conversely, increasing $\delta t(h)$ corresponds to analyzing a signal with a larger frequency band, which (asymptotically) grows linearly with $\delta t(h)$. In between these two extreme points, there always exists an optimal window size $\delta t_{\mathrm{opt}}(h)$, for which the frequency spread of the spectrogram attains a minimum $\delta \nu_{\mathrm{min}}(S_t)$.[20] It is easy to verify that

$$\delta t_{\mathrm{opt}}(h) = |\beta|^{-1/2} \quad \Longrightarrow \quad \delta \nu_{\mathrm{min}}(S_t) = (2|\beta|)^{1/2} \;. \qquad (3.80)$$

However, in the case of a blind analysis the existence of this optimal value is of minor use, as it assumes an *a priori* knowledge of the local slope of the modulation. In other words, this underlines the fact that a spectrogram is not defined intrinsically; rather, it requires the introduction of an external function (the arbitrary window function). This function is more closely associated with the way the signal is observed than with the signal itself. This amounts to an unavoidable interaction of object-measurement type. Accordingly, the result essentially represents this interaction, and it only becomes indicative of the signal when the interaction can be considered nondisruptive (which we denoted by a quasi-stationary behavior). In the opposite case, or when the valid region of a possible quasi-stationary behavior is unknown beforehand, the window function sets limits to the information about the structure of the analyzed signal that can be drawn from the representation. If two types of nonstationary quantities with very different rates coexist in the same signal, the analysis cannot account for both simultaneously. It proves blind to either of them depending on the selected window. Therefore, with an *a priori* fixed window, all short-lived events (i.e., of a much shorter duration than the window) lose their specific nature—one can even say, that in this case the signal analyzes the window rather than the opposite ... .

Chapter 3 Issues of Interpretation   221

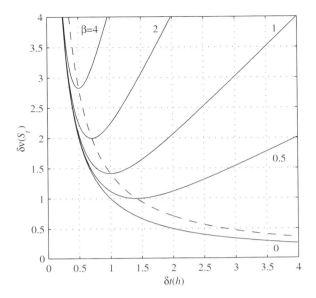

Figure 3.2. Resolutions of various spectrograms.

The time- and frequency-resolutions of a spectrogram of a linear chirp with a Gaussian window are shown. The curves drawn as solid lines correspond to different values of the slope of the frequency modulation ($\beta = 0, 0.5, 1, 2, 4$); the dashed line shows the positions of the minimal frequency spread for each fixed β (see the text).

Spectrogram and reassignment. The utility of the spectrogram faces two limitations. The first, as seen before, comes from poor properties of the representation (concerning its localization in the plane, for example). The second limitation is caused by the fact that the spectrogram is a nonunitary transformation of the signal. These bad properties result in a *loss in information*. One way to improve this situation is to use the *phase* information of the underlying short-time Fourier transform, which is completely ignored by a spectrogram. A possible way to do this, while staying within the setting of an energy distribution, was upheld by Kodera, Gendrin, and de Villedary. Their proposed method is called "Modified Moving Window Method."[21] Its argument is as follows: If we start over from the definition equation (3.70) of the spectrogram as a twice-smoothed version of the Wigner-Ville distribution, we can write

$$S_x(t,\nu) = \int_{-\infty}^{+\infty} I_x(s;t,\nu)\,ds$$

where we let

$$I_x(s;t,\nu) = \int_{-\infty}^{+\infty} W_x(s,\xi)\, W_h(s-t, \xi-\nu)\, d\xi \ .$$

At every frequency ν the spectrogram can thus be regarded as a superposition (in form of a time integral) of the terms $I_x(s;t,\nu)$; in other words, the resulting value of the spectrogram at an instant t is the *mean value* of these contributions. Here the important point is that the time average is not weighted by construction; that is, a constant weight is assigned to all values of I_x as a function of s, which itself can vary largely. Rather than associating the final result with the given time t of the evaluation, it seems wiser to *reassign* it to another instant, which gives a more representative picture of the time localization of I_x. One possible choice for the instant of the reassignment is the center of gravity of I_x, defined by

$$t_a(t,\nu) = \frac{1}{S_x(t,\nu)} \int_{-\infty}^{+\infty} s\, I_x(s;t,\nu)\, ds \ .$$

A direct computation shows that

$$t_a(t,\nu) = \mathrm{Re}\left\{ \frac{1}{F_x(t,\nu)} \int_{-\infty}^{+\infty} s\, x(s)\, h^*(s-t)\, e^{-i2\pi\nu s}\, ds \right\}$$

$$= -\frac{1}{2\pi} \mathrm{Im}\left\{ \frac{1}{F_x(t,\nu)} \frac{\partial F_x}{\partial \nu}(t,\nu) \right\} \ .$$

Consequently, if we represent the short-time Fourier transform in polar coordinates,

$$F_x(t,\nu) = |F_x(t,\nu)|\, \exp\left\{i \arg F_x(t,\nu)\right\} \ ,$$

we obtain the result

$$t_a(t,\nu) = -\frac{1}{2\pi} \frac{\partial}{\partial \nu} \arg F_x(t,\nu) \ . \tag{3.81}$$

Hence, the *group delay* of the observation of the signal through the short-time window turns out to be the result of the previous analysis.

In a dual fashion, we can write

$$S_x(t,\nu) = \int_{-\infty}^{+\infty} J_x(\xi;t,\nu)\, d\xi$$

Chapter 3 Issues of Interpretation

where

$$J_x(\xi; t, \nu) = \int_{-\infty}^{+\infty} W_x(s, \xi) W_h(s - t, \xi - \nu) \, ds \ .$$

Then we observe that the center of gravity of this last quantity, in the frequency-direction, is given by

$$\nu_{\mathrm{a}}(t, \nu) = \nu + \frac{1}{2\pi} \frac{\partial}{\partial t} \arg F_x(t, \nu) \ , \tag{3.82}$$

and this is the *instantaneous frequency* of the filtered signal.

Therefore, the computation of a modified spectrogram (in the sense of Kodera et al.) consists in:

(*i*) Computing a short-time Fourier transform;

(*ii*) Evaluating its local centers of gravity $t_{\mathrm{a}}(t, \nu)$ and $\nu_{\mathrm{a}}(t, \nu)$ in every point; and

(*iii*) Assigning the energetic contents to this new point of the time-frequency plane according to the reassignment rule

$$|F_x(t, \nu)|^2 \quad \rightarrow \quad S_x\left(t_{\mathrm{a}}(t, \nu), \nu_{\mathrm{a}}(t, \nu)\right) . \tag{3.83}$$

This operation has the advantage of "reallocating" the energy of the spectrogram at points that are close to the rule of the instantaneous frequency. Hence, if a linear frequency modulation

$$\nu_x(t) = \nu_0 + \beta t$$

is analyzed by a Gaussian window of width $1/\sqrt{\alpha}$ as in the preceding example (cf. Eq. (3.78)), we realize that

$$\arg F_x(t, \nu) = -\frac{\pi \beta}{\alpha^2 + \beta^2} \left[(\nu - \nu_0)^2 + 2\frac{\alpha^2}{\beta}(\nu - \nu_0) - \alpha^2 t^2 \right] \ .$$

It follows that the centers of gravity $t_{\mathrm{a}}(t, \nu)$ and $\nu_{\mathrm{a}}(t, \nu)$, at every point (t, ν) of the plane, verify the relation

$$\nu_{\mathrm{a}}(t, \nu) = \nu_0 + \beta \, t_{\mathrm{a}}(t, \nu) = \nu_x \left(t_{\mathrm{a}}(t, \nu) \right) \ , \tag{3.84}$$

and this signifies that they are perfectly lined up with the instantaneous frequency (cf. Fig. 3.3).

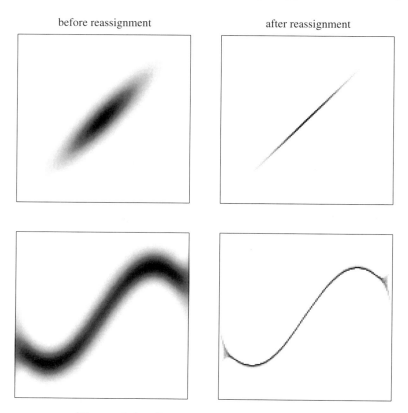

Figure 3.3. Spectrogram and reassignment.

The figure presents two examples of spectrograms before (left column) and after reassignment (right column). In the case of a linear frequency modulation (top), the reassignment yields a perfect alignment with the rule of the instantaneous frequency. Roughly the same happens for nonlinear modulations, assuming that the chosen short-time window permits a local linear approximation of the modulation (example of a sinusoidal instantaneous frequency is at the bottom).

We can therefore construct a modified spectrogram by making use of the phase information, which is supplied by the short-time Fourier transform. This can be done, so that the localization to linear chirps becomes as good as for the Wigner-Ville distribution. Moreover, Auger showed how one can circumvent an explicit computation of the phase of the short-time Fourier transform. A simple manipulation of the previous equations yields the result that the points for the reassignment are equivalently character-

Chapter 3 Issues of Interpretation

ized by

$$t_{\mathrm{a}}(t,\nu) = t + \operatorname{Re} \left\{ \frac{\langle x, \hat{t}\, h_{t\nu} \rangle}{\langle x, h_{t\nu} \rangle} \right\}, \tag{3.85}$$

$$\nu_{\mathrm{a}}(t,\nu) = \nu + \operatorname{Re} \left\{ \frac{\langle x, \hat{\nu}\, h_{t\nu} \rangle}{\langle x, h_{t\nu} \rangle} \right\}. \tag{3.86}$$

This procedure amounts to employing three window functions instead of one: $h(t)$, $th(t)$ and $(dh/dt)(t)$. In addition, this formal representation can be further generalized to other distributions as well.

Remark. For Eq. (3.82) one of the key principles of the reassignment method is the computation of an instantaneous frequency *of the filtered signal*, which is given by the short-time Fourier transform. Suppose we are interested in the fine analysis of the signal in a frequency band centered around ν. Then we can choose a window with an (adapted) width $T = 1/\nu$, so that

$$\frac{1}{2\pi} \frac{\partial}{\partial t} \arg F_x(t, 1/T) = \nu_{\mathrm{a}}(t, 1/T) - \frac{1}{T}$$

measures the *fluctuations* of the frequency around the given center. This falls in the category of another method, called *Differential Spectral Analysis*.[22]

Discretization.[23] We already mentioned the advantages of localization due to the special quadratic form of the Wigner-Ville transform. Another consequence, which can be annoying if not taken care of correctly, concerns the discretization of the transform. In fact, let us rewrite the definition of the Wigner-Ville distribution in the form

$$W_x(t,\nu) = 2 \int_{-\infty}^{+\infty} x(t+\tau)\, x^*(t-\tau)\, e^{-i4\pi\nu\tau}\, d\tau,$$

which leads to a more convenient way of discretization. We denote by

$$x[n] = x(nT_{\mathrm{e}})$$

the samples of the signal with a regular sampling period T_{e} (deliberately chosen to be 1). Then a possible discrete-time version of the previous expression is given by

$$W_x[n,\nu] = 2 \sum_{k=-\infty}^{+\infty} x[n+k]\, x^*[n-k]\, e^{-i4\pi\nu k}. \tag{3.87}$$

This expression turns out to be periodic in the frequency direction, with period 1/2. This is only half the period of the classical discrete Fourier transform of the same sequence of samples, which directly enters the discretization of the spectrogram

$$S_x[n,\nu] = \left| \sum_{k=-\infty}^{+\infty} x[k]\, h^*[k-n]\, e^{-i2\pi\nu k} \right|^2. \tag{3.88}$$

If the components of the analyzed signal are spread over the whole frequency band $[-1/2, +1/2]$, the discrete-time form of the Wigner-Ville distribution will be affected by *aliasing* in the bands $[-1/2, -1/4]$ and $[+1/4, +1/2]$. This occurs, in particular, for real-valued signals that are sampled with a rate close to the Shannon rate.

Another interpretation of this phenomenon of aliasing is based on the fact that the Wigner-Ville distribution is the Fourier transform of a bilinear kernel, which is the "cross-product" of the signal and its mirror image. Consequently, if the analyzed signal contains a spectrum-line at frequency $1/2 - \epsilon$, for example, then the cross-product maps it into a line at frequency $1 - 2\epsilon$. This *expansion of the frequency band* causes an effective undersampling of the cross-product in connection with the subsequent Fourier transform.

There are two principal ways to avoid this problem of aliasing (cf. Fig. 3.4). The first, of course, is an *oversampling* of the (real) signal by at least twice the minimal Shannon rate. Starting with a useful band $[-1/4, +1/4]$ of the signal, the cross-product occupies the larger band $[-1/2, +1/2]$, and this allows us to use a discrete Fourier transform with a unit sampling period. The second conceivable procedure is to use the *analytic signal* instead. By omitting the negative frequencies (cf. Eq. (1.22)), this amounts to a division by two of the useful band of the real signal, which can then be sampled at the minimal rate. In both cases (oversampled real or analytic signal) the number of samples must approximately be multiplied by two: The oversampling results in interlacing two sampling sequences at the Shannon rate, while the usual sampling rate for the analytic signal yields two samples per point, defined by the real and the imaginary part.

3.2.2. The Mechanism of Interferences

Because the Wigner-Ville distribution is a sesquilinear form of the signal, it cannot submit to the principle of linear superposition. This signifies that the Wigner-Ville distribution of the sum of two signals does not yield the sum of the individual distributions.

Chapter 3 Issues of Interpretation

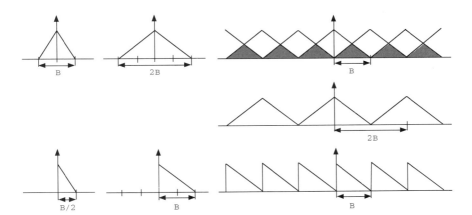

Figure 3.4. Wigner-Ville and sampling.

The figure gives a symbolic representation of the frequency behavior of the Wigner-Ville distribution of a continuous signal with a given spectrum (left). The frequency behavior of its Wigner-Ville distribution before sampling (center) and after sampling (right) is also shown. In the first line, the analyzed signal is real-valued and sampled at the minimal rate: This results in aliasing. A first solution to this problem is illustrated in the second row, where the analyzed signal is real as well, but the sampling rate is twice the Shannon rate. The bottom line presents a second solution, which uses a sampling at the minimal rate, and operates on the analytic signal (thus omitting the negative frequencies).

Just like $(a+b)^2 = a^2 + b^2 + 2ab$, one can easily verify that

$$W_{x+y}(t,\nu) = W_x(t,\nu) + W_y(t,\nu) + 2\operatorname{Re}\left\{W_{xy}(t,\nu)\right\} , \qquad (3.89)$$

with the definition of

$$W_{xy}(t,\nu) = \int_{-\infty}^{+\infty} x\left(t+\frac{\tau}{2}\right) y^*\left(t-\frac{\tau}{2}\right) e^{-i2\pi\nu\tau} d\tau . \qquad (3.90)$$

More generally, we obtain for any linear combination

$$x(t) = \sum_{n=1}^{N} a_n x_n(t) \qquad (3.91)$$

the expression

$$W_x(t,\nu) = \sum_{n=1}^{N} |a_n|^2 W_{x_n}(t,\nu) + 2\sum_{n=1}^{N-1}\sum_{k=n+1}^{N} \operatorname{Re}\left\{a_n a_k^* W_{x_n x_k}(t,\nu)\right\} . \qquad (3.92)$$

Hence, the Wigner-Ville distribution contains $N(N-1)/2$ additional components for a signal with N basic constituents. They result from the *interaction* of different components of the signal. They are called *interference terms* for two different reasons. First, the mechanism of their creation is analogous to the usual interference, which can be observed for physical waves. A second reason for this terminology lies in the *disturbing* effect that these terms introduce concerning the readability of a time-frequency diagram. So they interfere with the interpretation of the diagram, as they amount to a combinatorial proliferation of additional components.

The presence of interference terms (or cross-terms) in a Wigner-Ville distribution can be regarded as a natural consequence, which is characteristic of any bilinear structure. On the other hand, this very structure also leads to most of the good properties of the distribution (such as localization). Alhough it first looks like a drawback, the presence of interference terms is the price to pay for gaining other advantages. No matter what it is, advantage or disadvantage, it is important to understand fully the mechanism of their creation. It is indispensable for drawing the correct interpretation from the representation of an unknown signal, or for reducing the importance of these terms if desired, or even for gaining some useful information from them if needed.[24]

Construction principle. It might be useful to begin with a simple example, in order to understand the formation of the interference terms in the Wigner-Ville distribution. Let us consider a well-localized signal $x_0(t)$ (a time-frequency "atom," for example), and let us derive the quantities

$$x_+(t) = a_+ \, x_0\left(t - \frac{\Delta t}{2}\right) e^{i2\pi(\Delta\nu/2)t} \, e^{i\varphi_+} \, , \quad a_+ \geq 0 \, ,$$

$$x_-(t) = a_- \, x_0\left(t + \frac{\Delta t}{2}\right) e^{i2\pi(-\Delta\nu/2)t} \, e^{i\varphi_-} \, , \quad a_- \geq 0 \, ,$$

from it. They define two new signals, which are time-frequency translates of the original signal $x_0(t)$ including attenuations and phase factors. By construction, the signals $x_+(t)$ and $x_-(t)$ are shifted by an amount of $(+\Delta t/2, +\Delta\nu/2)$ and $(-\Delta t/2, -\Delta\nu/2)$, respectively. The covariance of the Wigner-Ville distribution relative to time-frequency shifts yields, for each signal separately, that

$$W_{x_+}(t,\nu) = a_+^2 \, W_{x_0}\left(t - \frac{\Delta t}{2}, \nu - \frac{\Delta\nu}{2}\right),$$

$$W_{x_-}(t,\nu) = a_-^2 \, W_{x_0}\left(t + \frac{\Delta t}{2}, \nu + \frac{\Delta\nu}{2}\right).$$

Let us next turn to the sum

$$x(t) = x_+(t) + x_-(t)$$

of these time-frequency atoms. Its Wigner-Ville distribution acquires the form

$$W_x(t,\nu) = W_{x_+}(t,\nu) + W_{x_-}(t,\nu) + I_{x_+x_-}(t,\nu) \ . \qquad (3.93)$$

Here the expression $I_{x_+x_-}(t,\nu)$ denotes the cross-term, which is given by

$$I_{x_+x_-}(t,\nu) = 2\operatorname{Re}\left\{\int_{-\infty}^{+\infty} x_+\left(t+\frac{\tau}{2}\right) x_-^*\left(t-\frac{\tau}{2}\right) e^{-i2\pi\nu\tau}\, d\tau\right\}. \qquad (3.94)$$

A simple computation shows that

$$I_{x_+x_-}(t,\nu) = 2\, a_+ a_-\, W_{x_0}(t,\nu)\, \cos[2\pi(t\Delta\nu - \nu\Delta t) + (\varphi_+ - \varphi_-)] \ . \qquad (3.95)$$

This means that the interference term essentially is a *modulated* version of the Wigner-Ville distribution of the unshifted signal. It thus has an *oscillating* structure, which can be better explained by looking at the previous equation more closely. In fact, the oscillations are determined by the argument of the cosine-function: they depend mainly on the time- and frequency-*distances* Δt and $\Delta\nu$ of the interacting components. Hence, we observe faster oscillations in time (or in frequency) for larger values of $\Delta\nu$ (Δt, respectively).

Regarded as a joint quantity in time and frequency, it is actually the *time-frequency distance* between the components that intervenes in this formula. The direction of the oscillations is perpendicular to the straight line connecting the centers $(t + \Delta t/2, \nu + \Delta\nu/2)$ and $(t - \Delta t/2, \nu - \Delta\nu/2)$ of the two distinct components (cf. Fig. 3.5).

The chosen example emerged from a symmetric shift in the plane; however, the covariance relative to time-frequency shifts will also yield a similar result for arbitrary displacements. In fact, let us consider the general case of two atoms

$$x_1(t) = a_1\, x_0(t - t_1)\, e^{i2\pi\nu_1 t}\, e^{i\varphi_1} \ , \qquad a_1 \geq 0,$$

$$x_2(t) = a_2\, x_0(t - t_2)\, e^{i2\pi\nu_2 t}\, e^{i\varphi_2} \ , \qquad a_2 \geq 0.$$

Then the signal $x(t) = x_1(t) + x_2(t)$ has the Wigner-Ville distribution

$$W_x(t,\nu) = a_1^2\, W_{x_0}(t-t_1, \nu-\nu_1) + a_2^2\, W_{x_0}(t-t_2, \nu-\nu_2) + I_{x_1 x_2}(t,\nu) \ , \qquad (3.96)$$

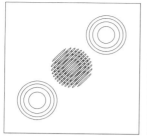  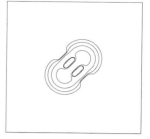

Figure 3.5. Wigner-Ville transform of two atoms.

If a signal is the superposition of two atoms, the Wigner-Ville distribution consists of two smooth contributions (related to both atoms) and one interference term (related to their interaction). The latter is located at the midpoint of the line connecting the time-frequency centers of the given atoms. Moreover, it has an oscillatory structure in a direction perpendicular to that line, and a frequency proportional to the distance between the atoms. The figure depicts this behavior for pairs of identical atoms at different distances.

with the cross-term given by

$$I_{x_1 x_2}(t,\nu) \tag{3.97}$$
$$= 2\, a_1\, a_2\, W_{x_0}(t - t_\mathrm{i}, \nu - \nu_\mathrm{i})\, \cos\left\{2\pi[(t - t_\mathrm{i})\Delta\nu - (\nu - \nu_\mathrm{i})\Delta t] + \Delta\varphi\right\}\,.$$

Here we employed the notations

$$t_\mathrm{i} = \frac{t_1 + t_2}{2}\,,\qquad \nu_\mathrm{i} = \frac{\nu_1 + \nu_2}{2}\,, \tag{3.98}$$

$$\Delta t = t_1 - t_2\,,\qquad \Delta \nu = \nu_1 - \nu_2\,, \tag{3.99}$$

$$\Delta\varphi = \varphi_1 - \varphi_2 + 2\pi\, t_\mathrm{i}\, \Delta\nu\,. \tag{3.100}$$

We thus recover the principle that the interference term is concentrated in a neighborhood of the point in the time-frequency plane, which is the (geometric) midpoint between the interacting components. We can also retain the idea that its oscillatory structure is governed by the time-frequency distance between the components.

Chapter 3 Issues of Interpretation

A different perspective from the ambiguity plane.[25] The same construction principle also admits a dual interpretation in the ambiguity plane. Indeed, for the signal $x(t) = x_1(t) + x_2(t)$ (from the preceding text) we obtain

$$A_x(\xi, \tau) = A_{x_1}(\xi, \tau) + A_{x_2}(\xi, \tau) + J_{x_1 x_2}(\xi, \tau)$$

where

$$J_{x_1 x_2}(\xi, \tau) = A_{x_1 x_2}(\xi, \tau) + A_{x_2 x_1}(\xi, \tau) .$$

From the Fourier relation of Eq. (3.17) between the Wigner-Ville distribution and the ambiguity function, we can conclude that

$$I_{x_1 x_2}(t, \nu) = \iint_{\mathbb{R}^2} J_{x_1 x_2}(\xi, \tau) \, e^{-i 2\pi (\nu \tau + \xi t)} \, d\xi \, d\tau .$$

Furthermore, let us recall that any ambiguity function is Hermitian (which is dual to the reality of the Wigner-Ville distribution); that is, we have

$$A_{x_2 x_1}(\xi, \tau) = A^*_{x_1 x_2}(-\xi, -\tau) .$$

Hence, the behavior of the interference term $I_{x_1 x_2}(t, \nu)$ can be understood by inspecting the cross-ambiguity function $A_{x_1 x_2}(\xi, \tau)$. For the previously considered example we find

$$A_{x_1 x_2}(\xi, \tau) = a_1 \, a_2 \, A_{x_0}(\xi + \Delta\nu, \tau - \Delta t) \, e^{i[2\pi(\xi t_i + \tau \nu_i) + \Delta\varphi]} . \qquad (3.101)$$

This relation implies that the Fourier transform of the cross-term of the two signals $x_1(t)$ and $x_2(t)$ reproduces the ambiguity function of the basic signal $x_0(t)$, from which they were derived by time-frequency shifts; the only change is its displacement from the origin of the ambiguity plane. The amount of the displacement is identical to the time-frequency distance of the two signals. Certainly, this result conforms to the interpretation of an ambiguity function in terms of time-frequency *correlations*: If two components are localized to neighborhoods of the points (t_1, ν_1) and (t_2, ν_2) in the time-frequency plane, the essential contribution to their cross-correlation involves those shift parameters, which put them in an overlay position; these correspond to the time- and frequency-separations Δt and $\Delta \nu$ between the two components.

By combining the cross-ambiguity function with its Hermitian copy and taking the localized and eccentric character of $A_{x_1 x_2}(\xi, \tau)$ into account, we recognize from a different angle, why the cross-term $I_{x_1 x_2}(t, \nu)$ has an oscillatory structure: It results from the properties of the Fourier transform operating on a "bandpass" function in the ambiguity plane. From this point of view, the interference term oscillates more rapidly, if the cross-ambiguity function of the two components is located in a region farther from the origin, and this coincides with the direct argument of a greater distance between the components.

Conversely, and opposite to the cross-term $I_{x_1 x_2}(t, \nu)$, the "signal" terms $W_{x_1}(t, \nu)$ and $W_{x_2}(t, \nu)$ are the Fourier transforms of the auto-ambiguity functions $A_{x_1}(\xi, \tau)$ and $A_{x_2}(\xi, \tau)$. Carrying on the forementioned interpretation in terms of correlations, these two functions essentially live close to the origin, which lends them a "lowpass" character in the ambiguity plane. This results in an *a priori* smooth structure of the signal terms, which are just their Fourier transforms.

Inner and outer interferences. [26] The construction principle, which was just established for distinct "atoms," can also serve to discuss more complex situations, which can be brought back into the form of a superposition of atoms. We will denote them as *outer interferences*, when they result from the interaction of components (atoms or sets of atoms) that are significantly separated in the time-frequency plane. This point of view, however, might seem artificial in certain cases, where it does not correspond to an objective or physically relevant decomposition of the signal. This case occurs, in particular, for modulated chirps. While such signals cannot be nicely described by considering an objective separation into distinct components, their Wigner-Ville distributions can still possess oscillating structures. We shall regard them as *inner interferences* (cf. Fig. 3.6).

Obviously, there is no clean borderline between these two concepts of inner and outer interferences. We adopted both notions only for the sake of convenience. One can observe that they overlap and complement each other by inspecting the so-called *interference formula of Janssen*. [27] For its demonstration, let us only consider the envelope of the cross-term of two signals $x(t)$ and $y(t)$, which we write as

$$|W_{xy}(t,\nu)|^2 = \left| \int_{-\infty}^{+\infty} \kappa_x(s;t,\nu)\, \kappa_y^*(-s;t,\nu)\, ds \right|^2$$

where

$$\kappa_x(s;t,\nu) = x\left(t + \frac{s}{2}\right) e^{-i2\pi\nu(t+s/2)} .$$

It immediately follows from this definition that the Wigner-Ville transform of $\kappa_x(s;t,\nu)$, considered as a function of s with parameters t and ν, is equal to

$$W_{\kappa_x}(s,\xi) = 2\, W_x\left(t + \frac{s}{2}, \nu + 2\xi\right) .$$

Chapter 3 Issues of Interpretation

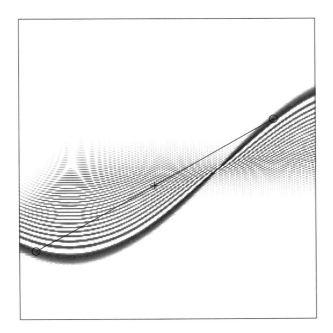

Figure 3.6. Wigner-Ville and inner interferences.

For a nonlinear frequency modulation (which is of sinusoidal type in the example), the Wigner-Ville distribution gives rise to an interference structure, called inner interference, whose construction principle is based on a pointwise application of the principle depicted in Fig. 3.5.

Likewise, the Wigner-Ville transform of $\widetilde{\kappa}_y(s;t,\nu) = \kappa_y(-s;t,\nu)$ is given by

$$W_{\widetilde{\kappa}_y}(s,\xi) = 2\,W_y\left(t - \frac{s}{2}, \nu - 2\xi\right)\ .$$

As a consequence, we infer from an application of Moyal's formula (cf. Eq. (2.95)) that

$$\left|\int_{-\infty}^{+\infty} \kappa_x(s;t,\nu)\,\kappa_y^*(-s;t,\nu)\,ds\right|^2 = \iint_{\mathbb{R}^2} W_{\kappa_x}(s,\xi)\,W_{\widetilde{\kappa}_y}(s,\xi)\,ds\,d\xi\ .$$

This leads to the searched result

$$|W_{xy}(t,\nu)|^2 = \iint_{\mathbb{R}^2} W_x\left(t + \frac{s}{2}, \nu + \frac{\xi}{2}\right) W_y\left(t - \frac{s}{2}, \nu - \frac{\xi}{2}\right) ds\,d\xi\ , \quad (3.102)$$

which is known as the *formula of outer interferences*.

This formula permits several interpretations. First, if the Wigner-Ville distributions of the components $x(t)$ and $y(t)$ are concentrated in regions around (t_1, ν_1) and (t_2, ν_2), their cross-term essentially exists around the midpoint (t_i, ν_i). On the other hand, we can look upon the existence of an interference term at a point (t, ν) as the result of the (infinite) superposition of *pointwise* interactions, which obey the "midpoint rule" of separated atoms as already established.

As long as the given components are reasonably separated, this qualitative description explains the existence of *outer* interferences. However, there is no reason why the same formula should not be applicable in the extreme case where $x(t) = y(t)$. Then one obtains the *formula of inner interferences*

$$|W_x(t,\nu)|^2 = \iint_{\mathbb{R}^2} W_x\left(t + \frac{s}{2}, \nu + \frac{\xi}{2}\right) W_x\left(t - \frac{s}{2}, \nu - \frac{\xi}{2}\right) ds\, d\xi . \quad (3.103)$$

This accounts for the interaction of a single component with itself, and is thus denoted as *inner* interference.

Approximation by the method of stationary phase. [28] The model of two distinct atoms was well-suited for the study of the mechanism of outer interferences. As far as inner interferences are concerned, a useful model of nonatomic type is given by a signal with a modulated amplitude and frequency.

Let us therefore consider a signal (supposed to be analytic) of the form

$$x(t) = a(t)\, e^{i\varphi(t)} , \quad (3.104)$$

where $a(t) \geq 0$ denotes its instantaneous amplitude and

$$\nu_x(t) = \frac{1}{2\pi} \frac{d\varphi}{dt}(t) \quad (3.105)$$

is the instantaneous frequency. Intuitively, a time-frequency representation should reduce to a "signal" term, which is localized to a vicinity of the curve of the instantaneous frequency. The fact that a distribution can be supported outside this curve (as seen in Fig. 3.6, for example) results from the inner interferences. We shall now describe their geometry more precisely.

The Wigner-Ville distribution of the chosen model signal of Eq. (3.104) has the form

$$W_x(t,\nu) = \int_{-\infty}^{+\infty} L(\tau;t)\, e^{i\Phi(\tau;t,\nu)}\, d\tau , \quad (3.106)$$

Chapter 3 Issues of Interpretation

with
$$L(\tau;t) = a\left(t+\frac{\tau}{2}\right) a\left(t-\frac{\tau}{2}\right),$$
$$\Phi(\tau;t,\nu) = \varphi\left(t+\frac{\tau}{2}\right) - \varphi\left(t-\frac{\tau}{2}\right) - 2\pi\nu\tau.$$
(3.107)

Suppose the variations of the amplitude L are slow in comparison with those of the phase Φ. (This corresponds to endowing the mathematical definition of the instantaneous frequency with its intuitive physical meaning.) Then the Wigner-Ville distribution attains the form of an *oscillatory integral*. It can thus be computed approximately by the method of *stationary phase*. As we recall from Subsection 1.2.1, the underlying principle of this method is rooted in the assumption that the significant contributions to the integral of a highly oscillating function emerge only from neighborhoods of the "stationary" points, for which the derivative of the phase vanishes.

By definition, the stationary points of the integral (Eq. (3.106)) under consideration are the solutions of the equation

$$\frac{\partial \Phi}{\partial \tau}(\tau;t,\nu) = 0 \qquad (3.108)$$

solved for the variable τ. Provided that N solutions exist, and that they meet the condition

$$\frac{\partial^2 \Phi}{\partial \tau^2}(\tau_n;t,\nu) \neq 0, \quad n = 1, \ldots, N,$$

the approximation by the method of stationary phase allows us to write

$$W_x(t,\nu) \approx (2\pi)^{1/2} \sum_{n=1}^{N} \left|\frac{\partial^2 \Phi}{\partial \tau^2}(\tau_n;t,\nu)\right|^{-1/2}$$
$$\times L(\tau_n;t) \, \exp\left\{i\Phi(\tau_n;t,\nu) + i\frac{\pi}{4}\,\mathrm{sgn}\,\frac{\partial^2 \Phi}{\partial \tau^2}(\tau_n;t,\nu)\right\}.$$

In view of the definition (Eq. (3.107)) of $\Phi(\tau;t,\nu)$, the explicit computation of these stationary points boils down to finding the solutions (in τ) of the equation

$$\nu = \frac{1}{2}\left\{\nu_x\left(t+\frac{\tau}{2}\right) + \nu_x\left(t-\frac{\tau}{2}\right)\right\}, \qquad (3.109)$$

under the side-condition

$$\frac{d\nu_x}{dt}\left(t+\frac{\tau}{2}\right) \neq \frac{d\nu_x}{dt}\left(t-\frac{\tau}{2}\right).$$

It is a simple fact that the stationary points appear in pairs; in other words, if τ_n is a solution, then $-\tau_n$ must be a solution as well. If we retain only the positive stationary points, we can once more rewrite the previous approximation as

$$W_x(t,\nu) \approx (8\pi)^{1/2} \sum_{n=1}^{N/2} \left| \frac{\partial^2 \Phi}{\partial \tau^2}(\tau_n; t, \nu) \right|^{-1/2} \quad (3.110)$$

$$\times\ L(\tau_n; t)\ \cos\left\{ \Phi(\tau_n; t, \nu) + \frac{\pi}{4}\operatorname{sgn}\frac{\partial^2 \Phi}{\partial \tau^2}(\tau_n; t, \nu) \right\}.$$

Hence, the stationary phase approximation yields a result, which coincides with the construction principle of the outer interferences: The Wigner-Ville distribution has non-negligible values at all points (t, ν) of the time-frequency plane, which are midpoints of a line connecting any two points of the curve of the instantaneous frequency.

Moreover, the approximation permits a further quantization of the importance of the inner interferences. In fact, let us consider a point (t, ν), for which the interference results from the interaction of exactly two points $(t_1, \nu_x(t_1))$ and $(t_2, \nu_x(t_2))$, so that

$$t = \frac{t_1 + t_2}{2}, \qquad \nu = \frac{\nu_x(t_1) + \nu_x(t_2)}{2}.$$

Hence, we look at a situation where only two stationary points $\tau = \pm(t_1 - t_2)$ exist. The Wigner-Ville distribution has the approximate value

$$W_x(t,\nu) \approx 4\, a(t_1)\, a(t_2) \left(\left| \frac{d\nu_x}{dt}(t_2) - \frac{d\nu_x}{dt}(t_1) \right| \right)^{-1/2}$$

$$\times\ \cos\left\{ \varphi(t_2) - \varphi(t_1) - 2\pi\nu(t_2 - t_1) \pm \frac{\pi}{4} \right\}.$$

This gives

$$W_x(t,\nu) \approx 4\, a(t_1)\, a(t_2) \left(\left| \frac{d\nu_x}{dt}(t_2) - \frac{d\nu_x}{dt}(t_1) \right| \right)^{-1/2} \cos\left\{ A(t_1, t_2) \pm \frac{\pi}{4} \right\}, \quad (3.111)$$

where $A(t_1, t_2)$ is the area enclosed by the curve of the instantaneous frequency and the line connecting the two interacting points. It follows that the oscillations of the inner interference at a point of the plane are faster, if this point has a greater distance from the instantaneous frequency.

Chapter 3 Issues of Interpretation   237

This local oscillating behavior can be determined more precisely when we look into small perturbations δt and $\delta \nu$ of the point (t, ν), at which the Wigner-Ville transform is evaluated. If the distance $t_1 - t_2$ remains constant, we find

$$A(t_1, t_2) \approx 2\pi \left[\Delta \nu \, \delta t - \Delta t \, \delta \nu\right] + \text{const} .$$

This reproduces our earlier result, according to which the *local* structure of $W_x(t + \delta t, \nu + \delta \nu)$ can be derived from a *pointwise* application of the principle of outer interferences.

It is clear that the approximation by the method of stationary phase is only valid, if the slopes of the instantaneous frequency at the interacting points are different. The closer these slopes are, the bigger is the amplitude of the interference term. Divergence occurs in the extreme case of equal slopes. This equality of the slopes may correspond to two different scenarios:

(*i*) The interacting points *coalesce* ($\tau = 0$, line of zero length). Hence, the rule of the instantaneous frequency itself turns out to be characteristic of a divergent behavior of the approximation: As we anticipated intuitively, the amplitude of the Wigner-Ville distribution will be significant at all points of this curve, and the distribution has a nonoscillating structure due to the relation $A(t, t) = 0$; and

(*ii*) The interacting points are *distinct* ($\tau \neq 0$, line of nonzero length). This defines a "phantom" curve, which is different from the curve of the instantaneous frequency. The Wigner-Ville distribution has a big amplitude along this curve. Its oscillating behavior is characteristic of the presence of interferences.

This behavior is depicted in Fig. 3.7, which displays an example of a segment of a sinusoidal frequency modulation.

Singularities and catastrophes. [29] The curve of the instantaneous frequency, along which the stationary phase approximation diverges, can locally be interpreted as a *borderline* between two regions of the time-frequency plane with radically different properties: For every point inside the convex hull there exist at least two stationary points, while outside there exist none. This abrupt change in the behavior when passing from one region to the other can be regarded as a *catastrophe* of the Wigner-Ville distribution, using the language of Thom.

In our special setting it suffices to recall as a result from catastrophe theory, that a *qualitative* description of an oscillatory integral is obtained from considering the structure of the singularities of the phase. This means

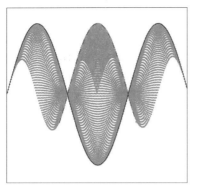 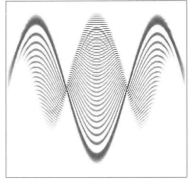

Figure 3.7. Wigner-Ville and stationary phase.

(Left: theoretical model; right: computed Wigner-Ville distribution.) The shadowed region represents the set of points, which are the midpoints of a line connecting two arbitrary points of the curve of the instantaneous frequency (which is sinusoidal in this example, drawn as a solid line). At these points, the stationary phase approximation predicts a significant value of the Wigner-Ville distribution. This approximation diverges in the subset of all points, which are midpoints of a line with endpoints on the curve of the instantaneous frequency, where the slopes of this curve are equal. This subset is composed of two lines, the instantaneous frequency (solid line) and a "phantom" curve (dashed line).

that we have to differentiate between several cases concerning the number of vanishing consecutive derivatives of the phase. The approach essentially reduces to employing a "generic" phase, which has the same structure as the real phase concerning its singularities. This leads to a *description* of the oscillatory integral, which remains valid even when the method of stationary phase breaks down. At the same time it amounts to a *classification* of the possible geometries that can materialize in a distribution. These reduce to a small number of typical cases called the *elementary catastrophes*. The number of possible cases depends on the dimension of the problem; that is, it is connected with the number of parameters that are needed to determine the phase. In case of the Wigner-Ville distribution, considered as an oscillatory integral, the phase $\Phi(\tau; t, \nu)$ is a function of the variable τ and two additional parameters, namely time t and frequency ν. A remarkable result emerging from this theory is that only *two* elementary catastrophes are dominant and structurally stable (in the sense that small perturbations of Φ do not change their qualitative nature). The following distinction of the conceivable cases renders this result understandable, although it lacks

Chapter 3 Issues of Interpretation

any validity of a proof.

There actually exist five families of points in the plane, as far as the qualitative behavior of the Wigner-Ville distribution regarding the consecutive derivatives of the phase is concerned. Only two of them correspond to stable singularities:

1. The points (t, ν) such that

$$\frac{\partial \Phi}{\partial \tau}(\tau; t, \nu) \neq 0 .$$

These are the points, at which the equation

$$\nu = \frac{1}{2} \left\{ \nu_x \left(t + \frac{\tau}{2} \right) + \nu_x \left(t - \frac{\tau}{2} \right) \right\}$$

has no solution for τ. Consequently, they correspond to the *areas*, where $W_x(t, \nu)$ is negligible.

2. The points (t, ν) such that

$$\frac{\partial \Phi}{\partial \tau}(\tau; t, \nu) = 0 , \quad \frac{\partial^2 \Phi}{\partial \tau^2}(\tau; t, \nu) \neq 0 .$$

They define the *areas*, where $W_x(t, \nu)$ is non-negligible and where the approximation by the method of stationary phase is appropriate. As explained before, these points correspond to the midpoints of lines connecting two arbitrary points of the instantaneous frequency, where the slopes of this curve do not coincide. Hence, they verify the relations

$$\nu = \frac{1}{2} \left\{ \nu_x \left(t + \frac{\tau}{2} \right) + \nu_x \left(t - \frac{\tau}{2} \right) \right\} ,$$

$$\frac{d\nu_x}{dt} \left(t + \frac{\tau}{2} \right) \neq \frac{d\nu_x}{dt} \left(t - \frac{\tau}{2} \right) .$$

The Wigner-Ville distribution has an oscillatory structure near these points, which is characterized by a circular function (cf. Eq. (3.111)).

3. The points (t, ν) such that

$$\frac{\partial \Phi}{\partial \tau}(\tau; t, \nu) = \frac{\partial^2 \Phi}{\partial \tau^2}(\tau; t, \nu) = 0 , \quad \frac{\partial^3 \Phi}{\partial \tau^3}(\tau; t, \nu) \neq 0 .$$

They correspond to the *curves* where $W_x(t, \nu)$ has large values. The approximation by the method of stationary phase diverges here. These curves are composed of the midpoints of lines with endpoints on the instantaneous frequency, where the slopes agree and the absolute values of the curvatures do not coincide, the curvatures themselves having opposite signs. More precisely, these points are characterized by

$$\nu = \frac{1}{2}\left\{\nu_x\left(t+\frac{\tau}{2}\right) + \nu_x\left(t-\frac{\tau}{2}\right)\right\},$$

$$\frac{d\nu_x}{dt}\left(t+\frac{\tau}{2}\right) = \frac{d\nu_x}{dt}\left(t-\frac{\tau}{2}\right),$$

$$\frac{d^2\nu_x}{dt^2}\left(t+\frac{\tau}{2}\right) + \frac{d^2\nu_x}{dt^2}\left(t-\frac{\tau}{2}\right) \neq 0.$$

This situation corresponds to a so-called *fold catastrophe*. We already mentioned (without referring to this name), that these points constitute two curves, one of which is the instantaneous frequency itself, and another is a "phantom" curve emerging from it (cf. Fig. 3.7).

In the neighborhood of a fold singularity, the approximation is governed by the *Airy function*

$$\mathrm{Ai}(x) = \int_{-\infty}^{+\infty} \exp\left\{i\left(xu - \frac{u^3}{3}\right)\right\} du.$$

By a third-order expansion of $\Phi(\tau; t, \nu)$ around this point and by ignoring the influence of the instantaneous amplitude (or, which is equivalent, considering it as locally constant and fixing this constant to 1), we obtain the *transitional approximation*

$$W_{e^{i\varphi}}(t,\nu) \approx \frac{1}{\epsilon(t)}\,\mathrm{Ai}\left(\frac{1}{\epsilon(t)}\left[\nu - \nu_x(t)\right]\right), \qquad (3.112)$$

with

$$\epsilon(t) = \left(\frac{1}{32\pi^2}\left|\frac{d^2\nu_x}{dt^2}(t)\right|\right)^{1/3}.$$

Remark. If we wish to take the instantaneous amplitude into account explicitly, it suffices to use the compatibility of the Wigner-Ville transform with products. Then the approximation

$$W_x(t,\nu) \approx \frac{1}{\epsilon(t)} \int_{-\infty}^{+\infty} W_a(t, \nu - \xi)\,\mathrm{Ai}\left(\frac{1}{\epsilon(t)}\left[\xi - \nu_x(t)\right]\right) d\xi$$

is justified.

The transitional approximation thus appears as a refinement of the method of stationary phase in a neighborhood of the points, where the

Chapter 3 Issues of Interpretation

latter diverges. For a nonlinear frequency modulation the form of the Airy function implies that the Wigner-Ville distribution has significant oscillations inside the convex hull of the instantaneous frequency and that it decays exponentially outside (cf. Fig. 3.7). (We can further observe that the instantaneous frequency at an instant t is the *center of gravity* in the frequency-direction of the distribution, and not its maximum.) The velocity of the oscillations inside the convex hull is controlled by the local *curvature* of the instantaneous frequency: The oscillations are slower, if the curvature is small, and vice versa. The extreme case of a linear modulation corresponds to letting the curvature tend to zero. Then the remarkable property of the Airy function

$$\lim_{\epsilon \to 0} \frac{1}{\epsilon} \operatorname{Ai}\left(\frac{x}{\epsilon}\right) = \delta(x)$$

yields the anticipated result related to linear chirps, which states that

$$W_x(t,\nu) \approx W_a(t, \nu - \nu_x(t)).$$

4. The points (t, ν) such that

$$\frac{\partial \Phi}{\partial \tau}(\tau; t, \nu) = \frac{\partial^2 \Phi}{\partial \tau^2}(\tau; t, \nu) = \frac{\partial^3 \Phi}{\partial \tau^3}(\tau; t, \nu) = 0, \quad \frac{\partial^4 \Phi}{\partial \tau^4}(\tau; t, \nu) \neq 0.$$

These points are midpoints of lines with endpoints on the curve of the instantaneous frequency, where the slopes are equal and the curvatures agree in absolute values and have opposite signs. Hence, the points are characterized by

$$\nu = \frac{1}{2}\left\{\nu_x\left(t + \frac{\tau}{2}\right) + \nu_x\left(t - \frac{\tau}{2}\right)\right\},$$

$$\frac{d\nu_x}{dt}\left(t + \frac{\tau}{2}\right) = \frac{d\nu_x}{dt}\left(t - \frac{\tau}{2}\right),$$

$$\frac{d^2\nu_x}{dt^2}\left(t + \frac{\tau}{2}\right) + \frac{d^2\nu_x}{dt^2}\left(t - \frac{\tau}{2}\right) = 0,$$

$$\frac{d^3\nu_x}{dt^3}\left(t + \frac{\tau}{2}\right) \neq \frac{d^3\nu_x}{dt^3}\left(t - \frac{\tau}{2}\right).$$

By construction, they must belong to a "phantom" curve of folds: In fact, they represent its *pucker points*. The corresponding catastrophe is called a *cusp*. Its characteristics are subsumed by the *Pearcey function*

$$\operatorname{Pe}(x, y) = \int_{-\infty}^{+\infty} \exp\left\{i\left(y u - x \frac{u^2}{2} + \frac{u^4}{8}\right)\right\} du.$$

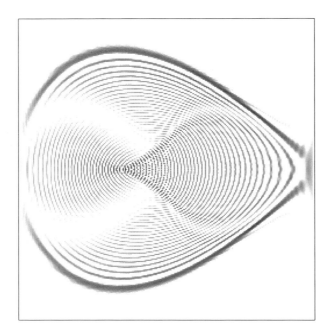

Figure 3.8. Wigner-Ville and cusp singularities.

The Wigner-Ville distribution shows a cusp singularity (which is locally described by a Pearcey function) at each midpoint of a straight line between two points of the instantaneous frequency, where the slopes agree and the curvatures have equal absolute values and opposite signs. A cusp corresponds to a pucker point of a fold curve.

Remark. We should mention that the appearance of a cusp, while staying in the setting of frequency modulations, requires a superposition of distinct instantaneous frequencies; it thus refers to a situation of outer interferences (cf. Fig. 3.8).

5. The points (t, ν), which are centers of a perfect symmetry or antisymmetry. These points are the midpoints of infinitely many straight lines. The local behavior has the form of a Dirac distribution (cf. Fig. 3.9).

However, this last scenario cannot be considered on the same level as the previous ones because it describes an "unstable" situation. This means that the slightest modification of the phase causes the exceptional character of the singularity to disappear and return to one of the previous cases. An analogous situation is known for the focus of a perfect lens: Every deformation of the lens, as small as it may be, transforms the ideal

Chapter 3 Issues of Interpretation

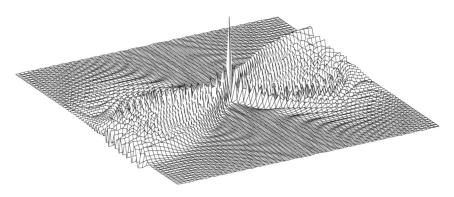

Figure 3.9. Wigner-Ville and higher-order singularities.

The Wigner-Ville distribution shows a singularity of higher order (which looks like a Dirac distribution) at each point, which is the midpoint of infinitely many straight lines connecting two points of the instantaneous frequency. The figure depicts an example of such a situation (case of a sinusoidal frequency modulation).

focal point into a caustic (whose sections are cusps). On the other hand, if we directly start from a lens with a certain aberration, a small modification affects the *quantitative* nature of its caustic, but not its *qualitative* feature.

The objective of catastrophe theory is a classification of structurally stable morphologies. Accordingly, we are able to conclude that only two possible structures can occur in a Wigner-Ville distribution in the time-frequency plane (for signals with amplitude and frequency modulations); these are fold curves and cusps. The corresponding construction principles were explained in cases 3 and 4 presented in the preceding text.

Interferences, localization, and symmetries. We can roughly summarize the construction principle of the (outer and inner) interferences of the Wigner-Ville distribution as follows: *Two points of the plane interfere, so that they create a contribution at a third point, which is the midpoint of the straight line connecting the two.* Although this principle is so simple, it still gives rise to several consequences and interpretations, which put a new light on the Wigner-Ville transform.

The first conclusion in connection with this "midpoint" rule can be drawn from its recursive application. Suppose we begin with two arbitrary points and pretend that the midpoint, which is generated in each step of the iteration, belongs to the signal itself. Then it can again interfere with already existing points. By an iteration of the same construction principle, we obtain a perfect alignment of infinitely many points on the straight line connecting the two starting points. Hence, we can argue just on the basis of the geometry of the interferences, that the Wigner-Ville distribution must be perfectly localized to linear chirps. Turning this argument upside down, we conclude that the Wigner-Ville distribution can only be localized to a curve in the time-frequency plane (viewed as an instantaneous frequency), if the midpoint of any two points of the curve lies on the same curve again; the only possible solution that meets this criterion is the straight line.

A second interpretation of the "midpoint" rule can be gained from an opposite perspective. If two points create a new point by interference, then they must be *symmetrical* about this third point. More specifically, let us rewrite the definition of the Wigner-Ville distribution as

$$W_x(t,\nu) = 2 \int_{-\infty}^{+\infty} \left[x(2t-\tau) \, e^{-i4\pi\nu(t-\tau)} \right] x^*(\tau) \, d\tau \; ,$$

so that it takes the form of an operator expectation value. We next introduce the *displacement operator* (following Glauber's terminology), which is defined by

$$\widehat{D}_{t\nu} = e^{i2\pi(\nu\hat{t} - t\hat{\nu})} \; . \tag{3.113}$$

A direct computation leads to

$$W_x(t,\nu) = 2 \, \langle \widehat{D}_{t\nu} \, \widehat{\Pi} \, \widehat{D}_{-t,-\nu} \rangle_x \; , \tag{3.114}$$

where $\widehat{\Pi}$ is the parity operator (or the reflection about the origin of the time-frequency plane), so that

$$(\widehat{\Pi} x)(t) = x(-t) \; , \quad (\widehat{\Pi} X)(\nu) = X(-\nu) \; . \tag{3.115}$$

It follows that the value of the Wigner-Ville distribution at one point of the time-frequency plane can be regarded as the expectation value of a parity operator, which is centered around this point by means of a composition with the displacement operator.[30] Hence, we can conclude that the existence of the Wigner-Ville transform in a point is determined by the interaction of components that are symmetrically distributed about this point. This conforms to the interpretation of Janssens's interference formula (Eq. (3.103)). Restated once again, the value of the Wigner-Ville distribution in one point is a measure of how much the signal is "centered" around this point.

Chapter 3 Issues of Interpretation

Generalization to the s-Wigner distributions. Several generalizations of the previous results can be addressed. The first deals with an extension of the geometry of the interferences to the class of s-Wigner distributions

$$W_x^{(s)}(t,\nu) = \int_{-\infty}^{+\infty} x\left(t - \left(s - \tfrac{1}{2}\right)\tau\right) x^*\left(t - \left(s + \tfrac{1}{2}\right)\tau\right) e^{-i2\pi\nu\tau}\, d\tau \, .$$

Recall that the Wigner-Ville distribution appears as the special case $s = 0$ (cf. Subsection 2.3.2).

Without giving all the details, we mention only some typical results pointing in this direction. When we look at the s-Wigner distribution of the previously discussed example (sum of two shifted atoms, cf. Eqs. (3.96)-(3.101)), we obtain

$$W_x^{(s)}(t,\nu) = a_1^2\, W_{x_0}^{(s)}(t-t_1, \nu-\nu_1) + a_2^2\, W_{x_0}^{(s)}(t-t_2, \nu-\nu_2) + I_{x_1 x_2}^{(s)}(t,\nu)\,.$$

This is the same structure as before, except that the interference comprises two terms

$$\begin{aligned}I_{x_1 x_2}^{(s)}(t,\nu) = a_1 a_2\, \big[&F^+\left(t - (t_\text{i} + s\Delta t), \nu - (\nu_\text{i} - s\Delta\nu)\right) \\ + &F^-\left(t - (t_\text{i} - s\Delta t), \nu - (\nu_\text{i} + s\Delta\nu)\right)\big]\,,\end{aligned}$$

with

$$F^\pm(t,\nu) = W_{x_0}^{(s)}(t,\nu)\, \exp\left\{i\left[2\pi(t\Delta\nu - \nu\Delta t) + \Delta\varphi - 2\pi\Delta\nu(\pm s\Delta t - t_\text{i})\right]\right\}\,.$$

Therefore, if the "signal" terms are localized to neighborhoods of the given points (t_1,ν_1) and (t_2,ν_2), the two cross-terms have significant values near the points $(t_\text{i} + s\Delta t, \nu_\text{i} - s\Delta\nu)$ and $(t_\text{i} - s\Delta t, \nu_\text{i} + s\Delta\nu)$. The location of these points depends on the parameter s. In case $s = 0$ both terms are localized to a region about the same point $(t_\text{i}, \nu_\text{i})$. One can easily verify that they are complex conjugates of each other. Hence, this restates the previous result, see Eq. (3.97), for the Wigner-Ville distribution. In all other cases, when $s \neq 0$, the distribution is complex-valued, and the cross-terms have a "distance" from their midpoint $(t_\text{i}, \nu_\text{i})$ which is proportional to s. In the extreme cases of the Rihaczek distribution or its complex conjugate ($s = \pm 1/2$), the signal terms and cross-terms are located at four corners of a time-frequency rectangle with sides parallel to the axes and with opposite corners (t_1,ν_1) and (t_2,ν_2) (cf. Fig. 3.10).

This behavior, which varies as a function of the parameter s, yields another geometric interpretation of the optimality of the Wigner-Ville distribution regarding its localization among all s-Wigner distributions. In

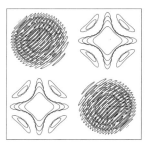  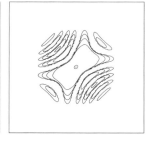

Figure 3.10. Rihaczek distribution of two atoms.

If a signal is the superposition of two atoms, the Rihaczek distribution contains two smooth contributions (corresponding to both atoms) and two cross-terms (relative to their interference). These are located at opposite corners of the rectangle with sides parallel to the time-frequency axes, whose other two corners are the centers of the given atoms. They show an oscillatory behavior with a frequency proportional to the distance of the atoms. The figure depicts this structure for pairs of identical atoms with different separation (only the real part of the distributions is shown).

fact, we explained the perfect localization of the Wigner-Ville distribution to linear chirps by conceptualizing the cross-terms of a linear frequency modulation as staying on the same curve as defined by the instantaneous frequency. For an s-Wigner distribution with $s \neq 0$ this result is no longer true: the relocation of the cross-terms amounts to an expansion of the support of the distribution. In fact, in the case of the Rihaczek distribution the whole rectangle supported by the modulation between (t_1, ν_1) and (t_2, ν_2) is occupied (cf. Fig. 3.11).

Just as in the case of the Wigner-Ville distribution, we can look at the geometry of the s-Wigner distributions and the creation of their interferences from a complementary perspective. This is expressed in terms of the *formula for the outer interferences*, which has the form

$$\left|W_{xy}^{(s)}(t,\nu)\right|^2 = \iint_{\mathbb{R}^2} W_x^{(s)}\left(t + \left(\frac{1}{2} - s\right)\tau, \nu + \left(\frac{1}{2} + s\right)\xi\right)$$

$$\times W_y^{(s)}\left(t - \left(\frac{1}{2} + s\right)\tau, \nu - \left(\frac{1}{2} - s\right)\xi\right) d\tau\, d\xi\ . \tag{3.116}$$

Likewise, we can deduce a similar formula for the *inner* interferences by letting $y(t) = x(t)$.

Chapter 3 Issues of Interpretation

Figure 3.11. Rihaczek distribution of linear frequency modulation.

For a linear frequency modulation the pointwise application of the construction principle shown in Fig. 3.10 leads to a Rihaczek distribution, which occupies the entire rectangle. Its sides are parallel to the time-frequency axes and it contains the line of the instantaneous frequency as its diagonal (only the real part of the distribution is shown).

Generalization to the affine distributions. Another generalization of our results deals with the localization of a distribution to a time-frequency curve by means of arguments of interference. For the time-frequency case, considered in the strict sense of Cohen's class, we have observed and justified in different ways that only the Wigner-Ville distribution admits a perfect localization to a time-varying frequency modulation; moreover, this can happen only if the modulation is *linear*. However, we saw in Subsection 2.3.2 that other types of perfect localization to more general curves can be achieved. For this purpose, however, we have to turn to distributions of another class, namely the affine class, and use their time-frequency interpretation. As the affine distributions have a bilinear character as well, their underlying principle of superposition also generates cross-terms. Their geometric interpretation, however, differs vastly from the corresponding principle of the Wigner-Ville distribution.

In order to explain the mechanism for the creation of these terms, we make use of the following geometric argument: If a distribution is known to be perfectly localized to a curve in the time-frequency plane, then the interference relative to any two points of this curve must be located at another point of the same curve. Therefore, we can compute the cross-term of two (undamped) pure frequencies in two consecutive steps. First, we determine its frequency location. In a second step we derive the corresponding location in the time domain from our *a priori* knowledge about the curve, on which the distribution lives.[31] Let us consider, for instance, the Unterberger distribution in its active form (cf. Table 2.3). Its time-frequency representation ($\nu = \nu_0/a$) has the form

$$U_x^A(t,\nu) = \nu \int_0^{+\infty} \left(1 + \frac{1}{\gamma^2}\right) X(\nu\gamma) X^* \left(\frac{\nu}{\gamma}\right) e^{-i2\pi\nu(\gamma - 1/\gamma)t} d\gamma .$$

It can easily be seen that the cross-term associated with two pure frequencies

$$x_1(t) = e^{i2\pi\nu_1 t} , \qquad x_2(t) = e^{i2\pi\nu_2 t}$$

is given by

$$I_{x_1 x_2}(t,\nu) = 2 \operatorname{Re} \{U_{x_1 x_2}^A(t,\nu)\}$$

$$= 2\nu \int_0^{+\infty} \left(1 + \frac{1}{\gamma^2}\right) \delta(\nu\gamma - \nu_1) X^* \left(\frac{\nu}{\gamma} - \nu_2\right) e^{-i2\pi\nu(\gamma - 1/\gamma)t} d\gamma$$

$$= 2(\nu_1 + \nu_2) \cos\{2\pi(\nu_1 - \nu_2)t\} \delta(\nu^2 - \nu_1 \nu_2) .$$

This yields its perfect localization to a frequency ν_i, which must be defined as the *geometric mean*

$$\nu_i = \sqrt{\nu_1 \nu_2} \tag{3.117}$$

of the two interacting frequencies.

We next use the fact (cf. Subsection 2.3.2) that the Unterberger distribution is perfectly localized to the curve of the group delay of the type ν^{-2}; hence, two interacting points (t_1, ν_1) and (t_2, ν_2) define a corresponding curve, whose parameters t_0 and α are determined by the system of equations

$$\begin{cases} t_1 = t_0 + \alpha \nu_1^{-2} \\ t_2 = t_0 + \alpha \nu_2^{-2} . \end{cases}$$

We can thus derive the values of t_0 and α and insert them into the equation

$$t_i = t_0 + \alpha \nu_i^{-2} .$$

Chapter 3 Issues of Interpretation

This last relation states that the interference point (t_i, ν_i) must also lie on the same curve. The previous expression for ν_i can now be used in order to find the value

$$t_i = \frac{t_1 + t_2}{2} + \frac{1}{2}(t_1 - t_2)\frac{\nu_1 - \nu_2}{\nu_1 + \nu_2}. \tag{3.118}$$

We have thus shown that the underlying geometric construction principle of the Wigner-Ville distribution does not immediately extend to other distributions in the affine class. While the former "midpoint rule" is given by

$$\begin{cases} t_i^W = \dfrac{t_1 + t_2}{2} \\ \nu_i^W = \dfrac{\nu_1 + \nu_2}{2}, \end{cases} \tag{3.119}$$

the (active) Unterberger distribution is governed by the modified rule

$$\begin{cases} t_i^U = \dfrac{t_1 + t_2}{2} + \dfrac{1}{2}(t_1 - t_2)\dfrac{\nu_1 - \nu_2}{\nu_1 + \nu_2} \\ \nu_i^U = \sqrt{\nu_1 \nu_2}. \end{cases} \tag{3.120}$$

It is possible, of course, to follow the same procedure for other localized distributions of the affine class. In particular, we obtain that the Bertrand- and the D-distributions (which are localized to the group delays by ν^{-1} and $\nu^{-1/2}$, respectively) obey the construction rules

$$\begin{cases} t_i^B = \dfrac{t_1 + t_2}{2} + \dfrac{1}{2}(t_1 - t_2)\left(\dfrac{\nu_1 + \nu_2}{\nu_1 - \nu_2} + \dfrac{2\nu_1\nu_2 \log(\nu_2/\nu_1)}{(\nu_1 - \nu_2)^2}\right) \\ \nu_i^B = \dfrac{\nu_2 - \nu_1}{\log(\nu_2/\nu_1)} \end{cases} \tag{3.121}$$

and

$$\begin{cases} t_i^D = \dfrac{t_1 + t_2}{2} + \dfrac{1}{2}(t_1 - t_2)\dfrac{\sqrt{\nu_1} - \sqrt{\nu_2}}{\sqrt{\nu_1} + \sqrt{\nu_2}} \\ \nu_i^D = \left(\dfrac{\sqrt{\nu_1} + \sqrt{\nu_2}}{2}\right)^2. \end{cases} \tag{3.122}$$

As an interesting feature of these constructions, we note that each of them can be associated with a different definition of a "midpoint" between two given points. This is laid out more clearly when the preceding equations are rewritten using the variable $\theta = t - t_0$, where t_0 denotes the vertical asymptote of the group delay (i.e., the curve on which the distribution is localized). Then we obtain

$$\begin{cases} \theta_i^W = \dfrac{1}{2}\left(\theta_1 + \theta_2\right) \\ \nu_i^W = \dfrac{1}{2}\left(\nu_1 + \nu_2\right), \end{cases} \tag{3.123}$$

$$\begin{cases} \left(\theta_i^D\right)^{-1} = \dfrac{1}{2}\left(\theta_1^{-1} + \theta_2^{-1}\right) \\ \left(\nu_i^D\right)^{1/2} = \dfrac{1}{2}\left(\nu_1^{1/2} + \nu_2^{1/2}\right), \end{cases} \tag{3.124}$$

$$\begin{cases} \log \theta_i^U = \dfrac{1}{2}\left(\log \theta_1 + \log \theta_2\right) \\ \log \nu_i^U = \dfrac{1}{2}\left(\log \nu_1 + \log \nu_2\right). \end{cases} \tag{3.125}$$

Thus the three rules correspond to as many ways of computing a midpoint after performing a nonlinear transformation of the coordinates. Furthermore, in the limit of narrowband signals the other two midpoints tend to the usual arithmetic mean, which is characteristic of the Wigner-Ville distribution.[32]

Similar to the Wigner-Ville distribution, we can interpret these midpoint rules as symmetries considered in a modified geometry (owing to the nonlinear transformation). Conversely, we can devise an explicit construction of a distribution based on a given midpoint rule. This procedure starts from rewriting the operator, whose expectation value is the Wigner-Ville distribution (cf. Eq. (3.114)), according to

$$\widehat{D}_{t\nu}\,\widehat{\Pi}\,\widehat{D}_{-t,-\nu} = e^{-i2\pi t\hat{\nu}}\left(e^{i2\pi\nu\hat{t}}\,\widehat{\Pi}\,e^{-i2\pi\nu\hat{t}}\right)e^{i2\pi t\hat{\nu}}.$$

In this form, the central operator

$$\widehat{\Pi}_\nu = e^{i2\pi\nu\hat{t}}\,\widehat{\Pi}\,e^{-i2\pi\nu\hat{t}} \tag{3.126}$$

is the operator of a *frequential parity* associated with the idea of an *arithmetic* mean. Its action is defined by

$$(\widehat{\Pi}_\nu X)(\xi) = X(2\nu - \xi). \tag{3.127}$$

Chapter 3 Issues of Interpretation

We can thus think of substituting another unitary operator for it, which is associated with the idea of a *geometric* mean. Its action can be defined by [33]

$$(\widehat{H}_\nu X)(\xi) = \frac{\nu}{\xi} X\left(\frac{\nu^2}{\xi}\right) U(\xi) . \tag{3.128}$$

An immediate computation shows that the action of the global operator, which can be derived as in Eq. (3.114), is given by

$$(e^{-i2\pi t\hat{\nu}} \, \widehat{H}_\nu \, e^{i2\pi t\hat{\nu}} X)(\xi) = (e^{-i2\pi t\hat{\nu}} \, \widehat{H}_\nu) X(\xi) \, e^{i2\pi \xi t}$$

$$= (e^{-i2\pi t\hat{\nu}}) \frac{\nu}{\xi} X\left(\frac{\nu^2}{\xi}\right) e^{i2\pi(\nu^2/\xi)t} U(\xi) \tag{3.129}$$

$$= \frac{\nu}{\xi} X\left(\frac{\nu^2}{\xi}\right) e^{i2\pi[(\nu^2/\xi)-\xi]t} U(\xi) .$$

Hence, the corresponding time-frequency representation has the form

$$\left\langle e^{-i2\pi t\hat{\nu}} \, \widehat{H}_\nu \, e^{i2\pi t\hat{\nu}} \right\rangle_x = \int_0^{+\infty} \frac{\nu}{\xi} X\left(\frac{\nu^2}{\xi}\right) e^{i2\pi[(\nu^2/\xi)-\xi]t} X^*(\xi) \, d\xi$$

$$= \nu \int_0^{+\infty} \frac{1}{\gamma} X(\nu\gamma) X^*\left(\frac{\nu}{\gamma}\right) e^{i2\pi\nu(\gamma-1/\gamma)t} \, d\gamma .$$

This expression, except for a factor 2, is identical to the definition of the Unterberger distribution in its "passive" form (cf. Table 2.3).

Analogously, it is conceivable to start from the rule of Eq. (3.122), which is connected with the *D*-distribution. Then an operator can be defined by

$$(\widehat{K}_\nu X)(\xi) = \left(2 - \sqrt{\xi/\nu}\right) X\left((2\sqrt{\nu} - \sqrt{\xi})^2\right) , \tag{3.130}$$

and we find in this case that

$$\left\langle e^{-i2\pi t\hat{\nu}} \, \widehat{K}_\nu \, e^{i2\pi t\hat{\nu}} \right\rangle_x = \frac{1}{2} D_x(t,\nu) .$$

3.2.3. Reduction of the Interferences

In the previous section we noticed the importance of the interference terms in a Wigner-Ville distribution. Their significance materializes in a 2-fold manner. First, their *number* is large, because N interacting components generate $N(N-1)/2$ cross-terms. Second, their *size* may exceed the signal terms by a factor 2, as two components of equal amplitude in the model equation (3.91) yield

$$\max |I_{x_1 x_2}(t,\nu)| = 2 \max |W_{x_0}(t,\nu)| \ .$$

Consequently, it is worthwhile to dwell on the problem of their reduction; this will once more refer to the analysis of the mechanism of their creation.

Analytic signal. Because the number of cross-terms grows as the number of interacting components of the signal increases, a first step to avoiding an excessive proliferation of these terms is a best possible reduction of the redundancy in the representation of the signal. Viewed from a frequential perspective, the Wigner-Ville distribution can be written as

$$W_x(t,\nu) = \int_{-\infty}^{+\infty} X\left(\nu + \frac{\xi}{2}\right) X^*\left(\nu - \frac{\xi}{2}\right) e^{i2\pi\xi t} \, d\xi \ . \tag{3.131}$$

The number of created interference terms depends on the number of components in the spectrum $X(\nu)$. For a real-valued signal the Hermitian symmetry of the spectrum is the reason for an excess of half the number of frequency-components, namely those that are located at negative frequencies. As an example, we consider a signal which is the sum of *three* (positive) frequencies. In its real representation it produces *fifteen* cross-terms, *nine* of which have positive frequencies. Because the negative frequencies of the signal stand only for a repetition of the same information of the positive ones, it is certainly profitable to omit them completely. This can be achieved by passing to the analytic signal (cf. Eq. (1.22)).[34] Then only the physically important components with positive frequencies interfere with each other: In the preceding example the number of cross-terms is thus reduced to *three* instead of *nine*, and this enhances greatly the readability of the diagram (cf. Fig. 3.12).

Wigner-Ville and atomic decompositions. A second approach to reduce the interference terms makes use of atomic decompositions. For this purpose let us begin with the general result, that the Wigner-Ville distribution of a multicomponent signal

$$x(t) = \sum_{n=1}^{N} a_n \, x_n(t) \tag{3.132}$$

Chapter 3 Issues of Interpretation

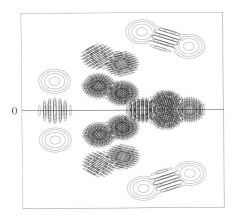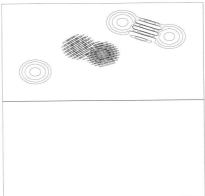

Figure 3.12. Wigner-Ville: real signal vs. analytic signal.

The Wigner-Ville distribution of a real-valued signal (left) contains interference terms, which are created by the interaction of components with positive and negative frequencies. These terms disappear, if the distribution is computed based on the corresponding analytic signal (right), since the components of the signal with negative frequencies are omitted.

is given by

$$W_x(t,\nu) = \sum_{n=1}^{N} |a_n|^2 W_{x_n}(t,\nu) + 2 \sum_{n=1}^{N-1} \sum_{k=n+1}^{N} \text{Re}\left\{ a_n\, a_k^*\, W_{x_n x_k}(t,\nu) \right\} .$$

An obvious way to suppress the cross-terms is just by retaining the first sum of the above expression, which corresponds to the signal terms, and omitting the second sum, in which the cross-terms are collected.

Unfortunately, while the $x_n(t)$ generally correspond to an objective decomposition of the signal into physical components, the observer can only dispose of their sum (Eq. (3.132)) as a whole. This renders a separation of the two series impossible. However, there is a trick by replacing the objective (and inaccessible) decomposition into *physical* constituents with an atomic decomposition into *mathematical* components. Hence, given a certain basis of functions $\varphi_n(t)$, the analyzed signal can be written as

$$x(t) = \sum_{n=1}^{N} \langle x\,,\,\varphi_n \rangle\, \varphi_n(t) . \tag{3.133}$$

Then a "cleared" form of the Wigner-Ville distribution, which suppresses the interactions of the different atoms of the decomposition, is given by [35]

$$\sum_{n=1}^{N} |\langle x, \varphi_n \rangle|^2 \, W_{\varphi_n}(t, \nu) \ . \tag{3.134}$$

This clever method is naturally limited by the (necessary) choice of the basis functions of the decomposition. When a sufficient a priori knowledge about the signal is unavailable, this problem can be remedied by a learning procedure that uses a large dictionary of admissible bases.

Smoothing. We explained in Subsection 3.2.2 that one of the characteristics of the interference terms is their *oscillatory* structure. This contrasts with the signal terms, which are more regular. This opposite behavior suggests the use of a *smoothing* operation as an appropriate tool for reducing the size of the interference terms. Such a smoothing can either be expressed in the time-frequency plane by means of a convolution with a function $\Pi(t, \nu)$, or equivalently, by a *weighting* or multiplication in the ambiguity plane by a function $f(\xi, \tau)$, which is the Fourier transform of $\Pi(t, \nu)$. In this regard, the reduction of the interferences results from replacing the Wigner-Ville distribution with a smoothed version

$$\iint_{\mathbb{R}^2} \Pi(s - t, \xi - \nu) \, W_x(s, \xi) \, ds \, d\xi = \iint_{\mathbb{R}^2} f(\xi, \tau) \, A_x(\xi, \tau) \, e^{-i2\pi(\nu\tau + \xi t)} \, d\xi \, d\tau \ .$$

We thereby recover the general definition of the distributions in Cohen's class (Eqs. (3.9)-(3.18)). It amounts to a new geometric interpretation of this class.

Therefore, the inherent problem of reducing the interferences boils down to a good choice for the weight function. The solution to this problem relies on the geometry of the interferences (explored in the previous Subsection). In fact, as the velocity of the oscillations is bigger for interacting components that are farther apart from each other, the required smoothing needs only a shorter window in such situation. Expressed in terms of the ambiguities, a big distance of the components of the signal manifests itself by a localization of the significant cross-ambiguities far away from the origin, and thus from the autoambiguities of the signal. We only want to keep the latter ones, which are the true images of the signal terms. In this case it suffices to choose a weight function with a large support, which in return corresponds to a smoothing by a small window in the time-frequency plane.

This very general interpretation of $f(\xi, \tau)$ as a *weight* function states that we should choose f so that it suppresses the cross-ambiguity terms

Chapter 3 Issues of Interpretation

(which are naturally eccentric) and preserves the autoambiguities (concentrated in a neighborhood of the origin) to their best possible integrity. It is thus justified to think of $f(\xi, \tau)$ as being maximal at the origin and having a support, which is determined by the time-frequency positions of the interacting terms.[36]

Coupled smoothing. A first example for the reduction of interferences by a fixed smoothing is realized by the spectrogram. Denoting its window function by $h(t)$, as before, we can write it as

$$S_x(t,\nu) = \iint_{\mathbf{R}^2} W_x(s,\xi)\, W_h(s-t, \xi-\nu)\, ds\, d\xi \ .$$

In case of a smooth window function, the associated parameter function

$$f(\xi, \tau) = A_h^*(\xi, \tau)$$

has a global lowpass character. This ascertains an effective reduction of the cross-terms that are created by components with a time-frequency distance, which puts them outside the region of influence of $A_h(\xi, \tau)$. Here we recover another example for the restrictions that apply to the spectrogram: Providing a better reduction of interferences in one direction of the plane (time or frequency) causes a loss in the same property relative to the other direction. This can be regarded as a direct consequence of the *invariant volume* property of the ambiguity function, which signifies that

$$\iint_{\mathbf{R}^2} |A_h(\xi, \tau)|^2\, d\xi\, d\tau = \iint_{\mathbf{R}^2} W_h^2(t, \nu)\, dt\, d\nu = E_h^2 \ . \tag{3.135}$$

Combined with the requirement

$$|A_h(\xi, \tau)| \leq |A_h(0, 0)| = E_h \ ,$$

this property implies that the reduction of the support of the ambiguity function in any direction is coupled with an enlargement of the support in the perpendicular direction.

As an illustration let us take the Gaussian window (with unit energy)

$$h(t) = (2\alpha)^{1/4} e^{-\pi \alpha t^2} \ . \tag{3.136}$$

Its ambiguity function is given by

$$A_h(\xi, \tau) = \exp\left\{-\frac{\pi}{2}\left(\frac{\xi^2}{\alpha} + \alpha \tau^2\right)\right\} \ . \tag{3.137}$$

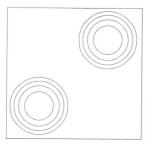

Figure 3.13. Spectrogram of two atoms.

The spectrogram of a signal, which is the superposition of two atoms, consists of two smooth contributions (related to the atoms) and a cross-term (relative to their interaction). As for the Wigner-Ville distribution, the latter is located at the midpoint of the straight line connecting the time-frequency centers of the given atoms. Its amplitude, however, decreases rapidly when the overlap of the "signal"-components gets smaller. The figure illustrates this behavior for identical pairs of atoms and different time-frequency distances.

For fixed α, the present *coupling* of the smoothing in the time- and frequency-directions is clearly underlined by the fact that

$$W_h(t,\nu) = 2\exp\left\{-2\pi\left(\alpha t^2 + \frac{\nu^2}{\alpha}\right)\right\} . \tag{3.138}$$

Even though a spectrogram shows a significant reduction of the cross-terms as compared to a Wigner-Ville distribution, it is not completely exempt from those terms.[37] Indeed, we can immediately see that

$$S_{x+y}(t,\nu) = S_x(t,\nu) + S_y(t,\nu) + 2\operatorname{Re}\left\{F_x(t,\nu)F_y^*(t,\nu)\right\} .$$

Hence a cross-term does generally exist. This is not surprising after all: It is related to the classical observation made when measuring an intensity in the presence of interfering physical waves. Nevertheless, this cross-term only arises when the short-time Fourier transforms of the interacting components overlap. Its importance decreases rapidly, if the distance between the components grows (cf. Fig. 3.13).

Separable smoothing. The use of the spectrogram as a smoothed version of the Wigner-Ville distribution faces two major drawbacks. The first is the loss in most theoretical properties that constitute the advantages of the

Chapter 3 Issues of Interpretation

Wigner-Ville transform. The second drawback of the spectrogram results from the restrictions of Heisenberg-Gabor type, which limit its smoothing capability. Heuristically, we can say that the time-frequency smoothing by a spectrogram is based on only one "degree of freedom," as it employs a unique short-time window $h(t)$: If the smoothing in time is of order $\delta t(h)$, by which we denote the width of $h(t)$, then the smoothing in frequency must be of order $1/\delta t(h)$.

If we take both dimensions of the time-frequency plane into consideration, we can use a smoothing with *two* "degrees of freedom," one related to time and one related to frequency. This yields an improvement over the spectrogram, as far as the forementioned second disadvantage is concerned. A natural way to proceed is to employ a *separable* parameter function

$$f(\xi,\tau) = G(\xi)\, h'(\tau) \ .$$

It is equivalent to a time-frequency smoothing by

$$\Pi(t,\nu) = g(t)\, H'(-\nu) \ ,$$

which is controlled in time (by g) and in frequency (by H') independently. The corresponding representation is called *smoothed pseudo-Wigner-Ville distribution* [38] and has the form

$$SPW_x(t,\nu) \tag{3.139}$$

$$= \int_{-\infty}^{+\infty} h'(\tau) \left[\int_{-\infty}^{+\infty} g(s-t) x\left(s+\frac{\tau}{2}\right) x^*\left(s-\frac{\tau}{2}\right) ds \right] e^{-i2\pi\nu\tau}\, d\tau \ .$$

Note that if no temporal smoothing is used, that is, $g(t) = \delta(t)$, the original definition equation (3.74) of the pseudo-Wigner-Ville distribution is recovered by putting

$$h'(\tau) = h^*\left(\frac{\tau}{2}\right) h\left(-\frac{\tau}{2}\right) \ .$$

Furthermore, if we also let the window function h tend to the constant 1, the resulting distribution tends to the Wigner-Ville distribution with no smoothing applied.

The use of separable functions permits a *continuous* and *independent* control (in time and frequency) of the employed smoothing. This contrasts with the spectrogram, which is based on a smoothing function whose time-frequency concentration cannot be pushed beyond the lower bound of the

Heisenberg-Gabor uncertainty principle. In fact, when we return to the example (Eq. (3.136)) of the Gaussian window, we find

$$\Pi_S(t,\nu) = W_h(t,\nu) = 2\exp\left\{-2\pi\left(\alpha t^2 + \frac{\nu^2}{\alpha}\right)\right\} \tag{3.140}$$

for the spectrogram. If we measure the extension of the time-frequency smoothing by the "duration-bandwidth product"

$$BT(\Pi) = 4\pi \left| \iint_{\mathbb{R}^2} t^2\nu^2\,\Pi(t,\nu)\,dt\,d\nu \right|^{1/2}, \tag{3.141}$$

we obtain

$$BT(\Pi_S) = 1 \tag{3.142}$$

for the spectrogram. This result holds regardless of the chosen parameter α; that is, it is not affected by the size of the short-time window.

However, in the separable case nothing hinders us from using a smoothing of the type

$$\Pi_{\text{SPW}}(t,\nu) = \frac{2}{\sqrt{\alpha\beta}}\,\exp\left\{-2\pi\left(\frac{t^2}{\beta} + \frac{\nu^2}{\alpha}\right)\right\}, \tag{3.143}$$

with $\alpha\beta \leq 1$. In this case we obtain

$$BT(\Pi_{\text{SPW}}) = \sqrt{\alpha\beta}, \tag{3.144}$$

and this enables us to pass *continuously* from the Wigner-Ville distribution ($\alpha\beta = 0$) to the spectrogram ($\alpha\beta = 1$) (cf. Fig. 3.14). While the spectrogram is confronted with an unavoidable trade-off between the time- and frequency-resolution of the analysis, this compromise is now located somewhere in between the *joint* time-frequency resolution of the signal terms and the size of the interference terms.

The actual choice of a pair of functions for a separable smoothing can be guided by the knowledge about the geometry of the interferences of the Wigner-Ville distribution. If we consider, for instance, the pseudo-Wigner-Ville distribution (Eq. (3.74)), which operates by a solely frequential smoothing, it becomes clear that it reduces mainly the oscillations in the frequency-direction. Such oscillations result from the interaction of components that are separated in time. Hence, we can regard their reduction equally as being gained from the introduction of a short-time

Chapter 3 Issues of Interpretation

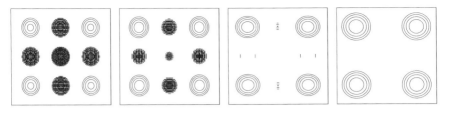

Figure 3.14. Transition between Wigner-Ville and spectrogram.

The spectrogram is a smoothed version of the Wigner-Ville distribution, which employs a time- and frequency-smoothing. By the use of a separable smoothing function, one can obtain a continuous transition between the Wigner-Ville transform (no smoothing, left figure) and the spectrogram (smoothing with respect to a Gabor cell, right figure). The transition depicts the resulting trade-off between the joint resolution of the analysis and the size of the interference terms.

window, which induces a temporal separation of the corresponding components (cf. Fig. 3.15). Only those interferences persist that are created by the interaction of components that are "seen" through the window *simultaneously*.

Alternatively, the use of a temporal smoothing essentially amounts to the reduction of interferences which oscillate in the time-direction. As we know, such oscillations result from the interaction of distinct spectral components, and the velocity of the oscillations is determined by their frequential distance. It is, thereby, quite easy to use this property as a guideline for choosing the duration of the time-smoothing properly (cf. Fig. 3.16).

While the use of a separable smoothing allows better control over the reduction of the cross-terms, it generally affects certain theoretical properties. However, we should retain the important observation that the introduction of a smoothing in one direction can only influence the theoretical properties related to this variable. This is another advantage over the spectrogram, as the latter disturbs the properties relative to both variables. As we already saw before, the conservation of the time-support of a signal is not affected in the pseudo-Wigner-Ville case by the introduction of a short-time window. Likewise, the property of the correct marginal distribution in time will be preserved if (cf. Table 2.4)

$$f_{\mathrm{PW}}(\xi, 0) = 1 \ .$$

Note that this condition is automatically satisfied, provided that we use a

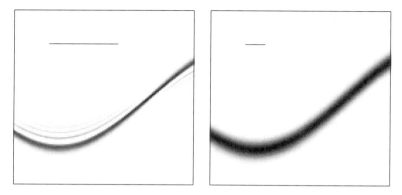

Figure 3.15. Pseudo-Wigner-Ville.

The pseudo-Wigner-Ville distribution uses a short-time window that performs a frequential smoothing of the Wigner-Ville distribution. Only the cross-terms created by components that overlap with the window simultaneously persist. The figure depicts two pseudo-Wigner-Ville distributions of the same signal, but computed with two different windows. Their respective width is indicated by the horizontal line segment.

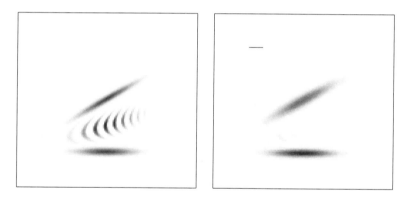

Figure 3.16. Smoothed pseudo-Wigner-Ville.

The short-time window of the pseudo-Wigner-Ville distribution leads to a reduction of the interferences of time-separated components. In a dual manner, the use of a second smoothing in the time-direction amounts to a reduction of the interferences of frequency-separated components. The figure illustrates this by showing two computed distributions of the same signal, which employ the same short-time window, but use a different smoothing along the time axis (left: no smoothing; right: smoothing with a duration indicated by the horizontal line segment).

normalized window h with
$$|h(0)| = 1 .$$

Moreover, the instantaneous frequency can be recovered as the local center of gravity if (cf. Table 2.4)

$$\frac{\partial f_{\text{PW}}}{\partial \tau}(\xi, 0) = 0 \quad \Longrightarrow \quad \text{Im}\left\{\frac{dh^*}{d\tau}(0)\, h(0)\right\} = 0 .$$

This condition is guaranteed if the window function is either real-valued or has a maximum at the origin.

Of course, we can collect a set of alternative conditions for the case of a pure time-smoothing. However, we must note that the mentioned properties are destroyed when the separable smoothing incorporates *both* directions. Consequently, in case a separable smoothing displays a noticeable *practical* improvement over the spectrogram, it is affected by the same loss in *theoretical* properties (although on a smaller scale).

Joint smoothing. By giving up the idea of separability, we can introduce other types of smoothing, so as to preserve some of the desirable theoretical properties and reduce the cross-terms at the same time. If, for instance, we wish to guarantee the correct marginal distributions of the modified representation, its parameter function must meet the condition (cf. Table 2.4)

$$f(\xi, 0) = f(0, \tau) = 1 .$$

If we want the same parameter function to yield a reduction of certain cross-terms, it must be decaying away from the origin so as to diminish the size of the cross-ambiguity of distinct components.

A simple way to achieve this double objective is to impose the special form

$$f(\xi, \tau) = \varphi(\xi\tau) \tag{3.145}$$

on the parameter function, where $\varphi(x)$ is decreasing and $\varphi(0) = 1$. The corresponding time-frequency distributions are generically called *reduced interference distributions*.[39] Their parameter functions of the type from Eq. (3.145) display a cross-like shape: The cross sections for $\xi = 0$ and $\tau = 0$ are constant, while the other contourlines ($\xi\tau =$ const) form a net of hyperbolas.

Choï-Williams distribution. A simple example (cf. Fig. 3.17) is obtained, when we choose a Gaussian for our function φ, hence

$$f_\sigma(\xi, \tau) = \exp\left(-\frac{1}{2\sigma^2}(\pi\xi\tau)^2\right) . \tag{3.146}$$

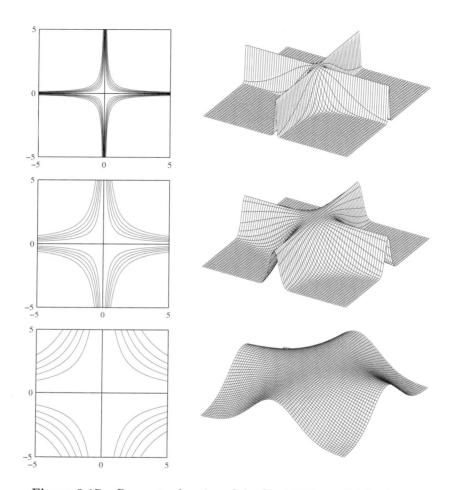

Figure 3.17. Parameter function of the Choï-Williams distribution.

The parameter function of the Choï-Williams distribution in the frequency-time domain (or ambiguity plane) is a Gaussian relative to the product of the two variables, with a free parameter σ denoting its spread. Its graph has the form of a "cross," because its contourlines build a net of hyperbolas and the cross sections along the two axes are constant. Its properties permit compatibility with the marginal distributions together with a certain reduction of the interference terms. This reduction is more pronounced for smaller values of σ. The figure presents three examples of such functions with parameter values $\sigma = 1$ (top), $\sigma = 5$ (center), and $\sigma = 25$ (bottom), respectively.

Chapter 3 Issues of Interpretation   263

Its variance σ^2 is used to fix its spread. We thus obtain

$$\lim_{\sigma \to +\infty} f_\sigma(\xi,\tau) = 1 ,$$

and this renders once more the Wigner-Ville distribution as a limiting case (with infinite variance). The formal application of the definition equation (3.141) of the duration-bandwidth product of a smoothing kernel leads to

$$BT(f_\sigma) = \frac{1}{\sigma} . \tag{3.147}$$

Hence, the smoothing gets more significant and the anticipated reduction of the interferences improves when we use smaller values of σ. Note that, strictly speaking, the Fourier transform of Eq. (3.146) does not define a "true" smoothing kernel in the sense of a lowpass filter.

The representation associated with the parameterization equation (3.146) is called *Choï-Williams distribution*. It has the form [40]

$$CW_x(t,\nu) = \sqrt{\frac{2}{\pi}} \iint_{\mathbb{R}^2} \frac{\sigma}{|\tau|} e^{-2\sigma^2(s-t)^2/\tau^2} x\left(s+\frac{\tau}{2}\right) x^*\left(s-\frac{\tau}{2}\right) e^{-i2\pi\nu\tau} \, ds \, d\tau. \tag{3.148}$$

The Choï-Williams distribution acts more like a *reduction* of the size of the interferences, combined with a *relocation*, rather than a true *deletion*. This occurs as a necessary counterpart of the requirement of the correct marginal distributions. It can better be explained by a simple example. Consider a signal composed of two pure frequencies

$$x(t) = e^{i2\pi\nu_1 t} + e^{i2\pi\nu_2 t} . \tag{3.149}$$

Its Wigner-Ville distribution has the form

$$W_x(t,\nu) = \delta(\nu-\nu_1) + \delta(\nu-\nu_2) + 2\cos[2\pi(\nu_1-\nu_2)t]\,\delta\left(\nu - \frac{\nu_1+\nu_2}{2}\right), \tag{3.150}$$

while a straightforward computation leads to

$$CW_x(t,\nu) = \delta(\nu-\nu_1) + \delta(\nu-\nu_2) \tag{3.151}$$

$$+ \, 2\cos[2\pi(\nu_1-\nu_2)t]\sqrt{\frac{2}{\pi}}\frac{\sigma}{|\nu_1-\nu_2|}\exp\left\{-\frac{2\sigma^2}{(\nu_1-\nu_2)^2}\left(\nu - \frac{\nu_1+\nu_2}{2}\right)^2\right\}.$$

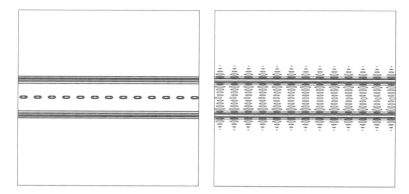

Figure 3.18. Pseudo-Wigner-Ville and Choï-Williams distribution of two pure frequencies.

If the analyzed signal is composed of two pure frequencies, the pseudo-Wigner-Ville distribution (left) yields two ridges with constant amplitude located at both frequencies, and an oscillating cross-term (of twice the amplitude) in between both frequencies. This interference structure is reduced by a Choï-Williams distribution (right). The case considered here ($\sigma = 1$) shows that the Choï-Williams distribution acts more like a *relocation* than a reduction of the interferences. This is needed for preserving the correct marginal distributions.

In both cases the cross-term is centered round the midpoint of the two interacting frequencies. It oscillates in the time-direction and its period is inversely proportional to the frequential distance. The difference between the two distributions results from the introduction of a finite parameter σ. It brings about a transformation of the normalized Dirac distribution (responsible for the perfect localization of the cross-term in the Wigner-Ville case) into a Gaussian with some nonzero spread and unit area. A small value of σ leads to a better reduction of the maximal absolute value of the interference term. However, it also widens its support, which (measured by the typical extension of Gaussians) attains the value $|\nu_1 - \nu_2|/2\sigma$ (cf. Fig. 3.18). Let us dwell on the connection with the correct marginal distributions once again. As already explained, when we use smaller values of the parameter σ, the interference term is reduced in size. It is also "spread" in the frequency-direction, so that the result of its integration parallel to the frequency axis remains constant. Obviously, this behavior is necessary for the conservation of the correct marginal distribution in time, as it involves the contributions of the interference term.

Chapter 3 Issues of Interpretation

Let us next return to the interpretation of the reduction of the interferences in the ambiguity plane. It can easily be seen that the efficacy of the Choï-Williams distribution depends crucially on the nature of the analyzed signal. By construction, it is most powerful if the cross-ambiguities are contained in the regions of the plane, where the attenuation by the weight function is maximal. On the other hand, the condition of correct marginal distributions requires the cross sections of the weight function at $\xi = 0$ and $\tau = 0$ to be constant. This hinders the Choï-Williams distribution from reducing the interference of terms that correspond to a cross-ambiguity, which is concentrated near one of the coordinate axes. (This case occurs when the interacting components have distinct frequencies, but are synchronous in time, or the other way around, when they are separated in time, but belong to the same frequency band.) Conversely, the efficacy of the distribution is best, when the cross-ambiguities are located near one of the diagonals of the plane (cf. Fig. 3.19). In this case the reduction is even more accentuated, when the interacting components are far apart.

Born-Jordan distribution. The compatibility with marginal distributions was only one conceivable restraint. We can add others and express them by further admissibility conditions on φ in the setting of parameterizations of the type from Eq. (3.145). For example, the conditions providing the instantaneous frequency and the group delay as local centers of gravity impose

$$\frac{d\varphi}{dx}(0) = 0 \,.$$

Obviously, the Choï-Williams distribution meets this condition.

However, the same distribution does not meet other desirable criteria such as the conservation of temporal support. As seen before (cf. Table 2.4), this corresponds to the requirement

$$F(t,\tau) = 0 \quad \text{for all} \quad |t/\tau| > 1/2 \,.$$

If we denote by $\Phi(y)$ the Fourier transform of the function $\varphi(x)$ in Eq. (3.145), we can establish the general relation

$$F(t,\tau) = \frac{1}{|\tau|} \Phi\left(\frac{t}{\tau}\right) \,.$$

Hence, the condition

$$\Phi(y) = 0 \,, \quad |y| > 1/2 \,,$$

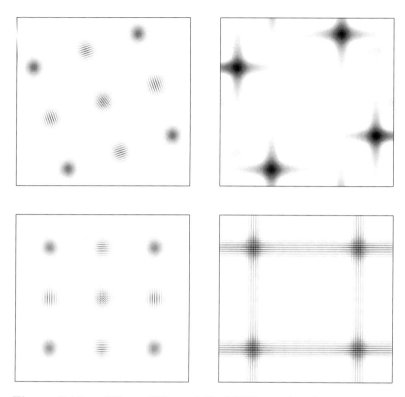

Figure 3.19. Wigner-Ville and Choï-Williams distribution of four atoms.

The figure shows two examples that demonstrate how large a reduction of interferences by a Choï-Williams distribution (right) can be as compared with the Wigner-Ville distribution (left). In both cases the analyzed signal is composed of four atoms, without (top) and with (bottom) overlapping time- or frequency supports. Evidently, the first situation proves more favorable for an analysis by the Choï-Williams distribution.

is sufficient for the conservation of temporal support. The simplest way to fulfill this condition within the class of distributions in Eq. (3.145) is given by the *cardinal sine function*. We thus let

$$f(\xi, \tau) = \frac{\sin \pi \xi \tau}{\pi \xi \tau}, \qquad (3.152)$$

Chapter 3 Issues of Interpretation

which defines the *Born-Jordan distribution*[41]

$$BJ_x(t,\nu) = \int_{-\infty}^{+\infty} \left[\frac{1}{|\tau|} \int_{t-|\tau|/2}^{t+|\tau|/2} x\left(s+\frac{\tau}{2}\right) x^*\left(s-\frac{\tau}{2}\right) ds \right] e^{-i2\pi\nu\tau} d\tau .$$
(3.153)

For the example of Eq. (3.149) (with $\nu_1 < \nu_2$) this distribution yields a similar reduction of the cross-term as the Choï-Williams distribution (3.151), as we obtain

$$\begin{aligned} BJ_x(t,\nu) &= \delta(\nu-\nu_1) + \delta(\nu-\nu_2) \\ &+ 2\cos[2\pi(\nu_1-\nu_2)t] \, \frac{1}{|\nu_1-\nu_2|} \, \mathbf{1}_{[\nu_1,\nu_2]}(\nu) . \end{aligned}$$
(3.154)

We should also note that the frequential support of the cross-term, in this special case, is restricted to the interval $[\nu_1, \nu_2]$. This results from the fact that the Born-Jordan distribution also guarantees the conservation of the frequency support in the wide sense (cf. Table 2.4).

Zhao-Atlas-Marks distribution. Next we allow a deviation from the class of parameterization equation (3.145). As a generalization of Eq. (3.152) we can consider every function of the form

$$f(\xi, \tau) = h_0(\tau) |\tau| \frac{\sin \pi \xi \tau}{\pi \xi \tau} ,$$

where h_0 may be any apodization function. This gives rise to the *Zhao-Atlas-Marks distribution*[42]

$$ZAM_x(t,\nu) = \int_{-\infty}^{+\infty} \left[h_0(\tau) \int_{t-|\tau|/2}^{t+|\tau|/2} x\left(s+\frac{\tau}{2}\right) x^*\left(s-\frac{\tau}{2}\right) ds \right] e^{-i2\pi\nu\tau} d\tau,$$
(3.155)

which is also called a distribution with "cone-shaped kernel." This nomenclature refers to the support of the parameter function in the time-time domain. A comparison of equations (3.153) and (3.155) yields that the Zhao-Atlas-Marks distribution is nothing but a smoothed version of the Born-Jordan distribution, the smoothing taken along the frequency axis. Due to the conservation of the temporal support of the signal, it cannot preserve the frequential support as well. Accordingly, the property of a correct marginal distribution in frequency is no longer satisfied, unless we require that $h_0(\tau) = 1/|\tau|$, which brings us back to the Born-Jordan case. Furthermore, we have to impose the condition

$$\lim_{|\tau| \to 0} h_0(\tau) |\tau| = 1$$

for preserving the correct marginal distribution in time (and for making it an energy distribution in the sense of Eq. (2.55)!).

Remark. By putting
$$h'(\tau) = h_0(\tau)|\tau|,$$
we see that the relation between the Zhao-Atlas-Marks and the Born-Jordan distribution is the same as between the pseudo-Wigner-Ville and the ordinary Wigner-Ville distribution. Hence, some evident computational advantages can be gained from the use of a function $h_0(\tau)$ of compact support. Then the evaluation of the distribution equation (3.155) is based only on those values of the product
$$q_x(s,\tau) = x\left(s+\frac{\tau}{2}\right)x^*\left(s-\frac{\tau}{2}\right)$$
that correspond to points (s,τ) of a *finite* cone (or "butterfly" area) in the time-delay plane.

Generalizations. There are many special cases as well as generalizations of the reduced interference distributions, which are defined by a parameter function of Eq. (3.145) type. As their main ingredient is a decreasing univariate function, we can employ the whole range of apodization functions of spectral analysis (Hamming, Hanning, Kaiser-Bessel, etc.). Moreover, we can draw our inspiration from the techniques of filter design, insofar as the resulting parameter function serves as a lowpass filter in the ambiguity plane (Butterworth, Chebyshev filter, etc.). Such techniques can be helpful for gaining control over the transition between "passband" and "attenuated band" of the filter.

In this regard, one can define a *Butterworth distribution* by choosing the weight function
$$f(\xi,\tau) = \frac{1}{1+(\pi\xi\tau/\sqrt{2}\sigma)^{2N}}, \quad N = 1, 2, \ldots .$$

A possible generalization of the Choï-Williams distribution is given by
$$f_{\sigma,N}(\xi,\tau) = \exp\left\{-\left(\frac{\pi\xi\tau}{\sqrt{2}\sigma}\right)^{2N}\right\}.$$

Even more generally, we can introduce an *asymmetrical* behavior of the time and frequency variables of the weight function. If we use the substitution
$$\left(\frac{\pi^2\xi^2\tau^2}{2\sigma^2}\right) \rightarrow \left(\frac{\pi\xi}{\sqrt{2}\sigma_\xi}\right)^{M_\xi}\left(\frac{\pi\tau}{\sqrt{2}\sigma_\tau}\right)^{M_\tau}$$

Chapter 3 Issues of Interpretation

in any of the previous definitions, the variances and exponents can be controlled for each variable independently. [43] From a practical point of view, however, these generalizations run into some difficulty: It is not clear how the corresponding distribution can be computed directly, that is, without involving the ambiguity function.

Remark. To bring the discussion of joint smoothing to an end, let us note that some constraints are incompatible with the idea of a reduction of interferences, if we stick to parameterizations of the type of Eq. (3.145). This is evident for the *unitarity* of the distribution (or *Moyal's formula*). Indeed, the latter requires the weight function to be *unimodular* (cf. Table 2.4), and this certainly rules out any idea of an *attenuation* in the ambiguity plane. Such examples are the *s*-Wigner distributions for which

$$f_s(\xi, \tau) = e^{i 2\pi s \xi \tau}$$

(cf. Table 2.1). On the other hand, this definition implies that every function of Eq. (3.145) can be written as

$$\varphi(\xi\tau) = \int_{-\infty}^{+\infty} \Phi(s) f_s(\xi, \tau) \, ds \ . \tag{3.156}$$

Hence, it can be viewed as a superposition of *s*-Wigner parameterizations. This last property carries over to the associated (reduced interference) distributions, which attain the form [44]

$$C_x(t, \nu; \varphi) = \int_{-\infty}^{+\infty} \Phi(s) W_x^{(s)}(t, \nu) \, ds \ . \tag{3.157}$$

Variable and/or adapted smoothing. For all the forementioned cases, which are the spectrogram, the pseudo-Wigner-Ville distribution and the "reduced interference" distributions, the employed smoothing has two properties in common that make it too restrictive in some situations: first, the smoothing is *fixed*, that is, the action is the same at every point of the time-frequency plane; and second, it prioritizes *fixed directions*, namely the time- or frequency-direction of the plane.

Affine smoothing. Regarding the first point, we can devise some variations of the foregoing forms. Let us return, for instance, to the spirit of Eq. (3.10), which transforms the Wigner-Ville distribution into a time-scale distribution by an *affine* smoothing. This leads to an equivalent form of the usual scalogram

$$|T_x(t, a)|^2 = \frac{1}{a} \left| \int_{-\infty}^{+\infty} x(s) h^* \left(\frac{s-t}{a} \right) ds \right|^2 \tag{3.158}$$

by means of the representation

$$|T_x(t,a)|^2 = \iint_{\mathbb{R}^2} W_x(s,\xi) W_h\left(\frac{s-t}{a}, a\xi\right) ds\, d\xi . \qquad (3.159)$$

As a simplification let us choose a Morlet wavelet for our function $h(t)$, which is defined by

$$h_{\nu_0}(t) = (2\alpha)^{1/4} e^{-\pi \alpha t^2} e^{i2\pi \nu_0 t} .$$

A straightforward calculation yields the formula

$$W_{h_{\nu_0}}\left(\frac{s-t}{a}, a\xi\right) = 2\exp\left\{-2\pi\left(\frac{\alpha}{a^2}(s-t)^2 + \frac{a^2}{\alpha}(\xi - \nu_0/a)^2\right)\right\} . \qquad (3.160)$$

By means of the formal identification $\nu = \nu_0/a$, one can say that the computation of the scalogram equation (3.159) is *locally* based on the same smoothing as the spectrogram equation (3.70). Here, however, the width of the smoothing window is linked to the scale and thus to the analyzed frequency. Let us denote by Δt and $\Delta \nu$ the *constant* widths (in time and frequency, respectively) of the smoothing window h of the spectrogram. Then we infer from Eq. (3.160), that the scalogram operates with varying widths according to the rule

$$\Delta t(\nu) = \frac{\nu_0 \Delta t}{\nu} \quad \text{and} \quad \Delta \nu(\nu) = \frac{\nu \Delta \nu}{\nu_0} .$$

Hence, these local quantities depend on the analyzed frequency, as we already observed in Eq. (2.30). We thus realize that the scalogram behaves like a spectrogram with a large time-window (and consequently a short frequency-window) at low frequencies. Conversely, at high frequencies it behaves like a spectrogram with a short time-window (and a large frequency-window). It is clear that this variation of the resolution of the analysis does not affect the *joint* degree of the smoothing, as the product of both widths yields a constant value

$$\Delta t(\nu)\, \Delta \nu(\nu) = \Delta t\, \Delta \nu .$$

As already presented, we can similarly reduce the effect of this type of smoothing, while maintaining its affine character, by having recourse to *separable* parameterizations. This defines the so-called *affine smoothed pseudo-Wigner-Ville distribution*[45]

$$ASPW_x(t,a) = \frac{1}{a}\iint_{\mathbb{R}^2} h\left(\frac{\tau}{a}\right) g\left(\frac{s-t}{a}\right) x\left(s+\frac{\tau}{2}\right) x^*\left(s-\frac{\tau}{2}\right) ds\, d\tau .$$

$$(3.161)$$

Chapter 3 Issues of Interpretation 271

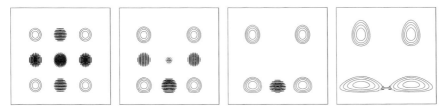

Figure 3.20. Transition between Wigner-Ville and scalogram.

The scalogram in its time-frequency form is a twice (affinely) smoothed version of the Wigner-Ville distribution. By means of a separable smoothing we can devise a continuous transition between the Wigner-Ville distribution (left, no smoothing) and the scalogram (right, smoothing by a Gabor atom). The transition illustrates the trade-off between the joint resolution of the analysis and the significance of the cross-terms.

It permits a smooth transition between the Wigner-Ville distribution and the scalogram (in an exact manner, if both h and g are Gaussians, and approximately otherwise) (cf. Fig. 3.20).

Signal-dependent smoothing. The "constant-Q" property of the affine smoothing materializes in a modification of the usual smoothing, which better fits to signals that are wideband at high frequencies and narrowband at low frequencies. More generally, in the case of an arbitrary signal we can say that a smoothing is "good," if it is *adapted* to the structure of the signal (in a possibly local way).

Let us also consider the example of a signal that is composed of two parallel linear chirps (Fig. 3.21). In this case, the Wigner-Ville distribution displays a *diagonal* structure, which does not support any of the directional smoothing methods with a preferably "rectangular" or "cross-like" shape. In fact, these latter smoothing operations are confronted with an inevitable trade-off between the reduction of the interferences and the loss in the resolution of the signal terms. Hence, one can only observe their poor efficacy. In view of the same directional preference of the signal terms and the cross-terms in the Wigner-Ville distribution, it would certainly be much better to use a directional smoothing along the straight line of the instantaneous frequency. However, this raises the practical problem of finding an automatic adaptation to the signal.[46]

Without dwelling on this issue in detail, we are satisfied with an indication that there exist *local* and *global* solutions to the problem of adaptation. From a global point of view, the easiest approach is to find a weight

Figure 3.21. Limits of fixed smoothing.

In this example the analyzed signal is the superposition of two linear and parallel frequency modulations. The resulting Wigner-Ville distribution is shown in the upper left, the spectrogram (optimized with respect to the slope of the modulation) in the upper right corner. The figures in the bottom row represent the smoothed pseudo-Wigner-Ville (left) and Choï-Williams distribution (right). This shows that the methods of fixed smoothing ("rectangular" or "cross-like") are inefficient in such a ("diagonal") situation, because they cannot furnish a reduction of the cross-terms without affecting the signal terms too much.

function in the ambiguity plane that depends on the signal. The usual interpretation of the interferences as cross-ambiguities associates them with an off-center position in the ambiguity plane. In order to reduce them, we can impose an immediate and natural requirement on the searched weight function that it be (real positive and) *radially decreasing*. By the use of

Chapter 3 Issues of Interpretation

polar coordinates, we can express this property in terms of the inequality

$$f(r_1 \cos\theta, r_1 \sin\theta) \geq f(r_2 \cos\theta, r_2 \sin\theta)$$

for any angle θ and radii $r_1 \leq r_2$. Given such a form, we thus search for the *best* weight function that solves the optimization problem

$$f_{\text{opt}}(\xi,\tau) = \max_f \iint_{\mathbb{R}^2} |f(\xi,\tau)\, A_x(\xi,\tau)|^2\, d\xi\, d\tau \qquad (3.162)$$

under the accompanying condition

$$\iint_{\mathbb{R}^2} |f(\xi,\tau)|^2\, d\xi\, d\tau \leq \alpha\,,$$

where α is fixed beforehand. In fact, for a given upper bound on the volume of f the property of being radially decreasing penalizes the eccentric contributions of $|A_x(\xi,\tau)|^2$ in the integral on the right-hand side of Eq. (3.162). An inclusion of these contributions would result in a "loss" in parts of the volume of f, which are allocated to the "empty" angular regions between the auto- and the cross-correlations. Thus the optimization prefers those contributions of the ambiguity function that are located near the origin; in other words, it essentially prefers the contributions of the signal terms. The choice of the upper bound α governs the trade-off between the joint resolution and the reduction of the interferences. As a reference value we use $\alpha = 1$, which compares with a spectrogram whose window has unit energy.

The actual solution to the optimization problem of Eq. (3.162) can take different forms. More details can be found elsewhere. One solution employs *radial Gaussians* as a model for the parameter function. In any case, it always involves the computation of the complete ambiguity function of the analyzed signal. Only then can one find the reduced interference distribution by a weighting (with the optimized f) and an inverse Fourier transform.

From a *local* point of view, we may express the degree of adaptation of a smoothing operation to a given representation by means of a measure of *concentration* in the plane. Such a measure can be, for instance, the local "kurtosis"

$$K(t,\nu;\Pi) = \frac{\iint_{\mathbb{R}^2} |C_x(\tau,\xi;f)\, w(\tau-t,\xi-\nu)|^4\, d\tau\, d\xi}{\left(\iint_{\mathbb{R}^2} |C_x(\tau,\xi;f)\, w(\tau-t,\xi-\nu)|^2\, d\tau\, d\xi\right)^2}\,, \qquad (3.163)$$

where $w(t,\nu)$ is a time-frequency function of lowpass-type that expresses the local character of the evaluation. The *locally best* smoothing is then defined by

$$\Pi_{\mathrm{opt}}(t,\nu) = \arg\max_{\Pi} \; K(t,\nu;\Pi) \; .$$

This optimization problem can actually be solved, if the smoothing is parameterized in a simple way (for example, by a Gaussian time-frequency function with unit energy, whose directionality and eccentricity may change). However, it is also true that such a procedure is difficult to implement and expensive as far as its computation time is concerned.

"Image" approaches. If we regard the result of the computation of a joint representation as an *image*, we can at last think of the reduction of the cross-terms as methods of image processing. In contrast to an image in the usual sense, however, a time-frequency representation possesses a very high degree of structured information (as can be deduced, for instance, from the formula for the inner interferences, Eq. (3.103)). These underlying structures must be taken into account in every post-processing of the image. We only wish to indicate that such methods exist. For an in-depth study we refer the reader to other sources. [47]

3.2.4. Usefulness of the Interferences

In general, the interference terms reduce the readability of a time-frequency diagram. Therefore it is often desirable to get rid of them. However, it is also true that the same terms constitute an integral part of the distribution, and they carry some information that can be useful in certain cases or from a certain perspective.

Unitarity. We mentioned before that the presence of the cross-terms is the price to pay for gaining the unitarity of the distribution. (Recall that the parameter function must be unimodular, which is incompatible with the idea of reducing the interferences by an apodization.) The important feature of this property (Eq. (2.95)) is that we can substitute the joint inner product in the time-frequency plane for the usual inner product (in time *or* frequency). As a consequence of this fact, which will be studied in more detail in Chapter 4, the comparison of two signals by means of their time-frequency "signatures" requires the consideration of the complete unitary representation *including the interference terms*.

The unitarity of a distribution always implies its *invertibility*; that is, it is possible to reconstruct the signal from its representation exactly, apart from a pure phase. Hence, the interference terms are essential, as they are needed for a complete reconstruction.

Chapter 3 Issues of Interpretation

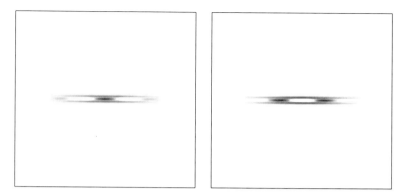

Figure 3.22. Pseudo-Wigner-Ville and phase displacement.

The (pseudo-)Wigner-Ville distribution is sensitive to a possible phase displacement between the different components of a signal. It codes it in the corresponding cross-term. The figure presents the pseudo-Wigner-Ville distributions of two superimposed pure frequencies with (right) or without (left) a phase displacement at the instant $t = 0$. The displacement materializes in a time-shift of the oscillating structure of the interference.

Phase information. The invertibility of a representation (with cross-terms) signifies that it contains all information about all components of the signal. This includes the modulus as well as the phase of the signal. Unlike the spectrogram, which neglects the phase information of the short-time Fourier transform by taking its squared absolute value, the Wigner-Ville distribution is a real quantity that codes these data in its cross-terms.

We have already seen such a coding of the phase in the example of Eq. (3.149), where the superposition of two pure frequencies was considered. The associated Wigner-Ville distribution (Eq. (3.150)) shows a direct sensitivity to a phase displacement of the sinusoidal oscillations at time $t = 0$ (cf. Fig. 3.22). In that example, the cross-term itself acquires a proper physical meaning: It can be regarded as a signature of the *beat frequency* resulting from the coexistence of two frequencies. Indeed, the instantaneous power of the signal of Eq. (3.149) is given by

$$|x(t)|^2 = 2\left[1 + \cos 2\pi(\nu_1 - \nu_2)t\right].$$

Hence, it is governed by fluctuations that have a longer period if the two frequencies are closer to each other. Owing to the correct marginal distribution in time, this value must coincide with the sum of all amplitudes of the signal terms *and* the cross-terms at this instant. Hence, it is the interference term that expresses the existence of the beat frequency and stands for its realization in the time-frequency plane.

Remark. Let us note that the phenomenon of the beat frequency can only be perceived physically, if the frequencies are very close to each other. In a time-frequency setting, this can be explained in terms of a *smoothing* of the distribution. For this purpose let us consider a post-processing filter, which performs a lowpass filtering relative to a nonzero time interval. Then only those interference terms subsist, which correspond to components with a small frequency distance: It must be smaller than the reciprocal of the response time of the filter (playing the role of the interval of the integration). For these components the effect of the interference can be perfectly understood as an *interference tone* in a physical sense. For other components, which are significantly farther apart, the smoothing destroys any coherence and leads to a time-frequency picture, which is simply the superposition of the spectral lines without interaction.

A second example, which explains how the phase enters a Wigner-Ville distribution, is an eventually existing *phase jump* of a spectral line. Let us consider the model of an undamped pure frequency, which has a phase discontinuity at the origin; hence

$$x(t) = e^{i(2\pi\nu_0 t + \varphi_1)} U(t) + e^{i(2\pi\nu_0 t + \varphi_2)} U(-t) \ . \qquad (3.164)$$

Here $U(t)$ denotes the normalized Heaviside step function, as before. An immediate computation of the Wigner-Ville distribution yields the result

$$W_x(t,\nu) = \cos(\varphi_1 - \varphi_2)\, \delta(\nu - \nu_0) \qquad (3.165)$$

$$+ \frac{2}{\pi(\nu - \nu_0)} \sin\left(\frac{\varphi_1 - \varphi_2}{2}\right) \cos\left\{ 4\pi(\nu - \nu_0)|t| - \frac{\varphi_1 - \varphi_2}{2} \right\} .$$

We can certainly verify that

$$W_x(t,\nu) = \delta(\nu - \nu_0) \qquad (3.166)$$

holds in case of a perfect alignment of the phases at the origin, and this corresponds to the Wigner-Ville distribution of the pure continuous frequency. At the breakpoint $t = 0$ Eq. (3.165) turns into

$$W_x(0,\nu) = \cos(\varphi_1 - \varphi_2)\, \delta(\nu - \nu_0) + \frac{1}{\pi(\nu - \nu_0)} \sin(\varphi_1 - \varphi_2) \ . \qquad (3.167)$$

It thus produces a Dirac distribution competing with a hyperbola, and both terms come with a factor depending on the phase displacement $\varphi_1 - \varphi_2$. Its

Chapter 3 Issues of Interpretation   277

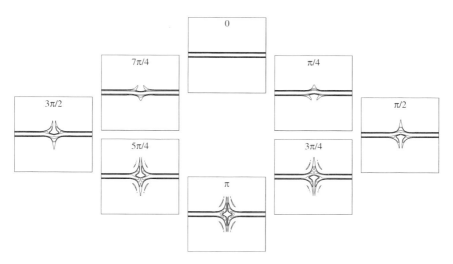

Figure 3.23. Pseudo-Wigner-Ville and phase jump.

For a signal that consists of only one pure frequency, the (pseudo-) Wigner-Ville distribution has a characteristic form near an eventual phase jump. The figure illustrates this fact for various values of the phase jump between 0 and 2π.

graph has a *peak*, whose amplitude and position tell how much the phase displacement $\varphi_1 - \varphi_2$ differs from π. In fact, if $\varphi_1 - \varphi_2$ is closer to π, then the peak is more accentuated; moreover, it is located at a lower (higher) frequency than ν_0, if $\varphi_1 - \varphi_2$ is smaller (larger, respectively) than π. This theoretical behavior is depicted in Fig. 3.23.

What is a component? We have approached the problem of the creation of interferences by considering general models of signals, which are formed by the linear superposition of *components* that interact. However, we never defined what this notion of components really means. Although it is difficult to give a definite answer to this question (what is a component?), it still deserves some special investigation.

Constructive interferences. In order to illuminate this issue, we should first recall that the distinction between "inner" and "outer" interferences was rather arbitrary. This is why the idea of decomposing a time-frequency distribution into "signal" terms related to the components of the signal and "interference" terms related to their interaction runs into difficulties. Let us illustrate this problem by referring to the example of Eq. (3.164) once more. When there is a phase jump at $t = 0$, the separation of the signal into two components has a physical meaning; however, if there is no phase jump at all, the splitting of the signal and sticking back together at the instant $t = 0$ is totally arbitrary. Let us think over what happens to the Wigner-Ville distribution in this borderline case. Each virtual "component" creates a distribution starting or terminating at $t = 0$. Then the interference term, which is superimposed on the individual distributions, causes the representation equation (3.166) to be continuous at this instant and look like a single component again.

In the opposite case, where a nonzero phase jump occurs at $t = 0$, the interference term becomes asymmetrical and does not yield the same perfect fit of the two parts. This results in the discontinuity of the representation at $t = 0$. In this sense, the mechanism of the creation of (inner) interferences renders the notion of a component meaningful. More generally, due to Eq. (3.103) of inner interferences, the value of the Wigner-Ville distribution in one point results from a "holographic" construction, which brings the values at all other points into play. We can thus say in a few words that *a component is a constructive interference of the Wigner-Ville distribution*.

Information logon. Another way to give a meaningful notion of a component relates to the definition of a *logon* in Gabor's sense; by this we denote an information "quantum" in the plane. Provided that we are able to define a *measure* for the information contents in the plane, a component can thus be associated with a *bit* of information. Unfortunately, as we know by now, a time-frequency representation is susceptible to having negative values, in general. Hence, it cannot be associated with a probability density in a straightforward manner, which would enable us to compute information contents in Shannon's sense. However, there exists a solution that employs a definition due to Rényi.[48] As a substitute of Shannon's information measure the quantity

$$R_\alpha(C_x) = \frac{1}{1-\alpha} \log_2 \iint_{\mathbf{R}^2} C_x^\alpha(t,\nu;f)\,dt\,d\nu\,, \quad \alpha > 1$$

is proposed. Recent studies have shown that this approach allows us to "count" the number of constituent logons of a signal, especially when $\alpha = 3$.

Independence. Finally, a third point of view relies on a stochastic interpretation. Suppose we analyze a *random* signal $x(t)$ that admits a decomposition

$$x(t) = \sum_{n=1}^{N} a_n x_n(t)\,,$$

Chapter 3 Issues of Interpretation

where the $x_n(t)$ are deterministic signals (not necessarily orthogonal) and the a_n are stochastically *independent* random coefficients. Under these assumptions we can conclude that

$$\mathbf{E}\left\{a_n a_m^*\right\} = 0 \quad \text{if} \quad n \neq m \ .$$

Hence, we can immediately deduce, due to Eq. (3.92), that

$$\mathbf{W}_x(t,\nu) = \sum_{n=1}^{N} |a_n|^2 \, W_{x_n}(t,\nu) \ . \tag{3.168}$$

This shows that a random signal, which is the superposition of stochastically independent components, has a Wigner-Ville spectrum which lacks any cross-terms. This can be compared with the method of "atomic" decomposition discussed earlier (see Eq. (3.132)); here, a simplified Wigner-Ville distribution is obtained by taking the sum of the distributions of the elements of the decomposition, provided that those are stochastically independent. [49] This last assumption contrasts with Eq. (3.132) where the atoms were fixed *ab initio*. However, this new approach requires the availability of as many linear combinations as there are components in the signal.

3.2.5. Statistical Estimation of the Wigner-Ville Spectrum

We already explained the importance of the Wigner-Ville *spectrum* of non-stationary random signals in Subsection 2.4.3. Let us first recall its definition

$$\mathbf{W}_x(t,\nu) = \int_{-\infty}^{+\infty} r_x\left(t + \frac{\tau}{2}, t - \frac{\tau}{2}\right) e^{-i2\pi\nu\tau} \, d\tau \ , \tag{3.169}$$

which is based on the autocovariance $r_x(t_1, t_2)$ of the analyzed random process.

Suppose we dispose of a *discrete-time* observation of the process. This is most often the case in practice and can be obtained, for instance, by sampling a realization of $x(t)$ at an arbitrary sampling rate, which we choose to be 1. Then the preceding definition can be put into another form, following the spirit of Eq. (3.87), by writing

$$\mathbf{W}_x[n,\nu] = 2 \sum_{k=-\infty}^{+\infty} r_x[n+k, n-k] \, e^{-i4\pi\nu k} \ , \quad |\nu| \leq 1/4 \ . \tag{3.170}$$

Now we face the problem of *estimating* this quantity, especially when only one realization of finite duration is assumed to be available. An estimator for the Wigner-Ville spectrum can be based on the Fourier transform of an estimator for the covariance. Under certain conditions on the random process, which describe the locally stationary behavior, we can thus devise a definition of a general class of estimators. They can be compared with the representations in Cohen's class of a single (certain) realization of the process. There are several motivations for defining such a unified conceptual framework: First, it puts some new light on Cohen's class; then it allows consideration of the properties of the different estimators from a very general perspective; and finally, it includes the treatment of deterministic and random signals in one and the same setting.[50]

Assumptions. Suppose an estimator for the Wigner-Ville spectrum is given by the Fourier transform

$$\widehat{\mathbf{W}}_x[n,\nu] = 2 \sum_{k=-\infty}^{+\infty} \hat{r}_x[n+k, n-k] \, e^{-i4\pi\nu k} \,, \qquad (3.171)$$

where $\hat{r}_x$ is an estimator for the autocovariance. In the stationary case, where only one realization of $x(t)$ is available, a commonly used assumption is the *ergodic hypothesis*, by which we can substitute a temporal mean value of the given data for the (inaccessible) stochastic expectation value of the theoretical covariance. Unfortunately, we cannot follow the same approach in the nonstationary case, in general, unless some additional conditions are met. Let us therefore confine ourselves to the class of *quasi-stationary* signals, which are characterized by stochastic properties that are steady enough, so that the instantaneous covariance at any instant may be approximated by a local time-average. This conforms to the interpretation that a quasi-stationary signal with instantaneous covariance $r_x[n+k, n-k]$ admits a *tangential stationary signal* at each instant n. The covariances r_x of the given signal and $\gamma_{x,n}$ of the tangential stationary signal are supposed to satisfy

$$|r_x[n+k, n-k] - \gamma_{x,n}[2k]| < \epsilon(T) \,, \quad |2k| \leq T \,. \qquad (3.172)$$

Here $\epsilon(T)$ is a measure for the gained approximation relative to a time-interval of length T. For fixed ϵ, the time of the stationary behavior T_s is defined as the largest value of T for which Eq. (3.172) holds. When we work with quasi-stationary signals, we can thus consider the average over the time interval T_s as an estimator for the instantaneous covariance.

Two further assumptions will be used in the course of the following analysis:

(*i*) The signals are given in their *analytic* form, so as to avoid all problems of spectral redundancy and to bestow a physical meaning on the notions of instantaneous amplitude and frequency; and

(*ii*) they are zero-mean *Gaussian* processes.

Chapter 3 Issues of Interpretation

Classes of estimators. Assuming a quasi-stationary character of the signal, a natural class of estimators for the covariance can be defined by

$$\widehat{r}_x[n+k, n-k] = \sum_{m=-\infty}^{+\infty} F[m-n, 2k]\, x[m+k]\, x^*[m-k]\ ,$$

where $F[n, k]$ is a free parameter function. Its form and duration determine the way of averaging the products $x[m+k]\, x^*[m-k]$. If we insert this expression into Eq. (3.171), we obtain a general class of estimators for the Wigner-Ville spectrum by

$$\widehat{\mathbf{W}}_x[n, \nu] = 2 \sum_{k=-\infty}^{+\infty} \sum_{m=-\infty}^{+\infty} F[m-n, 2k]\, x[m+k]\, x^*[m-k]\, e^{-i4\pi\nu k}\ . \quad (3.173)$$

Hence, we recover a discrete-time version of Eq. (3.11). We can therefore conclude that a general form of estimators for the Wigner-Ville spectrum of a random signal (with only one given realization) is provided by the representations in Cohen's class of the observed realization.

Remark. Cohen's class defines estimators based on a smoothing, which is independent of the frequency. Other general schemes for the estimation of the covariance may be based on the *affine* class, for example. We confine ourselves to the study of the forementioned class here. The second class of estimators will be used later, in Subsection 4.2.1, where it is employed for a spectral estimation of "$1/f$-noise" processes.

The complete set of results, which map the properties of the parameter function into associated properties of the representation (cf. Chapter 2), can be used here by an adaptation to the stochastic context. Moreover, the stochastic properties of the estimator equation (3.173) can be understood as a function of the parameterization F (or one of its equivalent forms) as well. In return, we are enabled to devise parameterizations, and thereby estimators, that satisfy some *a priori* imposed conditions.

Let us investigate the stochastic properties of first and second order. We begin with a consideration of the general form of the estimators, prior to specialization in some of the known cases. It is sometimes more convenient to work with the equivalent forms of F. By analogy with Subsection 3.1.1 we denote them by

$$\Pi[n, \nu] = 2 \sum_{k=-\infty}^{+\infty} F[n, 2k]\, e^{i4\pi\nu k}\ , \quad (3.174)$$

$$f(\xi, 2k) = \sum_{n=-\infty}^{+\infty} F[n, 2k]\, e^{-i2\pi\xi n}\ , \quad (3.175)$$

$$\psi(\xi,\nu) = 2 \sum_{n=-\infty}^{+\infty} f(\xi, 2k] e^{i4\pi\nu k} \;, \qquad (3.176)$$

which are discrete-time versions of the definitions given there.

Bias. Based on the general form of the estimator equation (3.173), we can immediately compute the expectation value

$$\mathbf{E}\left\{\widehat{\mathbf{W}}_x[n,\nu]\right\} = 2 \sum_{k=-\infty}^{+\infty} \sum_{m=-\infty}^{+\infty} F[m-n, 2k]\, r_x[m+k, m-k]\, e^{-i4\pi\nu k}$$

$$= 2 \sum_{k=-\infty}^{+\infty} \sum_{m=-\infty}^{+\infty} F[m-n, 2k]\, \frac{1}{2} \int_{-1/2}^{+1/2} \mathbf{W}_x[m, \zeta/2]\, e^{i2\pi\zeta k}\, d\zeta\, e^{-i4\pi\nu k}$$

$$= \sum_{m=-\infty}^{+\infty} \int_{-1/4}^{+1/4} \mathbf{W}_x[m, \xi)\, 2 \sum_{k=-\infty}^{+\infty} F[m-n, 2k]\, e^{i4\pi(\xi-\nu)k}\, d\xi \;,$$

and we thus obtain

$$\mathbf{E}\left\{\widehat{\mathbf{W}}_x[n,\nu]\right\} = \sum_{m=-\infty}^{+\infty} \int_{-1/4}^{+1/4} \mathbf{W}_x[m, \xi)\, \Pi[m-n, \xi-\nu)\, d\xi \;. \qquad (3.177)$$

This result is evidently comparable with that of Eq. (3.9). In the context of statistical estimation, it means that, in general, the estimators of Cohen's class are *biased* in time and frequency. The bias is controlled by the parameter function in the time-frequency form (and it grows with the "extension" of this parameterization). To have a correct normalization of the estimator, we must further impose the condition

$$\sum_{n=-\infty}^{+\infty} \int_{-1/4}^{+1/4} \Pi[n,\nu)\, d\nu = 1 \;,$$

which can be rewritten in a simpler form as

$$f(0, 0] = 1 \;. \qquad (3.178)$$

Chapter 3 Issues of Interpretation

Variance. For the determination of the stochastic properties of second order we must compute the covariance of the estimator at two fixed time-frequency positions. This yields the expression

$$\operatorname{cov}\{\widehat{\mathbf{W}}_x[n_1,\nu_1),\,\widehat{\mathbf{W}}_x[n_2,\nu_2)\}$$

$$= 4 \sum_{k_1=-\infty}^{+\infty} \sum_{k_2=-\infty}^{+\infty} \sum_{m_1=-\infty}^{+\infty} \sum_{m_2=-\infty}^{+\infty} F[m_1-n_1, 2k_1]\, F^*[m_2-n_2, 2k_2]$$

$$\times \operatorname{cov}\{x[m_1+k_1]x^*[m_1-k_1],\, x[m_2+k_2]x^*[m_2-k_2]\}\, e^{-i4\pi(\nu_1 k_1-\nu_2 k_2)}.$$

Under the given hypotheses we can further simplify the right-hand side. In fact, by the assumption of a Gaussian process we can express the moment of fourth order, which appears in the covariance of the cross-product, by a sum of products of the second-order moments. The assumption of an analytic signal allows us to reduce this sum to only one term, owing to the fact that real and imaginary parts of an analytic signal are uncorrelated. This leads to

$$\operatorname{cov}\{x[m_1+k_1]\,x^*[m_1-k_1],\, x[m_2+k_2]\,x^*[m_2-k_2]\}$$

$$= r_x[m_1+k_1,\, m_2+k_2]\, r_x^*[m_1-k_1,\, m_2-k_2].$$

By using the additional assumptions on quasi-stationary signals, we can employ the approximations

$$r_x[n_1+m_1\pm k_1,\, n_2+m_2\pm k_2] \approx r_x\left[\frac{n_1+n_2}{2}+k_\pm,\, \frac{n_1+n_2}{2}-k_\pm\right]$$

that are valid on the useful interval defined by F. If we next introduce the power spectrum $\Gamma_{x,n_i}(\nu)$ of the tangential stationary process at the moment $n_i = (n_1+n_2)/2$, we obtain as our final result (after some quite laborious computations)

$$\operatorname{cov}\{\widehat{\mathbf{W}}_x[n_1,\nu_1),\,\widehat{\mathbf{W}}_x[n_2,\nu_2)\}$$

$$\approx 2 \int_{-1/2}^{+1/2}\int_{-1/2}^{+1/2} \psi(2\xi, \nu-\nu_1)\, \psi^*(2\xi, \nu-\nu_2) \qquad (3.179)$$

$$\times\, \Gamma_{x,n_i}(\nu+\xi)\, \Gamma_{x,n_i}(\nu-\xi)\, e^{i4\pi\xi(n_1-n_2)}\, d\xi\, d\nu\,.$$

This shows that under given hypotheses the general estimators of Cohen's class are *correlated* in time and frequency. Let us add one further assumption, namely, that the radius of the local correlation of the tangential stationary process is small as compared to the time of the stationary behavior of the signal. (This amounts to the assumption that the Wigner-Ville spectrum is locally wideband relative to the frequency-band of the analysis.) Then we can derive an approximation of the *variance* of the estimator from the previous expression for the covariance. If we let

$$f'(\xi, k] = \sum_{m=-\infty}^{+\infty} f(\xi, m] f^*(\xi, m-k] ,$$

we obtain

$$\text{var}\left\{\widehat{\mathbf{W}}_x[n, \nu)\right\} \approx 2\, C_{\Gamma_{x,n}}[0, \nu; f') . \tag{3.180}$$

This shows that the variance depends on the frequency.

Examples. All distributions in Cohen's class can be viewed as estimators for the Wigner-Ville spectrum. As a first example let us consider the *spectrogram*. In the context of statistical estimation, it is identical with the *short-time periodogram* of classical spectral estimation, as it is defined by

$$S_x[n, \nu) = \left| \sum_{k=-\infty}^{+\infty} x[k]\, h^*[k-n]\, e^{-i2\pi\nu k} \right|^2 .$$

It can also be obtained by choosing the Wigner-Ville distribution of the short-time window $h[n]$ as the time-frequency weight $\Pi[n, \nu)$ for the estimator equation (3.173). Consequently, the bias in Eq. (3.177) attains the form

$$\mathbf{E}\left\{S_x[n, \nu)\right\} = \sum_{m=-\infty}^{+\infty} \int_{-1/4}^{+1/4} \mathbf{W}_x[m, \xi)\, W_h[m-n, \xi-\nu)\, d\xi . \tag{3.181}$$

We thus recover once more the classical trade-off between the time and frequency behavior, which materializes here in the form of bias.

As far as the correlation of the estimated values is concerned, the approximate form of Eq. (3.179) leads to

$$\text{cov}\left\{S_x[n_1, \nu_1), S_x[n_2, \nu_2)\right\} \approx |A_h(\nu_1 - \nu_2, n_1 - n_2]|^2\, \Gamma^2_{x, n_i}\left(\frac{\nu_1 + \nu_2}{2}\right). \tag{3.182}$$

Here the discrete-time ambiguity function is defined by

$$A_h(\xi, 2k] = \sum_{n=-\infty}^{+\infty} x[n+k]\, x^*[n-k]\, e^{i2\pi\xi n} , \quad |\xi| \leq 1/2 . \tag{3.183}$$

This signifies that the approximately uncorrelated behavior is subdued to a time-frequency compromise as well. With a correct normalization of the window (according to Eq. (3.178) it must have unit energy), we can obtain the approximation of the variance

$$\text{var}\{S_x[n,\nu]\} \approx \mathbf{W}_x^2[n,\nu] . \tag{3.184}$$

This follows from assimilating the Wigner-Ville spectrum locally (in time) to the spectral density of the tangential stationary process. In fact, this result is of the same type as the classical result in stationary spectral analysis, which states that the variance of a periodogram yields the square of the value to be estimated (which is the power spectrum).

Consequently, and by analogy with the stationary case, an improvement can be achieved, if we replace the raw periodogram with a *smoothed* or *averaged* version.[51] This amounts to the introduction of estimators of the type

$$\widetilde{S}_x[n,\nu] = \sum_{m=-\infty}^{+\infty} g[m]\, S_x[m-n,\nu] . \tag{3.185}$$

Their associated parameter function has the form

$$f(\xi, 2k] = G(\xi)\, A_h^*(\xi, 2k] .$$

In this case, the approximation of the variance in Eq. (3.184) becomes

$$\text{var}\left\{\widetilde{S}_x[n,\nu]\right\} \approx \mathbf{W}_x^2[n,\nu] \left| \sum_{m=-\infty}^{+\infty} g[m]\, A_h(0,m) \right|^2 . \tag{3.186}$$

Hence, if $g[m]$ is a rectangular filter with M taps and properly normalized so that

$$\sum_{n=-\infty}^{+\infty} |h[n]|^2 \cdot \sum_{m=-\infty}^{+\infty} g[m] = 1 ,$$

the factor in Eq. (3.186) is of order $1/M$. We can thus derive the approximation

$$\text{var}\left\{\widetilde{S}_x[n,\nu]\right\} \approx \frac{1}{M}\, \mathbf{W}_x^2[n,\nu] . \tag{3.187}$$

Note that enlarging the horizon M of the smoothing amounts to a reduction of the variance of the estimator. But then the bias in the time-direction increases owing to Eq. (3.181).

Let us reconsider the *geometric* point of view (this time in a *stochastic* setting), which was developed before and made us gain greater flexibility. In order to replace the spectrogram, which has "one degree of freedom," with a smoothed pseudo-Wigner-Ville distribution having "two degrees of freedom," it suffices to put

$$\Pi[n,\nu] = g[n]\, W_h[0,\nu] \ . \tag{3.188}$$

The corresponding estimator is defined by

$$SPW_x[n,\nu] \tag{3.189}$$

$$= 2 \sum_{k=-\infty}^{+\infty} |h[k]|^2 \left[\sum_{m=-\infty}^{+\infty} g[m-n]\, x[m+k]\, x^*[m-k] \right] e^{-i4\pi\nu k}.$$

As could be expected from the general setting, the bias is also a separable quantity with respect to the two variables, because we have

$$\mathbf{E}\left\{ SPW_x[n,\nu] \right\} = \sum_{m=-\infty}^{+\infty} \int_{-1/4}^{+1/4} \mathbf{W}_x[m,\xi]\, g[m-n]\, W_h[0,\xi-\nu]\, d\xi \ . \tag{3.190}$$

Hence, we can conclude in the pseudo-Wigner-Ville case (i.e., no smoothing in the time-direction is used), that the bias of the estimator affects only the frequency-direction; this becomes clear by examining the relation

$$\mathbf{E}\left\{ PW_x[n,\nu] \right\} = \int_{-1/4}^{+1/4} \mathbf{W}_x[n,\xi]\, W_h[0,\xi-\nu]\, d\xi \ . \tag{3.191}$$

The advantages of the separability also materialize in the properties of second order. Due to the general result of Eq. (3.179) and the specific form of the separable parameterization (Eq. (3.188)), we obtain

$$\operatorname{cov}\left\{ SPW_x[n_1,\nu_1],\, SPW_x[n_2,\nu_2] \right\}$$

$$\approx 2 \int_{-1/2}^{+1/2} W_h[0,\nu-\nu_1]\, W_h[0,\nu-\nu_2]\, C_{\Gamma_{x,n_i}}[n_1-n_2,\nu;f'']\, d\nu \ .$$

In this expression we put

$$f''(2\xi,k) = |G(\xi)|^2 \ .$$

Chapter 3 Issues of Interpretation 287

If we assume, as in the forementioned discussion, that h is wide enough, we obtain the approximate form

$$\text{cov}\left\{SPW_x[n_1,\nu_1)\,,\,SPW_x[n_2,\nu_2)\right\}$$
$$\approx 2\,C_{\Gamma_{x,n_i}}[n_1-n_2,(\nu_1+\nu_2)/2;f'')\int_{-1/2}^{+1/2} W_h[0,\nu-\nu_1)\,W_h[0,\nu-\nu_2)\,d\nu.$$

It follows that only the correlation in the frequency-direction enters this expression. The uncorrelated behavior in time can be understood as a result of the separation of the considered instants, whose distance is at least as large as the duration of the smoothing. Hence, under the given hypotheses it emerges that

$$\left.\begin{array}{l}|n_1-n_2|>M\\|\nu_1-\nu_2|>1/N\end{array}\right\}\quad\Longrightarrow\quad\text{cov}\left\{SPW_x[n_1,\nu_1)\,,\,SPW_x[n_2,\nu_2)\right\}\approx 0\;. \tag{3.192}$$

Here M and N denote the widths of the windows g and h, respectively. For such windows an approximate decorrelation of the estimated values at distinct points in the plane is guaranteed, if their distance is large enough. Conversely, for two given points in the plane an approximate decorrelation can be achieved, if the time-frequency "extension" of the smoothing is chosen to be sufficiently small, eventually smaller than for a spectrogram.

Finally, we assume that the normalization condition of Eq. (3.178) is satisfied and the window functions g and h are of rectangular shape (of length M and N, where N is supposed to be big). This leads to an approximation of the variance by

$$\text{var}\left\{SPW_x[n,\nu)\right\}\approx 2\,C_{\Gamma_{x,n}}[0,\nu;f'')\;. \tag{3.193}$$

If no smoothing is performed, we gain the result

$$\text{var}\left\{PW_x[n,\nu)\right\}\approx 2\,W_{\Gamma_{x,n}}[0,\nu)\;. \tag{3.194}$$

Hence, we can rewrite Eq. (3.193) in the case of a rectangular window function of length M as

$$\text{var}\left\{SPW_x[n,\nu)\right\}\approx \frac{2}{M}\sum_{m=-M}^{+M}\left(1-\frac{|m|}{M}\right)W_{\Gamma_{x,n}}[m,\nu)\;.$$

For sufficiently large M the property of the correct marginal distribution (Eq. (2.66)) allows us to write

$$\sum_{m=-M}^{+M} \left(1 - \frac{|m|}{M}\right) W_{\Gamma_{x,n}}[m, \nu] \approx \sum_{m=-\infty}^{+\infty} W_{\Gamma_{x,n}}[m, \nu] = \Gamma_{x,n}^2(\nu) .$$

(It suffices that M, being characteristic of the time of the stationary behavior, is big in comparison with the local radius of correlation of the tangential stationary signal.)

The variance of the estimator can thus be simplified to

$$\text{var}\left\{SPW_x[n, \nu]\right\} \approx \frac{2}{M} \mathbf{W}_x^2[n, \nu] . \qquad (3.195)$$

This result is comparable with Eq. (3.187). It is important to observe that the reduction of the variance due to the time-smoothing is of the same order of magnitude for the smoothed versions of the pseudo-Wigner-Ville estimator and the spectrogram. This reduction is accompanied by an enlargement of the bias in the time-direction, again for both estimators. Nevertheless, the advantage of the pseudo-Wigner-Ville estimator results from the fact that no other bias in the time-direction is introduced, because the non-smoothed estimator is truly unbiased in this direction (cf. Eq. (3.191)). This contrasts with the spectrogram: here the smoothing operation introduces a second bias, which is added to the bias related to the width of the analysis window. The smoothing thus gives rise to the same advantages and disadvantages over the nonsmoothed versions for both types of estimators. The difference, however, which expresses the superiority of the pseudo-Wigner-Ville estimator, is rooted in the nonsmoothed versions themselves.

Example. As an illustration let us consider the case of a white noise $b[n]$ with varying power. Its covariance can be written as

$$r_b[n+k, n-k] = (p[n]/2)\,\delta_{k0} . \qquad (3.196)$$

If the evolution of the power is sufficiently slow, we can think of the tangential stationary analytic signal at the instant n as being defined by the power spectrum

$$\Gamma_{b,n}(\nu) = \begin{cases} 0 & , \quad -1/2 \leq \nu < 0 , \\ 2p[n] & , \quad 0 \leq \nu < +1/2 . \end{cases}$$

Chapter 3 Issues of Interpretation 289

Hence, we infer

$$
W_{\Gamma_{b,n}}[m,\nu] = \begin{cases} 0 & , \quad -1/2 \leq \nu < 0 \ , \\ \dfrac{4p^2[n]}{\pi m} \sin 4\pi \nu m & , \quad 0 \leq \nu < 1/4 \ , \\ \dfrac{4p^2[n]}{\pi m} \sin 4\pi (1/2 - \nu) m & , \quad 1/4 \leq \nu < 1/2 \ . \end{cases}
$$

We further derive from Eqs. (3.184) and (3.194) that

$$\text{var}\left\{S_x[n,\nu]\right\} \approx 4p^2[n] \ , \qquad 0 \leq \nu < +1/2 \ , \qquad (3.197)$$

and

$$\text{var}\left\{PW_x[n,\nu]\right\} \approx \begin{cases} 4p^2[n]\, 4\nu & , \quad 0 \leq \nu < 1/4 \ , \\ 4p^2[n]\, 4(1/2 - \nu) & , \quad 1/4 \leq \nu < 1/2 \ . \end{cases} \qquad (3.198)$$

Therefore, in contrast to the spectrogram, the (nonsmoothed) pseudo-Wigner-Ville estimator has a variance that depends on the frequency; this is illustrated in Fig. 3.24, where the reducing effect of a time-smoothing on the variance is also shown.

§3.3. About the Positivity

The existence of negative values of the distribution sets limits on the analogy between a time-frequency representation and a probability density function. This also renders its interpretation more delicate. Evident problems arise when we attempt to bestow the rank of a local energy density on a negative value. Another important observation to be made concerns the rather vague justification of a *pointwise* interpretation of a time-frequency distribution, even if it happens to be positive everywhere. In spite of these restrictions, it is interesting to encompass the relations between time-frequency representations and positivity as well as some of their consequences.

3.3.1. Some Problems Caused by the Nonpositivity

The limitations caused by the nonpositivity have a *local* nature in the first place. This means that a negative value of the representation *at one point* renders any direct physical interpretation as an energy density at this particular point impossible.

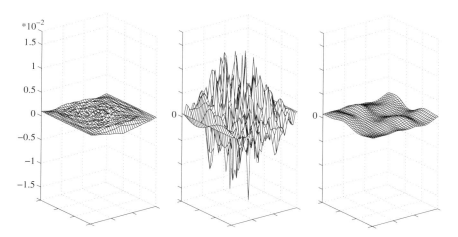

Figure 3.24. Wigner-Ville spectrum and its estimation.

The Wigner-Ville spectrum of a white noise (defined as the ensemble average of the Wigner-Ville distributions) is constant theoretically. This can be observed in the left figure, which shows the average of 100 computed Wigner-Ville distributions of independent realizations. In the case, where only one realization is available, the pseudo-Wigner-Ville distribution (center) yields an estimator with a big variation, as expected. An enhanced estimator is obtained by employing an additional smoothing in the time-direction (right).

They can also have a more *global* nature, insofar as the average quantities of a distribution with negative values can lose their specific meaning. If we wish to interpret a representation $\rho_x(t,\nu)$ as a joint *probability density* of the pair (t,ν), we should require, at least, that its marginal distributions be identical with the *individual probability densities* in t and ν (or the *a priori* densities)

$$\rho_x(t) = |x(t)|^2 \quad \text{and} \quad \rho_x(\nu) = |X(\nu)|^2. \qquad (3.199)$$

Within the setting of Cohen's class, we have seen that simple conditions on the parameterization guarantee this property (cf. Eq. (2.67)). Carrying on this analogy, we can also conceive definitions of *conditional probability densities* $\rho_x(t \mid \nu)$ and $\rho_x(\nu \mid t)$ (using Bayes' formula) according to

$$\rho_x(t,\nu) = \rho_x(t \mid \nu)\rho_x(\nu) = \rho_x(\nu \mid t)\rho_x(t). \qquad (3.200)$$

When we stick to Cohen's class, some additional conditions (cf. Eqs. (2.99) and (2.101)) turn out to be sufficient for an identification of the

Chapter 3 Issues of Interpretation

associated *conditional expectation values* with the instantaneous frequency and the group delay. Recall that under these hypotheses we have

$$\mathbf{E}\{\nu \mid t\} = \int_{-\infty}^{+\infty} \nu\, \rho_x(\nu \mid t)\, d\nu$$

$$= \frac{1}{|x(t)|^2} \int_{-\infty}^{+\infty} \nu\, C_x(t, \nu; f)\, d\nu = \nu_x(t) ,$$

(3.201)

$$\mathbf{E}\{t \mid \nu\} = \int_{-\infty}^{+\infty} t\, \rho_x(t \mid \nu)\, dt$$

$$= \frac{1}{|X(\nu)|^2} \int_{-\infty}^{+\infty} t\, C_x(t, \nu; f)\, dt = t_x(\nu) .$$

(3.202)

All conditions together (which are verified by the Wigner-Ville distribution, in particular) seem to provide a satisfactory (though not pointwise) interpretation, insofar as they suggest to define a localization of the representation around its local centers of gravity. It is therefore tempting to proceed with the stochastic analogy and to define a *dispersion* around these averages by means of *conditional variances*.[52] Let us consider the rule of the instantaneous frequency, for example. Then the expression

$$\sigma^2(\nu \mid t) = \mathbf{E}\{\nu^2 \mid t\} - (\mathbf{E}\{\nu \mid t\})^2$$

$$= \frac{1}{|x(t)|^2} \int_{-\infty}^{+\infty} \nu^2 C_x(t, \nu; f)\, d\nu - \nu_x^2(t)$$

(3.203)

is a suitable candidate for such a dispersion.

Unfortunately, simple computations in the Wigner-Ville case show that this "pseudo-variance" has the value

$$\sigma_{\mathrm{W}}^2(\nu \mid t) = -\frac{1}{8\pi^2} \frac{d^2 \log |x|}{dt^2}(t) ,$$

and this need not be nonnegative at all. The same happens for the "pseudo-variance" associated with the Rihaczek distribution. An analogous calculus leads to

$$\sigma_{\mathrm{R}}^2(\nu \mid t) = -\frac{1}{4\pi^2} \frac{1}{|x(t)|} \frac{d^2 |x|}{dt^2}(t) .$$

We can observe that in both cases a locally positive curvature of the envelope of the signal may lead, again locally, to a negative "pseudo-variance."

This, of course, rules out any legitimacy in terms of a *dispersion* round a mean value. One can show more generally that the "pseudo-variance" of an arbitrary distribution in Cohen's class (with correct marginal distributions and local first-order moments) can be written as

$$\sigma_f^2(\nu \mid t) = -\frac{1}{8\pi^2} \frac{d^2 \log |x|}{dt^2}(t) - \frac{1}{4\pi^2 |x(t)|^2} \int_{-\infty}^{+\infty} |x(s)|^2 \frac{\partial^2 F}{\partial \tau^2}(s-t, 0)\, ds \ . \tag{3.204}$$

If we consider, in particular, a parameterization of "product"-type as in Eq. (3.145), we obtain

$$\sigma_\varphi^2(\nu \mid t) = \frac{1}{8\pi^2} \left[\left(1 + \frac{d^2\varphi}{\pi^2 dt^2}(0)\right) \left(\frac{1}{|x(t)|} \frac{d|x|}{dt}(t)\right)^2 \right.$$
$$\left. + \left(1 - \frac{d^2\varphi}{\pi^2 dt^2}(0)\right) \left(\frac{1}{|x(t)|} \frac{d^2|x|}{dt^2}(t)\right) \right] . \tag{3.205}$$

Such a quantity can be ascertained to be nonnegative if it meets the condition [53]

$$\frac{d^2\varphi}{dx^2}(0) = \pi^2 \ . \tag{3.206}$$

Under this constraint the corresponding "pseudo-variance" attains the value

$$\sigma_\varphi^2(\nu \mid t) = \frac{1}{4\pi^2} \left(\frac{d \log |x|}{dt}(t)\right)^2 . \tag{3.207}$$

However, the foregoing condition is not satisfied by most classical parameterizations (in fact, it corresponds to a *positive* curvature at the origin of the ambiguity plane, which is inconsistent with the usual idea of a maximal weight at the origin). Nevertheless, one can propose examples of *ad hoc* constructions such as the "difference of two Gaussians"

$$\varphi(\xi\tau) = (1+a)\, e^{-\pi^2 \xi^2 \tau^2 / 2\sigma^2} - a\, e^{-\pi^2 \xi^2 \tau^2 / 2\sigma'^2} \tag{3.208}$$

(which actually corresponds to the difference of two Choï-Williams distributions), subject to the restriction

$$\frac{1}{\sigma'^2} = \frac{1}{a} + \frac{1}{\sigma^2}\left(1 + \frac{1}{a}\right).$$

Thinking more globally, we can use the quantity

$$\sigma_x^2(f) = \frac{1}{E_x} \iint_{\mathbb{R}^2} t^2 \nu^2\, C_x(t,\nu; f)\, dt\, d\nu \tag{3.209}$$

Chapter 3 Issues of Interpretation

as a measure for the *joint* dispersion of a representation. Again the nonpositive character of $C_x(t,\nu;f)$ can lead to a negative value in certain cases. Indeed, we know from Eqs. (3.44) and (3.45) that

$$\sigma_x^2(f) = \frac{1}{4\pi^2}\left[\frac{1}{E_x}\int_{-\infty}^{+\infty} t^2 \left|\frac{dx}{dt}(t)\right|^2 dt - \left(\frac{1}{2} - \frac{1}{4\pi^2}\frac{\partial^4 f}{\partial \xi^2 \partial \tau^2}(0,0)\right)\right].$$
(3.210)

A lower bound for the first term in brackets will be derived in Subsection 4.1.1. For the time being, we are satisfied with the consideration of Gaussian signals, for which a direct computation yields the result

$$\frac{1}{E_x}\int_{-\infty}^{+\infty} t^2 \left|\frac{dx}{dt}(t)\right|^2 dt = \frac{3}{4}.$$

Hence, the condition

$$\frac{\partial^4 f}{\partial \xi^2 \partial \tau^2}(0,0) \geq -\pi^2$$

is necessary for the positivity of Eq. (3.210). For the family of parameterizations of product type this condition simply turns into

$$\frac{d^2\varphi}{dt^2}(0) \geq -\frac{\pi^2}{2},$$

and in case of the Choï-Williams distribution (Eq. (3.146)) it is equivalent to $\sigma \geq 1$.

3.3.2. Positivity by the Signal

An immediate issue, which arises in conjunction with a fixed parameterization, is to find all signals whose distribution is nonnegative everywhere.

An example. Answering this question in full generality is too difficult. However, one can easily find one class of signals with a positive distribution. Let us consider the simplified form of a linear chirp as given by

$$x(t) = e^{-\pi(\gamma - i\beta)t^2}.$$
(3.211)

Straightforward computations show that its Wigner-Ville distribution has the form

$$W_x(t,\nu) = \sqrt{\frac{2}{\gamma}}\, e^{-2\pi v^+ A v} \geq 0 \ , \tag{3.212}$$

where we put

$$v^+ = (\sqrt{\gamma}\, t,\, \sqrt{\gamma}\, \nu/\beta)\ ,\quad A = \begin{pmatrix} 1+\alpha & -\alpha \\ -\alpha & +\alpha \end{pmatrix}\ ,\quad \alpha = \left(\frac{\beta}{\gamma}\right)^2. \tag{3.213}$$

Hence, the examples of Eq. (3.211) provide a class of solutions to the posed problem for the Wigner-Ville distribution. As a generalization the result remains true for all distributions in Cohen's class and the affine class, which are derived by a smoothing of the form of Eqs. (3.9) or (3.10) employing a positive time-frequency function $\Pi(t,\nu)$.

Hudson's Theorem. To be of the form of a linear chirp is *sufficient* for a signal to have a positive Wigner-Ville distribution. The *necessity* of this condition follows from a theorem of Hudson.[54] The idea of its proof is to introduce the signal

$$x_{ab}(t) = \exp\left\{-\frac{1}{2}(at^2 + bt + c)\right\},$$

which depends on arbitrary complex numbers a, b, and c, with the only assumption that Re $\{a\} > 0$. A generalization of Eq. (3.212) shows that the Wigner-Ville distribution of such a signal is nonnegative everywhere (in fact, it is the exponential of a quadratic form). An application of Moyal's formula (Eq. (2.95)) gives

$$|\langle x,\, x_{1z}\rangle|^2 = \iint_{\mathbb{R}^2} W_x(t,\nu)\, W_{x_{1z}}(t,\nu)\, dt\, d\nu\ .$$

This is a function of the complex variable z. If the signal $x(t)$ has a Wigner-Ville distribution, which is positive everywhere, the so-defined function has no zeros. Consequently, the analytic function

$$F(z) = e^{c/2}\, \langle x,\, x_{1z}\rangle \tag{3.214}$$

has no zeros as well. Furthermore, an application of the Cauchy-Schwarz inequality yields

$$|F(z)|^2 \leq \sqrt{\pi}\, E_x\, e^{(\mathrm{Re}\,\{z\})^2}\ .$$

Hence, $F(z)$ is a zero-free entire function of exponential type at most 2. It therefore suffices to invoke a theorem of Hadamard in order to ascertain the fact, that $F(z)$ is the exponential of a quadratic form of z. If we finally let $z = i2\pi\nu$, we can deduce from Eq. (3.214) that the unknown signal $x(t)$ itself is the exponential of a quadratic form of the variable t, and this completes the proof.

Random signals and positive spectra. The positivity is the exceptional case for deterministic signals; without being the rule, it can be found much more frequently for random signals. Due to Eq. (3.168), especially, every *independent* linear combination of (deterministic or random) signals with positive Wigner-Ville spectra has a positive spectrum as well. Moreover, there are plenty of examples of random signals with a positive spectrum.[55]

The positivity of the Wigner-Ville spectrum is evidently assured for the (already huge) class of *weakly stationary* signals, because their spectrum coincides with the power spectrum density. This is a nonnegative quantity, by definition. This feature extends easily to the class of signals, which are *locally stationary* in the sense of Eq. (1.44), as well as to a number of more particular cases. Let us consider, for example, the (*uniformly modulated*) discrete-time signal

$$x[n] = c[n]\,(e[n] + b\,e[n-2]) \quad \text{with} \quad \mathbf{E}\,\{e[n]e[m]\} = \delta_{nm}\;.$$

Then a straightforward calculation leads to

$$\mathbf{W}_x[n,\nu] = c^2[n]\left((1+b^2) + 2b\,\frac{c[n-1]\,c[n+1]}{c^2[n]}\,\cos 4\pi\nu\right).$$

Hence, the conditions

$$0 < c \le c[n] \le C < +\infty\;,\qquad \left(\frac{C}{c}\right)^2 \le \left(\frac{1+b^2}{2b}\right)$$

are sufficient for gaining the positivity of the Wigner-Ville spectrum. An analogous analysis can be carried out for the family of time-dependent MA(1) processes of the form

$$x[n] = e[n] + b[n-2]\,e[n-2]\;.$$

In this case we obtain

$$\mathbf{W}_x[n,\nu] = 1 + b^2[n-2] + 2b[n-1]\cos 4\pi\nu\;,$$

and this leads to the sufficient condition

$$2\,|b[n+1]| \le 1 + b^2[n]$$

for the positivity. This restraint is always satisfied in the stationary case, where $b[n+1] = b[n] = b$. It also holds for some special choices such as

$$b[n] = e^{-an}\,U[n]\;,\qquad a > 0\;,$$

or

$$b[n] = b + d\cos 2\pi\xi n, \quad 0 < b < 1, \quad d < \min\left\{b, (1-\sqrt{b})^2\right\}.$$

In all of these special cases, the positivity is closely linked to the structure of the signals. All of them stay close to the stationary case in certain ways. It is important to note, however, that the *quasi-stationary* properties are not needed, in general, in order for the Wigner-Ville spectrum to be positive: as an example we may refer to the Brownian motion, whose spectrum is positive (cf. Eq. (2.164)).

3.3.3. Positivity by the Distribution

A second question in connection with the positivity concerns the distributions and their capability of attaining only nonnegative values regardless of the analyzed signal. If such solutions exist (we have already encountered a few, such as the spectrogram or the scalogram), it is important to find out about the trade-off emerging from their realization (in theory or practice).

Positive distributions. We mentioned in Subsection 2.3.3 that there are no obstacles to gaining a nonnegative distribution with correct marginal values, if we accept representations outside Cohen's class, or if we let the parameter function of Cohen's class depend on the given signal. The simplest solution is given by

$$\rho_x(t,\nu) = \frac{1}{E_x} |x(t)|^2 |X(\nu)|^2. \qquad (3.215)$$

Cohen, Zaparovanny, and Posch showed that this solution can be extended to a larger class of distributions having the form [56]

$$CZP_x(t,\nu) = \frac{1}{E_x} |x(t)|^2 |X(\nu)|^2 \left[1 + c\,r\left(\alpha(t), \beta(\nu)\right)\right], \qquad (3.216)$$

with

$$\alpha(t) = \int_{-\infty}^{t} |x(s)|^2\, ds, \qquad \beta(\nu) = \int_{-\infty}^{\nu} |X(\xi)|^2\, d\xi,$$

and

$$r(\alpha,\beta) = h(\alpha,\beta) - \int_0^1 h(\alpha,\beta)\,d\beta - \int_0^1 h(\alpha,\beta)\,d\alpha + 1.$$

In the last expression, $h(\alpha,\beta)$ is an arbitrary positive function, whose integral over the square $0 \leq \alpha, \beta \leq 1$ equals 1. The constant c, by which the correlation term $r(\alpha,\beta)$ is multiplied, must be chosen so that

$$-\frac{1}{\max\left[r(\alpha,\beta)\right]} \leq c \leq -\frac{1}{\min\left[r(\alpha,\beta)\right]}.$$

Chapter 3 Issues of Interpretation 297

At first sight, this solution is satisfactory insofar as it enables us to overcome the constraint of Wigner's theorem. But it also incorporates certain disadvantages, which raise some doubts about its usefulness.[57] In fact, making the positivity compatible with the marginal distributions causes a loss in other important theoretical properties such as the attainment of the instantaneous frequency or the group delay. Moreover, there is no one-to-one correspondence between the given representation and the signal. In order to see this, it suffices to observe that any positive distribution, by definition, depends only on the squared absolute values $|x(t)|^2$ and $|X(\nu)|^2$ of the signal and its Fourier transform. Hence, such distributions cannot distinguish between signals that differ only by their phases. This happens, in particular, for two linear chirps whose slopes of the modulation have the same absolute value, but a different sign. The positive distributions of such signals coincide, no matter if the modulation is increasing or decreasing. This is unacceptable for methods that are used in order to determine the rule of the modulation, which is unknown *a priori*.

Positive smoothing. The close relationship between smoothing and positivity can be regarded as a consequence of our analysis of the reduction of the interference terms in Subsection 3.2.3. In fact, the negative values of a representation (the Wigner-Ville distribution, for example) are due mainly to the cross-terms and their oscillatory nature. A smoothing operation is therefore likely to reduce the importance of the negative values. This observation applies to both outer and inner interferences.

The simplest approach to gain positivity by a transformation of a nonpositive distribution (positive smoothing) is based on Moyal's formula (Eq. (2.95)). We recall for the reader's convenience, that

$$\iint_{\mathbb{R}^2} C_x(s,\xi;f) C_y^*(s,\xi;f)\, ds\, d\xi = \left| \int_{-\infty}^{+\infty} x(s)\, y^*(s)\, ds \right|^2$$

holds for every *unitary* distribution; that is, its parameter function f is unimodular (cf. Eq. (2.97)). By putting

$$y(s) = h(s-t)\, e^{i2\pi\nu s}$$

and using the covariance with respect to arbitrary time-frequency shifts (which is verified by every representation in Cohen's class), we can immediately derive

$$\iint_{\mathbb{R}^2} C_x(s,\xi;f) C_h^*(s-t,\xi-\nu;f)\, ds\, d\xi = S_x(t,\nu;h) \geq 0\ . \qquad (3.217)$$

Hence, we obtain a nonnegative distribution starting from any unitary representation of the given signal $x(t)$ and employing a certain smoothing operation. The smoothing kernel is the same type of distribution of an arbitrary signal h. (This observation extends, of course, to any linear combination of functions of this type.)

Remark. A result of the same type can certainly be obtained for the affine class, for which the "regularization" of a unitary distribution yields a scalogram.

This result, which was rediscovered several times,[58] places the spectrogram (and the scalogram) in the midst of the arguments about positivity. In particular, let us look at the case of the Wigner-Ville distribution again. Recall that its support cannot be limited to an arbitrarily small subset of the time-frequency plane. Equation (3.217) suggests that the smoothing, which is applied for gaining positivity, should occupy a region that is at least the size of the minimal time-frequency cell of the Heisenberg-Gabor uncertainty principle. Different results become effective in this sense. The first is related to the smoothing by *Gaussians*. It yields the assertion that

$$BT \geq \frac{1}{4\pi} \iff \iint_{\mathbf{R}^2} W_x(s,\xi) e^{-[(s-t)^2/2T^2+(\xi-\nu)^2/2B^2]} \, ds \, d\xi \geq 0 \ . \quad (3.218)$$

Let us demonstrate this equivalence in more detail. The condition is certainly sufficient. Indeed, if $BT = 1/4\pi$, and owing to Eq. (3.138), we have

$$e^{-[t^2/2T^2+8\pi^2 T^2 \nu^2]} = W_h(t,\nu) \quad \text{with} \quad h(t) = (8\pi T^2)^{-1/4} e^{-(t/2T)^2} \ .$$

If we now assume that $BT > 1/4\pi$, then with the same function $h(t)$ and

$$B' = B\sqrt{1 - (1/4\pi BT)^2}$$

it follows that

$$e^{-[t^2/2T^2+\nu^2/2B^2]} = 4\pi BT \int_{-\infty}^{+\infty} W_h(t,\xi) \frac{1}{\sqrt{2\pi}B'} e^{-(\xi-\nu)^2/2B'^2} \, d\xi \ .$$

Hence, for any $BT \geq 1/4\pi$, we can derive the relation

$$\iint_{\mathbf{R}^2} W_x(s,\xi) e^{-[(s-t)^2/2T^2+(\xi-\nu)^2/2B^2]} \, ds \, d\xi$$

$$= 4\pi BT \int_{-\infty}^{+\infty} S_x(t,\xi;h) \frac{1}{\sqrt{2\pi}B'} e^{-(\xi-\nu)^2/2B'^2} \, d\xi \geq 0 \ .$$

The last expression is positive as a convolution (in frequency) of a positive distribution (a spectrogram) and a Gaussian.

Chapter 3 Issues of Interpretation

Remark. The smoothing by Gaussians yields positive distributions as soon as the duration-bandwidth product exceeds the lower bound of the Heisenberg-Gabor uncertainty principle. However, a smoothing with an arbitrary window function that has the same effective time-frequency support as such a Gaussian need not furnish positivity. We will return to this issue in Subsection 4.1.2.

The condition $BT \geq 1/4\pi$ is also necessary for positivity. But the proof of this fact is more involved. In order to study the behavior of the smoothed representation in Eq. (3.218), we first make use of the fact that the Wigner-Ville distribution is invariant under time-frequency shifts. Then we can restrict our attention to the single term

$$E_x(B,T) = \iint_{\mathbf{R}^2} W_x(t,\nu)\, e^{-[t^2/2T^2 + \nu^2/2B^2]}\, dt\, d\nu \; . \tag{3.219}$$

Due to another invariance of the Wigner-Ville distribution with respect to changes of scale, one can easily see that this quantity depends on the duration-bandwidth product BT, but not on B and T independently. We thus find the equivalent relation

$$E_x(BT) = \iint_{\mathbf{R}^2} W_y(a,b) e^{-(a^2+b^2)/2BT}\, da\, db \; ,$$

by introducing the auxiliary signal (with a dummy variable a)

$$y(a) = (T/B)^{1/4}\, x((T/B)^{1/2} a) \; .$$

An expansion of $y(a)$ into a series of Hermitian functions $\psi_n(a)$ yields

$$E_x(BT) = \sum_{n=0}^{+\infty} \lambda_n(BT)\, |\langle y, \psi_n \rangle|^2$$

and

$$\lambda_n(BT) = (-1)^n \int_{-\infty}^{+\infty} L_n(2r)\, e^{-(1+1/4\pi BT)r}\, dr \; . \tag{3.220}$$

Here $L_n(r)$ is the nth Laguerre polynomial. Hence, we infer that the smoothed form in Eq. (3.219) is assured to be nonnegative, if and only if all $\lambda_n(BT)$ are nonnegative. An explicit calculation of the values in Eq. (3.220) gives

$$\lambda_n(BT) = \frac{4\pi BT}{4\pi BT + 1}\left(\frac{4\pi BT - 1}{4\pi BT + 1}\right)^n , \tag{3.221}$$

and this leads to the necessary condition $BT \geq 1/4\pi$.

A stochastic interpretation. The (eventually positive) smoothing of a time-frequency distribution admits a simple stochastic interpretation. For this purpose let us consider a random signal $x(t)$, which is induced by the *jittering* of a known deterministic signal $x_{\mathrm{d}}(t)$ in time and frequency. Then the (random) Wigner-Ville distribution of $x(t)$ can be written as

$$W_x(t,\nu) = W_{x_{\mathrm{d}}}(t+\tau, \nu+\xi) ,$$

where (τ, ξ) represents a pair of random variables associated with a probability density function $G(\tau, \xi)$. As a consequence, the Wigner-Ville *spectrum* of $x(t)$, defined in Eq. (2.158), has the form

$$\mathbf{W}_x(t,\nu) = \iint_{\mathbf{R}^2} W_{x_{\mathrm{d}}}(t+\tau, \nu+\xi)\, G(\tau, \xi)\, d\tau\, d\xi = C_{x_{\mathrm{d}}}(t,\nu; g) , \quad (3.222)$$

which is just a smoothed version of the Wigner-Ville distribution of $x_{\mathrm{d}}(t)$. The smoothing kernel is nothing but the probability density of the jitter. In the case of Gaussian fluctuations, the previously established result of positivity shows that a sufficient degree of disorder ascertains the positivity of the Wigner-Ville spectrum.

Although it was introduced based on a special example, this fact is rather general in nature. A second example pointing in the same direction concerns combinations of the type "signal + white noise"

$$x(t) = s(t) + b(t) .$$

On the assumption that the noise $b(t)$ has zero mean and variance σ^2, we obtain

$$\mathbf{W}_x(t,\nu) = W_s(t,\nu) + \sigma^2 . \quad (3.223)$$

Furthermore, the inequality

$$|W_s(t,\nu)| \leq 2E_s \quad (3.224)$$

can easily be verified for any finite energy signal $s(t)$. Consequently, we find that the condition

$$\frac{E_s}{\sigma^2} \leq \frac{1}{2}$$

is sufficient for the positivity of Eq. (3.223). The term on the left-hand side of the foregoing inequality can be regarded as a signal-to-noise ratio (SNR). We thus recover the forementioned interpretation, which states that positivity is gained from a minimal degree of disorder.

Chapter 3 Issues of Interpretation 301

Chapter 3 Notes

3.1

[1] See also Hlawatsch (1991; 1992).

3.1.1.

[2] The classical spectral estimation (correlogram, periodogram, etc.) is covered in many books. Possible references are, for example, Kunt (1984), Kay (1987), or Marple (1987). A first introduction to time-frequency methods considered from this angle is given in Schrœder and Atal (1962).

[3] The concept of cyclostationary signals was especially explored by Gardner and collaborators. Their investigation is based on a "signal" perspective. An overview can be found in Gardner (1988). We should also mention the work by Hurd (1969), who explicitly used the frequency-frequency interpretation given here.

3.1.2.

[4] The main work devoted to the functional calculus of operators in connection with correspondence rules is contained in the literature on quantum mechanics. The first approach that establishes the link to the notion of joint representations was given by Cohen (1966). There are more recent papers on this subject such as the one by Springborg (1983).

[5] See, for instance, Cohen-Tannoudji, Diu, and Laloé (1973).

[6] This interpretation is borrowed from Cohen (1970).

[7] A detailed interpretation of the correspondence rules from an angle of pseudo-differential calculus is presented in the monograph by Folland (1989) (with the needed mathematical rigor). The "Weyl symbol" owes its name to the work by Weyl (1928), who proposed the respective correspondence rule. Some examples of applications of this concept to signal theory are explained in the work by Kozek (1992).

[8] The notion of the "twisted product" (or "left product," or "∗-product") is discussed in Folland (1989) and Bayen et al. (1978a,b).

[9] This is the correspondence rule that was contained in the original article by Born and Jordan (1925).

[10] The example of the dilation operator is discussed in L. Cohen (1992), for instance. There it leads to an introduction of the Mellin transform. A mark of its appearance can also be found (as far as the Weyl symbol is concerned) in Flandrin and Escudié (1981a) and Folland (1989).

[11] The transition relation mentioned here, and the way of its proof, are taken from Flandrin and Escudié (1981a). A similar relation appeared in Altes (1971).

[12] There is no unique definition of a wideband ambiguity function (or 'in compression'), see the discussion in Shenoy (1991). The symmetrical form of Eq. (3.53) is due to Altes (1990). The forms mentioned here do not exhaust the infinity of mathematically conceivable definitions, which are discussed in Shenoy (1991), Papandreou, Hlawatsch, and Boudreaux-Bartels (1993), or Cohen (1993). Considered from an angle of physical significance, there should exist a most appropriate form in the following sense: The ambiguity function measures the correlation of a copy of an emitted signal and an echo, which has undergone the combined actions of a delay and a Doppler effect; hence, the most appropriate model for the ambiguity function should correspond to a unique modeling of the echo. This approach, which is discussed in Mamode (1981), for instance, tends to favor Eq. (3.50). It corresponds to a factorization of the type "propagation emitter-target, dilation/compression, propagation target-emitter," which is used for the echo. Another discussion of this issue of modeling of echoes for sonar/radar is contained in Blahut, Miller, and Wilcox (1991).

[13] For a different approach to the same problem, see L. Cohen (1992).

[14] The basic properties of the Mellin transform are described in Bracewell (1978), for example. Ovarlez (1992) offers a very compelling discussion about its application in signal processing.

[15] See Marinovic (1986).

[16] The quoted definition of the Q-distribution is taken from Altes (1990). Let us note that the transformation can be defined in two different ways: One of them is based on the signal and the other one employs its spectrum. The latter possibility seems to be more natural, as it does not require the introduction of an (arbitrary) point that plays the role of the temporal origin. A further generalization of these concepts of scale-invariant Wigner distributions to random signals can be found in Flandrin (1990). This work also contains examples, which explain the use of these concepts for the description of self-similar (stochastic) processes.

[17] The hyperbolic class was introduced by Papandreou, Hlawatsch, and Boudreaux-Bartels (1992; 1993).

3.2.1.

[18] The concept and terminology of the pseudo-Wigner(-Ville) distribution is due to Claasen and Mecklenbräuker (1980a).

Chapter 3 Issues of Interpretation

[19] This example essentially follows the analysis found in Flandrin and Escudié (1984).

[20] This minimal width in frequency is also called *Storey band*, referring to the discussion in Storey (1953). It equally determines the best achievable resolution for the separation of two parallel linear chirps.

[21] The idea of reassignment was introduced to the spectrogram by Kodera, Gendrin, and de Villedary (1976; 1978). Hence, it is a rather old idea, and we can contemplate the reasons why it almost fell into oblivion in spite of its promising results. In the original publications its introduction was based on both the Rihaczek distribution and the method of stationary phase. This is different from our presentation here, which uses the Wigner-Ville distribution. The reintroduction of the reassignment method in this new form by Auger and Flandrin (1995) is not limited to the spectrogram alone. It can be used with any distribution of "window-type" in the bilinear classes, and this greatly extends its possible impact.

[22] The concept of the differential spectral analysis was introduced in Gibiat *et al.* (1982). It was mostly applied to audio-acoustic applications for music.

[23] The discretization of the Wigner-Ville distribution has been addressed in very little of the existing literature. Nevertheless, current understanding exceeds our (deliberately) elementary discussion by far. The definition equation (3.87) is the simplest and most natural one. It goes back to Claasen and Mecklenbräuker (1980b). In the same publication the use of the analytic signal is proposed in order to avoid spectral aliasing. A more exhaustive discussion can be found in the works of Peyrin and Prost (1986) or O'Neill (1997).

3.2.2.

[24] The discussion of the interference terms of the Wigner-Ville distribution and the mechanism of their creation relies on the results of Flandrin and Escudié (1981b), Flandrin (1984; 1987), Hlawatsch (1984), Flandrin and Hlawatsch (1987), and Hlawatsch and Flandrin (1998).

[25] Looking at the interferences from an angle of ambiguities was proposed in Flandrin (1984).

[26] The distinction of inner and outer interferences was introduced by Hlawatsch (1984).

[27] See Janssen (1982).

[28] It seems that, historically, the first use of the approximation by the method of stationary phase (for investigating the structure of the Wigner-Ville distribution) is due to Berry (1977). There it was presented in the

context of semiclassical mechanics. Similar approaches in a "signal" framework can be found in Flandrin and Escudié (1981b) and Janssen (1982), and later in Flandrin and Hlawatsch (1987), Flandrin (1987), and Hlawatsch and Flandrin (1998).

[29] Berry (1977) was the first to propose the use of catastrophe theory for a description and classification of the possible geometric features in a Wigner-Ville transform. For a general description of catastrophe theory one should consult the fundamental book by Thom (1972). The special application that we have in mind here is a characterization of singularities stemming from oscillatory integrals. This part is very clearly explained in the book by Poston and Stewart (1978).

[30] The connection between the Wigner distribution and the parity operator was made evident by Grossmann (1976) and Royer (1977).

[31] This argument is borrowed from Ovarlez (1992).

[32] A more comprehensive treatment of generalized midpoint rules of affine distributions can be found in Flandrin and Gonçalvès (1996).

[33] This approach, which brings us back to the Unterberger distribution, was proposed by Grossmann and Escudié (1991). We should mention that Paul (1985) proposed another construction, which also relies on the idea of a "midpoint" in a modified geometry. However, it leads to a different definition.

3.2.3.

[34] Boashash (1982) (former Bouachache) was one of the first to realize the usefulness of the analytic signal in this context.

[35] Such a form was proposed by Qian and Morris (1992). It was based on a Gabor decomposition. A generalization by means of (Gaussian) atoms, which are indexed by time, frequency, and scale, was first discussed by Mallat and Zhang (1993).

[36] The general philosophy of reducing the interferences by working in the ambiguity plane first appears in Flandrin (1984). Since then it has served as a guideline for most of the considered methods. In Hlawatsch et al. (1995) one can find a comparison of different methods, which are based on this approach.

[37] We refer to Gendrin and de Villedary (1979) and the more recent work by Williams and Jeong (1992), where the existence of interferences in the very nature of a spectrogram is exposed. A similar discussion with respect to the scalogram is contained in Kadambe and Boudreaux-Bartels (1992).

Chapter 3 Issues of Interpretation   305

[38] We explicitly introduced the smoothed pseudo-Wigner-Ville distribution for the reduction of interferences in Flandrin (1984). It also appeared shortly before in Martin and Flandrin (1983), where its advantages for the statistical estimation were emphasized (we will come back to this in Subsection 3.2.5), and earlier in Escudié and Flandrin (1980), where a purely formal point of view of separable parameterizations was considered. One should also note that a definition of the same type was introduced in parallel by Jacobson and Wechsler (1983), who used the name of *composite Wigner distribution*.

[39] The concept of "Reduced Interference Distributions" (RID) is explicitly described in Williams and Jeong (1992). It formalizes the program that was sketched in Flandrin (1984).

[40] See Choï and Williams (1989).

[41] As we stated earlier, the Born-Jordan distribution is implicitly rooted in Born and Jordan (1925), but its explicit expression first appears in Cohen (1966). It was rarely used, however, and only a few examples of its application along with a justification of its introduction in terms of "reduced interference distribution" are contained in Flandrin (1984).

[42] See Zhao, Atlas, and Marks (1990).

[43] This generalization was proposed, discussed, and illustrated by Papandreou and Boudreaux-Bartels (1992).

[44] See Hlawatsch and Flandrin (1998), for example.

[45] The terminology of *affine smoothed pseudo-Wigner-Ville distribution* was introduced in Flandrin and Rioul (1990) and Rioul and Flandrin (1992). In these references one can also find an explanation of the continuous transition between spectrograms and scalograms via the Wigner-Ville distribution and its modified versions, which employ a separable smoothing.

[46] The idea of a directional smoothing, which automatically adapts to the special structure of the signal, was proposed in Flandrin (1984). But there was no efficient algorithmic solution included. This latter point was investigated in Andrieux *et al.* (1987) or Riley (1989), but the most successful and efficient approaches were developed by Jones and Parks (1990) and Baraniuk and Jones (1993).

[47] One can find a survey of "image" methods used for the post-processing of time-frequency distributions in Auger (1991).

3.2.4.

[48] The Rényi measure of information is defined in Rényi (1961). Its application for estimating the dimension of a signal in the time-frequency plane was proposed in Williams, Brown, and Hero (1991).

[49] This approach is due to Duvaut and Jorand (1991). We refer to Comon (1991) for a general presentation of the principles of an analysis by stochastically independent components.

3.2.5.

[50] This whole paragraph recapitulates the results in Flandrin and Martin (1983b; 1984; 1998), Flandrin (1987; 1989a), and Martin and Flandrin (1983; 1985b).

[51] See Note 2, this chapter.

3.3.1.

[52] The idea of measuring a local dispersion by the value of a conditional variance of a joint distribution was introduced in Flandrin (1982). The subsequent analysis is also taken from this source.

[53] The issue of the (non)positivity of the "variance" of a joint distribution was copiously discussed by Cohen and Lee (1988). Another useful reference is the work by Poletti (1993).

3.3.2.

[54] The Theorem of Hudson was proved in Hudson (1974). A generalization is given in Janssen (1984a).

[55] The quoted examples of random signals with a nonnegative Wigner-Ville spectrum are mostly taken from Flandrin (1986a).

3.3.3.

[56] The positive distributions were first introduced in the context of quantum mechanics by Cohen and Zaparovanny (1980). Later they were considered in signal theory by Cohen and Posch (1985). Interest has been revived in recent years. The main impact came from the design of efficient methods for their construction under certain constraints (Loughlin, Pitton, and Atlas, 1994).

[57] This question was raised by Altes (1984) and initiated a vivid discussion (see Janssen (1987) and the response by Cohen).

[58] As far as gaining positivity by a "regularization" of the Wigner-Ville distribution is concerned, we refer to Bopp (1956), Kuryshkin (1972; 1973), Srinivas and Wolf (1975), O'Connell and Wigner (1981), Janussis et al. (1982), and P. Bertrand et al. (1983), among others. More specifically, de Bruijn (1967) may have been the first to formalize the notion of a positive (Gaussian) smoothing.

Chapter 4
Time-Frequency as a Paradigm

The time-frequency representations do more than offer an arsenal of adaptive methods for nonstationary signals: They manifest a new paradigm. This last chapter attempts to illustrate by some typical examples, how an explicitly joint description can lead to a new vision of several problems in signal analysis and signal processing, and how it amounts to finding solutions that have "natural" interpretations.

Section 4.1 deals with the first of these issues, which is related to the questions of Chapter 1. It is concerned with the joint *localization* of a signal in time and frequency. More precisely, we introduce several definitions of mixed descriptions (based on the representations in Cohen's class), which reflect some of the inherent limitations. First, in Subsection 4.1.1, we examine different forms in which the Heisenberg-Gabor uncertainty principle carries over to bilinear distributions. This leads to new types of time-frequency inequalities regarding the minimal spread of a distribution in the plane.

Second, we follow the idea of a maximal energy concentration in a fixed time-frequency region in Subsection 4.1.2. This is related to the problem of Slepian-Pollak-Landau. The joint perspective amounts to maximizing the integral of a time-frequency distribution over a given bounded domain. This leads to a new eigenvalue problem. We are able to find its explicit solution for the special case, when the Wigner-Ville distribution is considered on ellipsoidal regions. We further discuss some generalizations and conjectures.

Finally, Subsection 4.1.3 describes other possible time-frequency inequalities. One construction is based on the use of different norms to quantify the local character of the Wigner-Ville distribution. A second approach relies on arguments of the method of stationary phase. It leads to a better description of the geometry of the Rihaczek distribution of frequency-modulated signals.

Some problems of signal *analysis* are addressed in Section 4.2. In each case we gain advantage of the two variables of the description for estimating the searched characteristics of the signal. The first example (studied in Subsection 4.2.1) deals with the spectral estimation of stationary random signals and its time-frequency interpretation. We show, in particular, how the "constant-Q" paving of the plane, which is associated with the time-scale representations, leads to an efficient spectral analysis of "$1/f$-noise" processes. More generally, we can use the same approach for self-similar stochastic processes.

Next, we discuss several issues in connection with the characterization of explicitly nonstationary quantities in Subsection 4.2.2. Among these are measures for the distance from the stationary case and estimators for the corresponding modulations by means of time-frequency tools. We also deal with the estimation of local or evolutionary singularities using time-scale tools. In each case the very nature of the representational space (time-frequency or time-scale plane) reveals the underlying (local) structure of the signal.

In the last section, Section 4.3, we focus on applications of time-frequency tools going beyond the purpose of analysis. Here we investigate how certain reference patterns in the transformed plane can be defined in order to formalize statistical *decision* problems (such as detection or classification). The first possibility (in Subsection 4.3.1) consists of employing a known (deterministic) reference and defining a matched time-frequency filter. This gives rise to a new interpretation of the optimality of some distributions (such as the Wigner-Ville or Bertrand distribution), which relies on their unitarity.

Then we investigate the classical binary detection problem, hypothesizing a Gaussian random process, in Subsection 4.3.2. It turns out that the unitarity of the representation is the key to finding optimal solutions in the sense of a maximum likelihood estimation.

Finally, we dwell on interpretations of these solutions in Subsection 4.3.3. They are intimately associated with the formulation of several problems of signal processing in a time-frequency context, such as the detection at a noisy output channel, the tolerance to the Doppler effect, or the matched filtering with an incompletely known reference. The chosen perspective enables us to collect a whole family of suboptimal receptors into one class (which again turns out to be Cohen's class). Their performance can be controlled directly by the choice of the smoothing function of a joint representation.

§4.1. Localization

As seen in Chapter 1, there are fundamental obstructions to an arbitrary localization of a signal in both time and frequency. A mixed description in the time-frequency plane cannot be exempt from an analogous restraint. But by its mere structure it should offer a new vision of these limitations, which is better suited to the posed problem. We illustrate this by considering several examples in the present section. At this point we focus on the two major issues of *minimal time-frequency spread* and *maximal energy concentration*, limiting ourselves to the representations in Cohen's class and some time-frequency variants of the corresponding Heisenberg-Gabor and Slepian-Pollak-Landau problems.

4.1.1. Heisenberg-Gabor Revisited

The primal restriction of the time-frequency localization is certainly expressed by the Heisenberg-Gabor inequality (Eq. (1.4)). It tells that the product of the time- and frequency extension of a signal, as measured by the second-order moments of the instantaneous power (temporal energy distribution) and the spectral energy density (frequential energy distribution) is bounded from below by a strictly positive constant. Intuitively, a similar inequality should exist for a time-frequency energy distribution, provided that we have suitable means to measure a joint dispersion in the time-frequency plane.

Example 1. As a first measure one can propose the quantity [1]

$$\Delta_x(f) = \frac{1}{E_x} \iint_{\mathbb{R}^2} \left(\frac{t^2}{T^2} + T^2 \nu^2 \right) C_x(t, \nu; f)\, dt\, d\nu , \qquad (4.1)$$

where T is an arbitrary nonzero duration and $f(\xi, \tau)$ denotes the (frequency-time) parameter function of the distribution $C_x(t, \nu; f)$ in Cohen's class. (As in Subsection 1.1.2, we always assume that the analyzed signal, and thus its distribution, have zero mean.)

A straightforward computation based on definition equation (4.1) leads to

$$\Delta_x(f) = \frac{\Delta t^2}{T^2} + T^2 \Delta \nu^2 + \delta_x(f) ,$$

where we put

$$\delta_x(f) = -\frac{1}{4\pi^2} \left[\frac{1}{T^2} \frac{\partial^2 f}{\partial \xi^2}(0,0) + T^2 \frac{\partial^2 f}{\partial \tau^2}(0,0) \right] . \qquad (4.2)$$

The general relation

$$\frac{\Delta t^2}{T^2} + T^2 \Delta \nu^2 = \left(\frac{\Delta t}{T} - T\Delta\nu\right)^2 + 2\,\Delta t\,\Delta\nu \geq 2\,\Delta t\,\Delta\nu$$

and an application of Eq. (1.4) imply

$$\Delta_x(f) \geq \frac{1}{2\pi} + \delta_x(f) \ . \tag{4.3}$$

If $f(\xi,\tau)$ is real and attains a local maximum at the origin, we infer from Eq. (4.2) that $\delta_x(f) \geq 0$. Hence, the time-frequency extension as defined by Eq. (4.1) is minimal for all parameter functions that satisfy

$$f(\xi,0) = f(0,\tau) = 1 \ .$$

As the Wigner-Ville distribution ($f_W(\xi,\tau) = 1$) obviously verifies these relations, we find that $\delta_x(f_W) = 0$ and

$$\Delta_x(f_W) = \frac{1}{E_x} \iint_{\mathbf{R}^2} \left(\frac{t^2}{T^2} + T^2 \nu^2\right) W_x(t,\nu)\,dt\,d\nu \geq \frac{1}{2\pi} \ . \tag{4.4}$$

The same property remains true for the s-Wigner and the Choï-Williams distributions.

On the other hand, we obtain for a spectrogram with window $h(t)$ (and associated parameter function $f_S(\xi,\tau) = A_h^*(\xi,\tau)$)

$$\delta_x(f_S) = \Delta_h(f_W) \geq \frac{1}{2\pi} \ .$$

This implies

$$\Delta_x(f_S) = \frac{1}{E_x} \iint_{\mathbf{R}^2} \left(\frac{t^2}{T^2} + T^2 \nu^2\right) S_x(t,\nu)\,dt\,d\nu \geq \frac{1}{\pi} \ . \tag{4.5}$$

Therefore, the minimal time-frequency spread of a spectrogram is twice as big as the one of a Wigner-Ville distribution. We can regard this result as a consequence of the regularization equation (3.70), due to which a spectrogram corresponds to the smoothing of the Wigner-Ville distribution of the signal by the same type of distribution of the window. This amounts to an accumulation of the individual extensions.

Chapter 4 Time-Frequency as a Paradigm

Remark 1.   The time-frequency inequality of Eq. (4.4) was established by an explicit application of the usual Heisenberg-Gabor inequality. The lower bound is therefore attained for Gaussian signals. As a corollary, the time-frequency inequality of Eq. (4.5) for the spectrogram turns into an equality when both the signal and the short-time window are Gaussians.[2]

Remark 2.   By Eq. (4.2) one could formally reduce the lower bound $(1/2\pi)$ in Eq. (4.4) by taking a parameter function $f(\xi,\tau)$ that has a local minimum at the origin of the plane. This would make both second derivatives in Eq. (4.2) negative. However, the physical meaning of such a parameterization becomes questionable.

Example 2.   Definition equation (4.1) presents one way to measure the time-frequency extension. There are, of course, other possibilities. We can use, for instance, the quantity

$$\sigma_x^2(f) = \frac{1}{E_x} \iint_{\mathbb{R}^2} t^2 \nu^2 \, C_x(t,\nu;f) \, dt \, d\nu \;, \tag{4.6}$$

which was already mentioned in Chapter 3 (see Eq. (3.209)). The computation in Eq. (3.210) showed that

$$\sigma_x^2(f) = \frac{1}{4\pi^2} \left[\frac{1}{E_x} \int_{-\infty}^{+\infty} t^2 \left| \frac{dx}{dt}(t) \right|^2 dt - \left(\frac{1}{2} - \frac{1}{4\pi^2} \frac{\partial^4 f}{\partial \xi^2 \partial \tau^2}(0,0) \right) \right] .$$

When we proceed as in Subsection 1.2.1, we can further prove that the first term in brackets is bounded from below. Indeed, the introduction of the intermediary quantity

$$I \equiv \int_{-\infty}^{+\infty} t \, \frac{dx}{dt}(t) \, x^*(t) \, dt$$

and integration by parts yield the result

$$I = \left[t \, |x(t)|^2 \right]_{-\infty}^{+\infty} - E_x - \int_{-\infty}^{+\infty} t \, x(t) \, \frac{dx^*}{dt}(t) \, dt \; .$$

Hence, for all signals that decrease sufficiently fast (especially those with finite energy) we have

$$\operatorname{Re}\{I\} = -\frac{E_x}{2} \; .$$

Consequently, we can write

$$\frac{1}{4} E_x^2 = (\operatorname{Re}\{I\})^2 \leq |I|^2 \leq \int_{-\infty}^{+\infty} t^2 \left| \frac{dx}{dt}(t) \right|^2 dt \cdot \int_{-\infty}^{+\infty} |x(t)|^2 \, dt$$

by an application of the Cauchy-Schwarz inequality. This leads to

$$\frac{1}{E_x} \int_{-\infty}^{+\infty} t^2 \left|\frac{dx}{dt}(t)\right|^2 dt \geq \frac{1}{4} . \tag{4.7}$$

It should be noted that this lower bound cannot be attained by any finite energy signal. In particular, we obtain for a Gaussian $g(t)$

$$\frac{1}{E_x} \int_{-\infty}^{+\infty} t^2 \left|\frac{dg}{dt}(t)\right|^2 dt = \frac{3}{4} .$$

Hence, the time-frequency *equality*

$$\sigma_g(f) = \frac{1}{4\pi} \left(1 + \frac{1}{\pi^2} \frac{\partial^4 f}{\partial \xi^2 \partial \tau^2}(0,0)\right)^{1/2} \tag{4.8}$$

follows for all distributions of a Gaussian g that use a parameter function $f(\xi, \tau)$ with

$$\frac{\partial^4 f}{\partial \xi^2 \partial \tau^2}(0,0) \geq -\pi^2 .$$

We thus find an identical result for the time-frequency extension of a Gaussian as in Eq. (1.4), if the respective partial derivative of the parameter function vanishes at the origin. This truly happens to be the case for the Wigner-Ville distribution. It is worthwhile to mention, however, that in case of a usual parameter function (i.e., maximal at the origin) the right-hand side of Eq. (4.8) may be less than $1/4\pi$, and even zero can be attained. This causes some evident problems concerning the interpretation of this measure.

On the other hand, if we consider the spectrogram of a signal $x(t)$ with a window $h(t)$, the mere structure of the ambiguity functions leads to the dispersion

$$\sigma_x^2(f_S) = \sigma_x^2(f_W) + \sigma_h^2(f_W) . \tag{4.9}$$

As a consequence of the positive character of the spectrogram we can now refine the lower bound in Eq. (4.7). Indeed, for every finite energy signal $x(t)$ we obtain

$$\frac{1}{E_x} \int_{-\infty}^{+\infty} t^2 \left|\frac{dx}{dt}(t)\right|^2 dt \geq \frac{1}{2} . \tag{4.10}$$

Moreover, in case we analyze a Gaussian signal with a Gaussian window, the minimal time-frequency extension in the sense of Eq. (4.6) attains the bound

$$\sigma_g(f_S) = \sqrt{2}\,\frac{1}{4\pi} = \sqrt{2}\,\sigma_g(f_W) .$$

4.1.2. Energy Concentration

The inequalities of Heisenberg-Gabor type characterize a first family of time-frequency restrictions by installing certain measures for the spread of a time-frequency function. A second type of restrictions concerns the fact that the total energy of a signal cannot be totally concentrated on finite intervals both in time and frequency, even if we allow the supports to be arbitrarily large. In this regard the time-frequency approach consists of defining a measure for the portion of the total energy of a time-frequency distribution, which is allocated to a bounded domain. This measure should operate *directly in the plane* and should permit a determination of the distributions with the best possible energy concentration.

Problem formulation. The classical problem (cf. Subsection 1.1.3) of the energy concentration of a signal is posed in "two times one dimension." Given a fixed duration, for example, we seek a signal that concentrates as much as possible in a certain frequency band. In this way we define implicitly a time-frequency "rectangle," to which the signal allocates most of its energy. Hence, a joint time-frequency perspective should certainly offer an adequate and more general setting for this problem. It should be adequate in the sense that the time- and frequency variables appear in the formulation of the problem; and being more general means that the copiousness of the time-frequency plane should be incorporated by passing to "one times two dimensions," which eliminates the restrictions to rectangular domains.[3]

Let us begin with the basic property of an energy distribution, which assures that

$$E_x = \iint_{\mathbb{R}^2} C_x(t, \nu; f) \, dt \, d\nu \ . \tag{4.11}$$

Then an appropriate definition of the *partial energy* is given by

$$E_x(D; f) = \iint_D C_x(t, \nu; f) \, dt \, d\nu \ , \tag{4.12}$$

where D may be any subset of the time-frequency plane. The new problem thus consists of finding the (largest possible) energy portion of a given type of distribution when restricted to this set D.

In a more relaxed version, we can replace the strict concentration on D, as expressed by the characteristic function of this set, with a (possibly smooth) function $G(t, \nu)$ whose effective support is equivalent to D. Then the problem is changed into the evaluation of

$$E_x(D; f) = \iint_{\mathbb{R}^2} G(t, \nu) \, C_x(t, \nu; f) \, dt \, d\nu \ . \tag{4.13}$$

Due to Eq. (3.27) the so-defined partial energy is nothing but the expectation value of the operator, which is associated with the function $G(t, \nu)$ by means of the correspondence rule based on f. Computing the maximal value of the partial energy is thereby equivalent to finding the signal that belongs to the largest eigenvalue of this "projection" onto D.

The general eigenvalue equation. The projection operator relative to the set D, in its completely general form, has the kernel (cf. Eq. (3.29))

$$\gamma_f(s, t) = \int_{-\infty}^{+\infty} F\left(\frac{t+s}{2} - \theta, t - s\right) \gamma(\theta, t - s) \, d\theta \;,$$

with

$$\gamma(t, \tau) = \int_{-\infty}^{+\infty} G(t, \nu) \, e^{-i 2\pi \nu \tau} \, d\nu \;.$$

The corresponding eigenvalue equation reads as

$$\int_{-\infty}^{+\infty} \gamma_f(t, s) \, x(s) \, ds = \lambda x(t) \;. \tag{4.14}$$

Every signal that constitutes an eigenfunction to the largest eigenvalue $\lambda_{\max}$ yields the best possible energy concentration of the distribution $C_x(t, \nu; f)$ to the domain D. This is easily inferred from the relation

$$\frac{E_x(D; f)}{E_x} \leq \lambda_{\max} \;.$$

It is too difficult, of course, to solve the eigenvalue problem (Eq. (4.14)) in the most general case (any distribution and arbitrary domain). However, we shall develop explicit solutions to the foregoing problem for some special cases.

Restriction to ellipsoidal domains. Let us consider the class of functions $G(t, \nu)$, which are elliptically symmetric in the sense

$$G(t, \nu) = H\left[\left(\frac{t}{T}\right)^2 + \left(\frac{\nu}{B}\right)^2\right] \;,$$

where H can be any function subject to the condition

$$\int_0^{+\infty} |H(s)|^2 e^{-bs} \, ds < +\infty$$

Chapter 4 Time-Frequency as a Paradigm

for all $b > 0$. In this case, the eigenvalue equation for the Wigner-Ville distribution can simply be written as

$$\int_{-\infty}^{+\infty} \gamma\left(\frac{t+s}{2}, t-s\right) x(s)\, ds = \lambda x(t) . \tag{4.15}$$

This renders an explicit computation of the solutions possible. In fact, by introducing the reduced coordinates

$$\tilde{x}(a) = \left(\frac{T}{B}\right)^{1/4} x\left(a\sqrt{\frac{T}{B}}\right) , \qquad R_{BT}(r) = H\left(\frac{r}{2\pi BT}\right) ,$$

one can show that [4]

$$E_x(D; f_W) = \sum_{n=0}^{+\infty} (-1)^n \int_0^{+\infty} e^{-r} R_{BT}(r)\, L_n(2r)\, dr \; \left|\langle \tilde{x}, \tilde{\psi}_n\rangle\right|^2 .$$

In the preceding formula, we employ the Hermitian functions

$$\psi_n(t) = (2^{n-1/2} n!)^{-1/2} H_n(\sqrt{2\pi}\, t)\, e^{-\pi t^2}$$

defined in terms of the Hermitian polynomials

$$H_n(t) = (-1)^n e^{\pi t^2} \left(\frac{d}{dt}\right)^n (e^{-\pi t^2}) ,$$

and the Laguerre polynomials

$$L_n(r) = \frac{1}{n!} e^r \left(\frac{d}{dt}\right)^n (e^{-r} r^n) .$$

It follows that the Hermitian functions are eigenfunctions of Eq. (4.15) with corresponding eigenvalues

$$\lambda_n(BT) = (-1)^n \int_0^{+\infty} e^{-r} R_{BT}(r)\, L_n(2r)\, dr . \tag{4.16}$$

Moreover, these values only depend on the (effective) duration-bandwidth product BT of the domain D. Next, we appeal to the generating function of the Laguerre polynomials

$$\sum_{n=0}^{+\infty} w^n L_n(r) = \frac{1}{1-w} \exp\left\{\frac{-rw}{1-w}\right\} .$$

Hence, the eigenvalues in Eq. (4.16) satisfy

$$\sum_{n=0}^{+\infty} \lambda_n(BT) = \frac{1}{2} \int_0^{+\infty} R_{BT}(r)\,dr \ . \tag{4.17}$$

As, by construction, this sum has the same order of magnitude as BT, the last equation conforms to the interpretation that the number of predominant eigenvalues close to 1 is directly proportional to the duration-bandwidth product BT.

In case of a strict confinement to the ellipse, where

$$H(s) = \begin{cases} 1, & 0 \le s \le 1/4, \\ 0, & \text{otherwise,} \end{cases} \tag{4.18}$$

the sum of Eq. (4.17) has the precise value

$$\sum_{n=0}^{+\infty} \lambda_n(BT) = \frac{\pi}{4} BT \ .$$

One can further show by some more involved analysis that

$$\lambda_n(BT) = 1 - e^{-\pi BT/2} \sum_{k=0}^{n} (-1)^k\, a_k\, L_k(\pi BT) \ , \tag{4.19}$$

with $a_k = 2$ for $0 \le k \le n-1$ ($n \ge 1$) and $a_n = 1$. The graphs of several of these eigenvalues as functions of BT are drawn in Fig. 4.1.

As a first consequence of Eq. (4.19) we conclude that

$$\lim_{BT \to \infty} \lambda_n(BT) = 1 \ ,$$

which was to be expected due to Eqs. (4.11) and (4.12). A second and less trivial observation is

$$\lambda_n(BT) \le \lambda_{\max} = \lambda_0(BT) = 1 - e^{-\pi BT/2} \le 1 \ , \tag{4.20}$$

and the maximal eigenvalue is obtained for the Gaussian signal. Indeed, based on Eq. (4.19) and the identity

$$\int_0^s L_n(r)\,dr = L_n(s) - L_{n+1}(s) \tag{4.21}$$

Chapter 4 Time-Frequency as a Paradigm

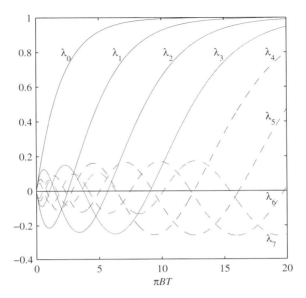

Figure 4.1. Eigenvalues (Eq. (4.19)).
Each of these different eigenvalues, as a function of the product BT, measures the energy portion of a Wigner-Ville distribution of a Hermitian function, which is strictly contained in an ellipse.

for the Laguerre polynomials, we obtain the relation

$$\lambda_n(BT) = \lambda_{n-1}(BT) + (-1)^n e^{-\pi BT/2}[L_{n-1}(\pi BT) - L_n(\pi BT)] \; .$$

A recursive application of this equality gives

$$\lambda_n(BT) = \lambda_0(BT) + e^{-\pi BT/2} \sum_{k=1}^{n}(-1)^k \left[L_{k-1}(\pi BT) - L_k(\pi BT)\right] \; .$$

Using Eq. (4.21) once more yields the result

$$\lambda_n(BT) = \lambda_0(BT) - e^{-\pi BT/2} \int_0^{\pi BT} \left[\sum_{k=1}^{n-1}(-1)^k L_k(t)\right] dt \; , \quad n \geq 1 \; .$$

If we finally realize that the sum in brackets is nonnegative [5] for every n, we obtain the first inequality in Eq. (4.20)

$$\lambda_n(BT) \leq \lambda_0(BT) \; .$$

Equality is attained for the function $\widetilde{\psi}_0(a)$, that is, for the Gaussian. Hence, the partial energy of a Wigner-Ville distribution contained in an ellipsoidal domain is maximal for a Gaussian signal.

The second inequality in Eq. (4.20) follows from Eq. (4.19) and the definition of the Laguerre polynomials, which imply

$$\lambda_0(BT) = 1 - e^{-\pi BT/2} ,$$

and, therefore,

$$\lambda_0(BT) \leq 1 .$$

As another consequence of the previous result, the integration of the Wigner-Ville distribution over an ellipsoidal domain always yields a value that cannot exceed the total energy of the signal. This fact is remarkable, as it is assured in spite of the possibly negative values of the distribution. This result looks quite satisfactory from a physical perspective; but we should keep in mind that it was by no means clear *a priori*.

Finally, as a last consequence of the explicit form of Eq. (4.19), we can consider the integration over very large domains. This signifies the cases where $BT \gg 1/2$. Intuitively, we could be tempted to expect that the result will always be positive. However, this conclusion is false, because for every finite value of BT, no matter how large, there exists an index n so that $\lambda_n(BT) < 0$. This contradicts our naive belief and ascertains that the smoothing of a Wigner-Ville distribution by a function, whose (effective) support in the plane exceeds the Heisenberg-Gabor limit, need not furnish, in general, a positive distribution. [6]

On the other hand, the anticipated conclusion about the positivity, which proved false in the situation of a strict confinement, becomes true, if the smoothing kernel is a Gaussian. This is related to the results of Subsection 3.3.2. In fact, when we replace the characteristic function in Eq. (4.18) with a Gaussian

$$H(s) = e^{-s^2/2} ,$$

we find

$$\lambda_n(BT) = \frac{4\pi BT}{4\pi BT + 1} \left(\frac{4\pi BT - 1}{4\pi BT + 1} \right)^n , \qquad (4.22)$$

and, consequently,

$$\sum_{n=0}^{+\infty} \lambda_n(BT) = 2\pi BT .$$

Chapter 4 Time-Frequency as a Paradigm 321

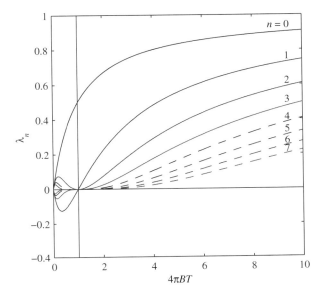

Figure 4.2. Eigenvalues (Eq. (4.22)).

Each eigenvalue is drawn as a function of BT. It measures the energy portion of the Wigner-Ville distribution of a Hermitian function, which is contained in an ellipsoidal domain associated with a Gaussian smoothing.

The result of Eq. (4.22) further leads to

$$BT \geq \frac{1}{4\pi} \implies \lambda_n(BT) \geq 0 \quad \text{for every} \quad n \ .$$

A graphical illustration of this fact is given in Fig. 4.2, where we draw the graphs of the eigenvalues of Eq. (4.22) as functions of the product BT.

As in the case of a strict restriction to an ellipsoidal domain, the maximal energy concentration with respect to a Gaussian weight function is obtained for a Gaussian signal. In this case it attains the upper bound

$$\lambda_n(BT) \leq \lambda_{\max} = \lambda_0(BT) = \frac{4\pi BT}{4\pi BT + 1} \ . \tag{4.23}$$

Remark 1. One can easily observe that in Fig. 4.2 the eigenvalues with odd indices are negative for all values of the duration-bandwidth product below the threshold $BT = 1/4\pi$. We can realize by a simple computation that the local extrema of Eq. (4.22) in the interval $0 \leq BT \leq 1/4\pi$ (for even and odd n) are

$$\lambda_n\left(\frac{1}{(2n+1)4\pi}\right) = (-1)^n \frac{n^n}{2(n+1)^{n+1}} .$$

Hence, the negative eigenvalue with the largest absolute value is λ_1. Its minimal value of $-1/8$ corresponds to a duration-bandwidth product $BT = 1/12\pi$, which is one-third of the lower bound in the Heisenberg-Gabor uncertainty relation.

Remark 2. For the same case of a Gaussian weight function, the Gaussian signal can be shown to be an eigenfunction of the equation

$$\int_{-\infty}^{+\infty} \gamma\left(\frac{t+u}{2} + s(t-u), t-u\right) x(u)\, du = \lambda x(t)$$

as well. This is the eigenvalue equation associated with an s-Wigner distribution. Its largest eigenvalue has the form

$$\lambda_0(BT;s) = \frac{4\pi BT}{4\pi BT[1+(s/2\pi BT)^2]^{1/2}+1} ,$$

from which we infer

$$\lambda_0(BT;s) \leq \lambda_0(BT;0) .$$

Hence, among the whole class of s-Wigner distributions the Wigner-Ville case (which corresponds to the value $s = 0$) yields a maximal energy concentration relative to a given ellipsoidal domain. This result presents a new justification of the minimal dispersion, which is characteristic of the Wigner-Ville distribution.[7]

Remark 3. The presented approach, which was restricted to ellipses, could possibly be generalized to other domains. However, it turns out that finding exact solutions to the eigenvalue problem becomes rather difficult. An approximate solution can be motivated by the observation, that maximizing Eq. (4.13) is equivalent to minimizing the distance

$$d^2(G, C_x) = \iint_{\mathbf{R}^2} [G(t,\nu) - C_x(t,\nu;f)]^2\, dt\, d\nu .$$

This transfers the considered problem to the *synthesis* of a signal, which has a Wigner-Ville distribution with a best least squares fit to a given time-frequency function.[8]

Chapter 4 Time-Frequency as a Paradigm

Interpretations and conjecture. The idea of measuring the energy concentration by integrating a time-frequency distribution over a bounded subset of the plane is confronted with the known difficulty, that the distribution may attain negative values. In fact, if we would regard a time-frequency distribution (such as Wigner-Ville) as an arbitrary function subject to the mere condition of Eq. (4.11), there is no reason why it should not have large values *outside* the domain of integration. We can easily devise functions that have arbitrarily large values with alternating signs and whose contribution to Eq. (4.11) is negligible. Moreover, there exists no obstruction, in general, against a local integration providing *greater* values than the total energy of the signal, because the excess could possibly be counterbalanced by negative values outside the domain of integration.

However, it seems that all this does not happen for the Wigner-Ville distribution, on convex domains at least. We want to justify this by a purely geometric argument, which is based on the structure of the interferences in a Wigner-Ville distribution as it was discussed in Subsection 3.2.2. Suppose, in fact, the integration of a Wigner-Ville distribution over a convex region yields a result that is arbitrarily close to the total energy of the signal. Assume further that the distribution has non-negligible values outside this domain. Due to the formula of inner interferences (Eq. (3.103)), the value at each point results from the interaction of contributions, which are symmetrically distributed about this point. Consequently, as the domain of the integration is supposed to be convex, the non-negligible values outside the domain can only result from an interaction of pairs of two contributions, at least one of which is also located outside the domain. On the other hand, the same principle of interferences together with the almost perfect conformity of total and partial energy imply that all significant contributions of the signal lie inside the convex domain. This contradicts the previous hypothesis.

Without having the quality of a mathematical proof, the forementioned reasoning supports the conjecture that

$$E_x(D; f_W) = \iint_D W_x(t, \nu) \, dt \, d\nu \leq E_x \qquad (4.24)$$

for every convex domain D.

4.1.3. Other Time-Frequency Inequalities

Apart from the generalization of the Heisenberg-Gabor and Landau-Slepian-Pollak approaches, the localization of the energy of a signal to a bounded domain in the time-frequency plane can be measured by other means, each having its own interpretation as a time-frequency inequality.

L_p-**norms.** A measure for the concentration in the plane, which is sensitive to the existence of large values in a small support, is given by the L_p-norm of the Wigner-Ville distribution. Recall that Moyal's formula (Eq. (2.95)) implies

$$\left(\iint_{\mathbb{R}^2} |W_x(t,\nu)|^2 \, dt \, d\nu \right)^{1/2} = E_x \ . \tag{4.25}$$

It therefore suggests itself to consider the more general quantity

$$\|W_x\|_p = \left(\iint_{\mathbb{R}^2} |W_x(t,\nu)|^p \, dt \, d\nu \right)^{1/p} . \tag{4.26}$$

First, we can show by a simple argument that this p-norm is finite. Indeed, let us write

$$\iint_{\mathbb{R}^2} |W_x(t,\nu)|^n \, dt \, d\nu = \iint_{\mathbb{R}^2} |W_x(t,\nu)| \, |W_x(t,\nu)|^{n-1} \, dt \, d\nu \ ,$$

and use the Hölder inequality, which gives

$$\|W_x\|_n^n \leq \|W_x\|_p \cdot \| \, |W_x|^{n-1} \|_q \ , \qquad q = \frac{p}{p-1}, \quad p \geq 1 \ .$$

We next employ the general relations

$$\lim_{p \to \infty} \|W_x\|_p = \max_{(t,\nu)} |W_x(t,\nu)| \leq 2 E_x \qquad \text{and}$$

$$\lim_{p \to \infty} \| \, |W_x|^{n-1} \|_q = \|W_x\|_{n-1}^{n-1} \ .$$

A recursive application of the estimate by Hölder's inequality leads to

$$\|W_x\|_n^n \leq 2 E_x \|W_x\|_{n-1}^{n-1} \leq (2 E_x)^2 \|W_x\|_{n-2}^{n-2} \leq \ldots \leq (2 E_x)^{n-2} \|W_x\|_2^2 \ ,$$

and this gives

$$\|W_x\|_n \leq 2^{1-2/n} E_x \ . \tag{4.27}$$

The foregoing estimate is not sharp. Lieb proved that a more accurate estimate is furnished by [9]

$$\begin{cases} \|W_x\|_n \leq n^{-1/n} \, 2^{1-1/n} E_x \ , & n > 2, \\ \|W_x\|_n \geq n^{-1/n} \, 2^{1-1/n} E_x \ , & 1 \leq n < 2. \end{cases} \tag{4.28}$$

Chapter 4 Time-Frequency as a Paradigm 325

The bounds on the right-hand side are actually attained for Gaussian signals (and only for them). These inequalities supply a new form of a time-frequency uncertainty, which renders an arbitrary concentration of the Wigner-Ville distribution of a finite energy signal impossible.

Localization and stationary phase. At last, let us consider another form of the localization, which was also mentioned in Subsection 3.2.2. It is related to frequency- or phase-modulated signals. We have already seen that certain pairs of the type "distribution/phase rule" permit a *perfect* localization; that is, the distribution behaves like a Dirac mass localized to a curve in the time-frequency plane, provided that this curve is infinitely extended. For dealing with more realistic models of signals it looks rather natural to measure this property of localization by means of a *dispersion* of the distribution about the considered time-frequency curve. This must be compared with the total spread of the signal in the time-frequency plane as measured by Eq. (4.6), for example. Hence, we arrive at the definition

$$\rho_x(f) = \frac{\sigma_x(f \mid \nu_x)}{\sigma_x(f)} \tag{4.29}$$

as a new measure for the localization, where

$$\sigma_x^2(f \mid \nu_x) = \frac{1}{E_x} \iint_{\mathbb{R}^2} t^2 \left[\nu - \nu_x(t)\right]^2 C_x(t,\nu;f) \, dt \, d\nu . \tag{4.30}$$

According to this definition a perfect localization (i.e., the ideal energy concentration on the curve of the instantaneous frequency) corresponds to a vanishing measure of Eq. (4.29). (We should note that dual definitions with respect to the group delay are possible as well.)

As we infer from previous discussions, the Wigner-Ville distribution is perfectly localized to a linear instantaneous frequency, provided that this line is extended infinitely. Let us next consider the model signal (cf. Eq. (3.211))

$$x(t) = e^{-\pi(\gamma - i\beta)t^2} ,$$

which is closer to the physical reality. A simple computation yields the result

$$\rho_x(f_\mathrm{W}) = [3(BT)^2 - 2]^{-1/2} , \qquad BT = \left(1 + \frac{\beta^2}{\gamma^2}\right)^{1/2} .$$

This implies

$$\lim_{BT \to \infty} \rho_x(f_\mathrm{W}) = 0 .$$

Hence, the localization improves, when the value of the duration-bandwidth product gets larger.

Remark 1. The same situation can be expressed differently, when we appeal to the explicit form of Eq. (3.212) that is associated with the preceding model. Because the contourlines of Eq. (3.212) define ellipses, we can use as a measure for the concentration in the plane the eccentricity $e(BT)$ of these ellipses. The latter is given by

$$e^2(BT) = \frac{1 + 2\alpha - \sqrt{1 + 4\alpha^2}}{1 + 2\alpha + \sqrt{1 + 4\alpha^2}}, \qquad \alpha = (BT)^2 - 1 = \frac{\beta^2}{\gamma^2}.$$

Consequently, we have

$$\lim_{BT \to \infty} e(BT) = 0.$$

Remark 2. The concentration on a curve in the time-frequency plane can be regarded as the analogue of the "classical" limit of quantum mechanics (existence of localized trajectories in phase space). The reciprocal of the duration-bandwidth product formally plays the role of the Planck constant in this context.

The previous result has a rather general significance, although it was based on a special case: The localization of a time-frequency representation of a modulated signal tends to be enhanced, when we enlarge the duration-bandwidth product. Here the notion of the duration-bandwidth product must be understood in a wider sense. The important point to be required is that the approximation based on the method of stationary phase be admissible. Then we can develop other relations to express the same property of concentration in the plane in a new form.

Let us begin with the approximation of the spectrum of a signal by the method of stationary phase, as it was laid out in Subsection 1.2.2. Given an (analytic) signal in its generic form

$$x(t) = a_x(t)\, e^{i\varphi_x(t)},$$

we suppose that the law of the instantaneous frequency $\nu_x(t)$ is strictly monotone. If the oscillations of the phase are fast as compared to the variations of the modulus, the spectrum of $x(t)$ admits the approximation

$$X(\nu) \approx \left|\frac{d\nu_x}{dt}(t_s)\right|^{-1/2} a_x(t_s)\, \exp\{i\, \Phi_x^s(\nu)\}$$

with

$$\Phi_x^s(\nu) = \varphi_x(t_s) - 2\pi\nu t_s + \frac{\pi}{4}\,\mathrm{sgn}\,\frac{d\nu_x}{dt}(t_s), \qquad \nu_x(t_s) = \nu.$$

Chapter 4 Time-Frequency as a Paradigm   327

Next we introduce the *Legendre transform*[10] of a (monotonic) function $\varphi(t)$ by

$$\mathcal{L}\{\varphi\}(\nu) = \nu t_s - \varphi(t_s), \qquad \nu = \frac{d\varphi}{dt}(t_s).$$

Then we immediately obtain

$$\frac{1}{2\pi}\Phi_x^s(\nu) = -\mathcal{L}\left\{\frac{1}{2\pi}\varphi_x\right\}(\nu) + \frac{1}{8}\operatorname{sgn}\frac{d\nu_x}{dt}(t_s), \qquad \nu_x(t_s) = \nu. \quad (4.31)$$

This implies, under the given hypotheses, that the phases of the signal and its spectrum are linked to each other by a Legendre transform. The involutive character of this operation permits a new justification of the fact, that the instantaneous frequency and the group delay are inverse functions. Another consequence is obtained by making use of *Young's inequality* in connection with the Legendre transform, which gives

$$\varphi(t) + \mathcal{L}\{\varphi\}(\nu) \begin{cases} \geq t\nu & \text{if } \varphi(t) \text{ is convex,} \\ \leq t\nu & \text{if } \varphi(t) \text{ is concave,} \end{cases}$$

with equality for $t = t_s$. As far as the Rihaczek distribution (cf. Table 2.1) is concerned, we infer from (4.31) that

$$\arg\{R_x(t,\nu)\} \begin{cases} \geq -\dfrac{\pi}{4} & \text{if } \nu_x(t) \text{ is increasing,} \\ \leq +\dfrac{\pi}{4} & \text{if } \nu_x(t) \text{ is decreasing.} \end{cases} \quad (4.32)$$

Equality holds for all points on the curve of the instantaneous frequency $\nu = \nu_x(t)$. This furnishes another form of a time-frequency inequality.

§4.2. Signal Analysis

Speaking of "good" properties of a representation draws its motivation from the goal of gaining a "good" *analysis* tool, no matter if this is expressed explicitly or implicitly. Such a tool should enable us to *reveal* the underlying structure of a signal and to *estimate* its characteristics. Viewed from this perspective and regarding most signals in our environment (in both their own nature and our modes of perception), the time-frequency approach offers a natural nomenclature for their description and thus facilitates their analysis.

4.2.1. Time-Frequency, Time-Scale, and Spectral Analysis

Although the time-frequency methods (in a wider sense) are designed to deal with nonstationary signals, they also provide a natural setting for the *spectral estimation* of stationary signals and signals with stationary increments.

Paving and marginal distributions. The most classical procedures of nonparametric spectral estimation are based on the smoothing or averaging of a *periodogram*.[11] This term is used for the same transform as a *spectrogram*, since its operation can be understood as a "squared absolute value of the Fourier transform of (independent) consecutive segments of the signal." Therefore, the spectral estimation by an averaged periodogram acquires a natural time-frequency interpretation. The way of partitioning the signal prior to taking its Fourier transform implicitly defines a *paving* of the time-frequency plane. The time-resolution of the corresponding cells is fixed by the short-time window, and the frequency-resolution is determined by the spectrum of this window.

From a point of view of time-frequency analysis, taking the *time-average* of a periodogram is the same as finding the marginal distribution of the spectrogram. We can clearly realize the *bias-variance trade-off*, which is inherent in this type of spectral estimation. Indeed, one can only reduce the variance of the estimation by increasing the number of (independent) mean values, which in turn results in reducing the length of the short-time window. At the same time, the bias is aggravated by (3.70),

$$\mathbf{E}\left\{\lim_{T\to+\infty}\frac{1}{T}\int_{(T)} S_x(t,\nu)\,dt\right\}$$

$$=\lim_{T\to+\infty}\frac{1}{T}\int_{(T)}\iint_{\mathbf{R}^2}\mathbf{W}_x(s,\xi)\,W_h(s-t,\xi-\nu)\,ds\,d\xi\,dt$$

$$=\lim_{T\to+\infty}\frac{1}{T}\int_{(T)}\iint_{\mathbf{R}^2}\Gamma_x(\xi)\,W_h(s-t,\xi-\nu)\,ds\,d\xi\,dt$$

$$=\int_{-\infty}^{+\infty}\Gamma_x(\xi)\,|H(\xi-\nu)|^2\,d\xi\,.$$

This shows that the resulting estimator yields a smoothed version of the searched power spectrum $\Gamma_x(\nu)$ as its expectation value. The paving associated with the considered procedure has a *uniform* and *rectangular* geometry, all cells having the same size and shape. This geometric structure renders the spectral estimation by an averaged periodogram suited for spectral densities with a *constant* rate of change in the frequency direction.

Due to the interpretation of a spectral estimation in terms of marginal distributions of a time-frequency representation, we can relax the restraint of identical time-frequency cells. This is useful for an adaptation to more

Chapter 4 Time-Frequency as a Paradigm

general variations of the power spectrum density. As an example, let us consider the case of a paving with *constant quality factor*, as it arises in connection with a time-scale analysis. In this case the frequency-resolution diminishes, when we turn to high frequencies, while the time resolution improves at the same rate. Hence, the trade-off between bias and variance is unevenly distributed over the frequencies. Indeed, at high frequencies the constant-Q paving provides a large number of segments, which (generally) leads to a smaller variance; however, this is only achievable by the simultaneous loss in the frequency-resolution. Conversely, the frequency-resolution is good at low frequencies, while the availability of few independent segments results in a bigger variance. Owing to the structure of this paving, the spectral estimation based on a scalogram fits to those cases where the power spectrum is governed by a law of the type of a *power of the frequency*, such as the "$1/f$-noise" signals.

The example of "$1/f$-noise." [12] The notion of "$1/f$-noise" processes is commonly used in a very general way. It encompasses all processes, which have a density spectrum of the form

$$\Gamma_x(\nu) = \frac{\sigma^2}{|\nu|^\alpha} \tag{4.33}$$

in a large frequency range, especially at low frequencies. [13] The problem of spectral estimation of such processes reduces to the estimation of the exponent α. This is often realized by a linear regression in a "log-log" scale, as we know

$$\log \Gamma_x(\nu) = \log \sigma^2 - \alpha \log |\nu| \,. \tag{4.34}$$

We want to estimate Eq. (4.33) by a marginal distribution of a representation in Cohen's class, that is,

$$\widehat{\Gamma}_x(\nu) = \lim_{T \to +\infty} \frac{1}{T} \int_{(T)} C_x(t, \nu; f) \, dt \,. \tag{4.35}$$

Provided that the process is stationary (which is needed in order that Eq. (4.33) is well defined), we obtain

$$\mathbf{E}\left\{\widehat{\Gamma}_x(\nu)\right\} = \Gamma_x(\nu) \int_{-\infty}^{+\infty} |1 + \xi/\nu|^{-\alpha} \, \psi(0, \xi) \, d\xi \,,$$

where we use the notation ψ from Subsection 3.1.1 for the parameterization. Hence, we observe that the power spectrum density is affected by a *bias factor* that depends on the analyzed frequency. This bias carries over to the estimation of the exponent α, if we insert the estimator of Eq. (4.35) into the linear regression model of Eq. (4.34). This consideration applies especially to the spectrogram (and its estimation by an averaged periodogram), because in case of a short-time window $h(t)$ we have $\psi(0,\xi) = |H(\xi)|^2$.

Let us next employ a representation of the affine class (used in its time-frequency form, Eq. (2.64)) in order to derive the estimator

$$\widehat{\Gamma}_x(\nu) = \lim_{T \to +\infty} \frac{1}{T} \int_{(T)} \widetilde{C}_x(t,\nu;f)\, dt \,. \qquad (4.36)$$

Then a computation, which is similar to the preceding one, provides

$$\mathbf{E}\left\{\widehat{\Gamma}_x(\nu)\right\} = \Gamma_x(\nu) \int_{-\infty}^{+\infty} |\xi|^{-\alpha}\, \psi(0,\xi)\, d\xi \,.$$

This shows that the bias (factor) becomes independent of the frequency. Inserting this new estimator into the regression model of Eq. (4.34) leads to an unbiased estimator for the exponent α. The time-frequency interpretation of the spectral estimation thus justifies the conclusion that the procedures based on constant-Q representations (especially the scalograms) are well suited for the study of "$1/f$-noise" processes.

Analysis of self-similar processes. There is another obvious situation, in which the time-scale analyses are naturally motivated and reveal very clearly the structure of the signal. This is the case for processes with an underlying *scaling operation* as expressed, for example, by a self-similarity.

By definition, a stochastic process is *self-similar* [14] (with exponent H), if it verifies the relation

$$\{x(kt)\} = \{k^H x(t)\}$$

for all $k > 0$, where the equality is to be understood in distribution. The simplest example of a self-similar stochastic process is a *fractional Brownian motion* [15] $B_H(t)$, which represents a (zero mean) Gaussian process with an autocovariance of the form

$$r_{B_H}(t,s) = \frac{1}{2}\sigma^2 \left[|t|^{2H} + |s|^{2H} - |t-s|^{2H}\right], \quad 0 < H < 1\,. \qquad (4.37)$$

The structure of the covariance renders its *nonstationary* character visible. It is self-similar with exponent H due to the identity

$$r_{B_H}(kt,ks) = k^{2H}\, r_{B_H}(t,s) = r_{(k^H B_H)}(t,s)\,.$$

Chapter 4 Time-Frequency as a Paradigm 331

It actually represents a generalization of the ordinary Brownian motion, which corresponds to the special case $H = 1/2$. Note that in this case Eq. (4.37) coincides with the known result

$$r_{B_{1/2}}(t,s) = \sigma^2 \min(t,s) \quad \Longrightarrow \quad \text{var}\left\{B_{1/2}(t)\right\} = \sigma^2 t \,, \qquad t, s > 0 \,.$$

While being a nonstationary process, the fractional Brownian motion has the property that its *increments* are stationary. Indeed, if we put

$$\delta B_H(t, \tau) = B_H(t + \tau) - B_H(t) \,,$$

we find by an application of Eq. (4.37)

$$\mathbf{E}\left\{\delta B_H(t,\tau)\,\delta B_H(t,\theta)\right\} = \frac{1}{2}\sigma^2 \left[|\tau|^{2H} + |\theta|^{2H} - |\tau - \theta|^{2H}\right] \,.$$

This expression is independent of the considered instant t. Furthermore, if we take *symmetric* increments (i.e., so that $\tau = -\theta$), the value of the normalized autocorrelation is

$$\rho_{B_H}(\tau) = -\frac{\mathbf{E}\left\{\delta B_H(t,\tau)\,\delta B_H(t,-\tau)\right\}}{\text{var}\left\{\delta B_H(t,\pm\tau)\right\}} = 2^{2H-1} - 1 \,.$$

This yields a *constant* value for every time-horizon of the increment (no matter how large). In case of the ordinary Brownian motion ($H = 1/2$) the correlation between disjoint increments is zero. The other cases reflect a *long-range correlation* of the process, which can have either positive (persistence) or negative (antipersistence) sign depending on H being greater or less than $1/2$, respectively. In the first case, the trajectories of the process become smoother when H gets closer to 1. Conversely, they become more and more irregular when H approaches 0. This behavior manifests itself in a *fractal dimension*,[16] which can be shown to be equal to $D = H - 2$. This dimension traverses the interval $[1, 2]$ when H passes from 1 to 0. The existence of long-range correlations, which decay slowly (typically by a rational power of the time-distance rather than exponentially as is the case for the ARMA processes), is also responsible for the eventual spectral divergence at the zero frequency (by a power of ν). Hence, it fits into the general picture of a "$1/f$- noise" process.

The analysis of self-similar processes, including the fractional Brownian motion as their prototype, runs into several difficulties. These can be reduced by assuming an adaptive viewpoint and introducing the time-frequency or time-scale context explicitly. For example, it is difficult a *priori* to attribute a meaning to the notion of the power spectrum of a fractional Brownian motion, because we are dealing with a nonstationary process. Nevertheless, it often serves as a model for the "1/f-noise" from a practical perspective. This disjunction can be taken care of naturally by using a time-frequency context and following the ideas of spectral estimation already developed here. In fact, we can make use of the stationary increments of the fractional Brownian motion together with the general admissibility condition equation (2.26) of a wavelet ($\psi(t)$ having zero mean). Then the expression for the covariance between two points of the time-scale plane ($a > 0$) can be simplified to

$$\mathbf{E}\left\{T_{B_H}(t,a)\,T_{B_H}^*(s,b)\right\} = -\frac{1}{2}\sigma^2\,a^{2H+1}\int_{-\infty}^{+\infty}|\theta|^{2H}\,T_\psi\left(\frac{t-s}{a}-\theta,\frac{a}{b}\right)d\theta. \quad (4.38)$$

This result is both simple and important. It indicates that a wavelet transform *makes* a fractional Brownian motion *stationary*.[17] Hence, one can define a scale-dependent stationary autocorrelation

$$\gamma_{B_H}(\tau,a) = \mathbf{E}\left\{T_{B_H}\left(t+\frac{\tau}{2},a\right)T_{B_H}^*\left(t-\frac{\tau}{2},a\right)\right\}$$

$$= -\frac{1}{2}\sigma^2\,a^{2H+1}\int_{-\infty}^{+\infty}|\theta|^{2H}\,T_\psi\left(\frac{\tau}{a}-\theta,1\right)d\theta\ .$$

This function reflects the self-similarity of the process, as we further obtain

$$\gamma_{B_H}(k\tau,ka) = k^{2H+1}\,\gamma_{B_H}(\tau,a) = \gamma_{(k^{H+1/2}B_H)}(\tau,a)\ .$$

Moreover, the existence of a *stationary* autocorrelation gives rise to a scale-dependent power spectrum, which is defined by the Fourier transform

$$\Gamma_{B_H}(\nu,a) = \int_{-\infty}^{+\infty}\gamma_{B_H}(\tau,a)\,e^{-i2\pi\nu\tau}\,d\tau$$

$$= C\,\frac{\sigma^2}{|\nu|^{2H+1}}\,a\,|\Psi(a\nu)|^2\ . \quad (4.39)$$

By the observation that $\sqrt{a}\Psi^*(a\nu)$ is the frequency response of the "wavelet filter" at scaling level a, this relation describes the spectral behavior of the fractional Brownian motion as a "1/f-noise" process, in which the frequency variable is raised to the usually assigned power $2H+1$.

As another simple consequence of Eq. (4.38) we obtain

$$\mathbf{E}\left\{|T_{B_H}(t,a)|^2\right\} = a^{2H+1}\,V_\psi(H) \quad (4.40)$$

where
$$V_\psi(H) = -\frac{1}{2}\sigma^2 \int_{-\infty}^{+\infty} |\tau|^{2H} T_\psi(-\tau, 1)\, d\tau \ .$$

This leads to a possible way of estimating H via

$$\log \mathbf{E}\left\{|T_{B_H}(t,a)|^2\right\} = (2H+1)\log a + \log V_\psi(H) \ . \tag{4.41}$$

The use of the scalogram thus admits a natural and particularly well-suited description of the structure of the fractional Brownian motion. The stationary increments find their counterpart in the stationary behavior of the representation on each scaling level; loosely speaking, the "wavelet filter" acts like a differentiator. The self-similarity materializes in the homogeneity of the expectation value of the scalogram relative to the scale parameter, as we infer from Eq. (4.40)

$$\mathbf{E}\left\{|T_{B_H}(t,ka)|^2\right\} = k^{2H+1}\, \mathbf{E}\left\{|T_{B_H}(t,a)|^2\right\} \ .$$

The flexibility of the wavelet analysis amounts to a further advantage regarding the estimation of the exponent (in the spectral or the self-similar form) of the process. For this purpose we can use Eq. (4.41) together with an empirical variance. The wavelet analysis not only leads to the stationary behavior on each scale of the initially nonstationary process, but it also reduces its long-range dependencies. We can give a qualitative explanation of this fact by rewriting Eq. (4.39) as

$$\gamma_{B_H}(\tau, a) = C\, \sigma^2\, a^{2H+1} \int_{-\infty}^{+\infty} \frac{|\Psi(\nu)|^2}{|\nu|^{2H+1}}\, e^{i2\pi\nu(\tau/a)}\, d\nu \ .$$

Hence, we observe that the behavior of the autocorrelation at infinity is determined by the behavior of the ratio $|\Psi(\nu)|^2/|\nu|^{2H+1}$ at zero. Consequently, the frequency response of the wavelet at the zero frequency can counterbalance the divergence of the "$1/f$-spectrum" of the process. The determining factor in control of this behavior is the *number of vanishing moments* (or degree of "cancellation") of the analyzing wavelet. More precisely, for a wavelet with at most R vanishing moments the ratio $|\Psi(\nu)|^2/|\nu|^{2H+1}$ behaves like $|\nu|^{2(R-H)-1}$ near the origin. Hence, the condition $R \geq H + 1/2$ must be satisfied in order that we can expect a reduction of the long-range dependencies in the transform of the signal. This requires the wavelet to have at least *two* vanishing moments, a priori, given that we limit ourselves to the cases $0 < H < 1$. Provided that R is large enough, indeed, the asymptotic decay of the autocovariance is $|\tau|^{2(H-R)}$. Hence, the correlation between adjacent segments of the scalogram becomes less important for wavelets with a large number of vanishing moments.[18]

4.2.2. Nonstationary Characteristics

We have already explained, why purely temporal or frequential methods are often unsatisfactory for the analysis of nonstationary signals. In such situations the time-frequency methods deploy their full capability. Moreover, they can serve to define new quantities that are well-suited for nonstationary situations and allow simple physical interpretations.

Distance from the stationary case. The notion of stationary signals in the wider sense (or weakly stationary signals) is a well-defined concept. This is no longer true for notions of a nonstationary behavior, because many different types can be imagined (see Subsection 1.2.2). Instead of defining what a nonstationary signal *is* in a very general sense, we can identify more easily what it *is not*. We thus try to explain in which sense its properties differ from those of a stationary signal. This approach can be considered as defining a *distance from the stationary case*. Assuming a time-frequency perspective turns out to be quite profitable in this regard.

According to the definitions of Subsection 2.4.3, a signal is (weakly) stationary if and only if its Wigner-Ville spectrum coincides with the power spectrum at each instant; that is, we have

$$\mathbf{W}_x(t,\nu) = \Gamma_x(\nu) .$$

For a fixed frequency ν and a time-horizon T, a natural measure for the distance from the stationary case can therefore be defined by

$$d_x^2(\nu;T) = \frac{1}{T} \int_{(T)} \left[\mathbf{W}_x(t,\nu) \, dt - \frac{1}{T} \int_{(T)} \mathbf{W}_x(s,\nu) \, ds \right]^2 dt . \quad (4.42)$$

This quantity is zero if and only if the signal is weakly stationary, and it increases as we depart from this situation.

The measure of Eq. (4.42) can obviously be used by itself (e.g., for problems of classification). However, it can also serve to find an optimal time-smoothing for the estimation of the Wigner-Ville spectrum. [19] Intuitively, an optimal smoothing should be more pronounced for processes that are almost stationary; then one can expect a reduced variance of the estimation without the penalizing effect of an enlarged bias. Hence, we should be able to find an optimal smoothing by an *adaptation* to the local time of the stationary behavior. A possible solution can be found in the design of an adapted piecewise stationary model for the considered signal. The underlying idea comprises three consecutive steps, namely:

(*i*) Start from an estimated spectrum without smoothing;

(ii) Estimate the instants at which the stationary properties change; and finally

(iii) Replace the nonsmoothed estimator with the sequence of averaged estimators on each interval between two such instants.

Let us consider the situation of a discrete-time estimation with an initial estimator $PW_x[n, \nu]$, $n = 1, \ldots, N$, of pseudo-Wigner-Ville type. Then the first problem consists in finding, for each selected frequency ν, a partition of the interval $[1, N]$ into p blocks of lengths N_i so that

$$\sum_{i=1}^{p} N_i = N \ .$$

The smoothed estimator is then defined by the sequence

$$\overline{PW}_x[i, \nu] = \frac{1}{N_i} \sum_{n=n_{i-1}+1}^{n_i} PW_x[n, \nu] \ , \qquad i = 1, 2, \ldots, p, \qquad (4.43)$$

where

$$n_0 = 0 \ , \qquad n_i = \sum_{j=1}^{i} N_j \ .$$

The N_i are the estimated times of the stationary behavior, and n_i are the estimated points in time where we pass from one stationary segment to the next. By intuition, a model using fewer segments than needed (i.e., a value of p which is too small) results in a significant bias and a reduced variance. Conversely, if the value of p is chosen too large, the bias becomes small, but the variance increases. This supports the idea of choosing p according to a compromise of bias-variance type, which is the result of a minimization of the total mean square error (MSE). (Note that for an *a priori* fixed number of segments the optimal partition, in the sense of a minimal MSE, can be constructed by a dynamic programming technique, because the problem resembles the approximation of a curve by a step function.[20])

For each stationary part the analytical Wigner-Ville spectrum coincides with the power spectrum $\Gamma_{x,n}(\nu)$ of the tangential stationary process. Therefore, we can compute the distance between the smoothed estimator on this segment and the power spectrum by taking the sum of squares of their differences. Taking the different variances of the estimators on different pieces into account (which results from a possibly unequal partition), a proper choice for a confidence measure of the estimator in Eq. (4.43) is

$$M_p(\nu) = \sum_{i=1}^{p} \frac{1}{V_i} \sum_{n=n_{i-1}+1}^{n_i} \left[\overline{PW}_x[i, \nu] - \Gamma_{x,n}(\nu)\right]^2 \ . \qquad (4.44)$$

Here V_i denotes the variance of the estimator in Eq. (4.43). In order to minimize the *expectation value* of Eq. (4.44) one can use the statistics (of Akaike type [21])

$$C_p(\nu) = \sum_{i=1}^{p} \frac{N_i}{V_i} d_x^2(\nu; N_i) - N + 2p ,\qquad(4.45)$$

with

$$d_x^2(\nu; N_i) = \frac{1}{N_i} \sum_{n=n_{i-1}+1}^{n_i} \left[PW_x[i,\nu] - \overline{PW}_x[i,\nu]\right]^2 .\qquad(4.46)$$

Note that Eq. (4.46) is a discrete-time analogue of Eq. (4.42). We thus see that the distance from the stationary case, as defined here, occurs as a natural ingredient in the solution of the underlying optimization problem.

Demodulation. The demodulation of signals is certainly among the first and most important applications of time-frequency representations. Concerning the quantitative and qualitative aspects of a representation, its success in applications depends not only on its theoretical capability of rendering the rules of the modulations accessible, but also requires it to lay these rules open in the time-frequency plane to the best possible degree.

Example 1. When we consider a frequency-modulated signal, we expect that a representation displays a "ridge" that marks the curve of the instantaneous frequency. [22]

Example 2. The group delay carries the useful information in case of a *dispersive* propagation of different wavelengths. This notion describes a situation, where the synchronous frequency components of an initial pulse become more and more asynchronous by a dispersion during their propagation. In a time-frequency terminology, the different frequencies propagate with different (group) velocities; this maps a straight line, which is parallel to the frequency axis initially, into a *curve* describing the group delay. A time-frequency representation should display the information about the law of this type of dispersion *directly*. This appears all the better as the estimation of the group delay gets more accurate and the localization to its curve is more pronounced. [23]

Let us recall the two conditions

$$f(\xi, 0) = 1 \qquad(4.47)$$

$$\frac{\partial f}{\partial \tau}(\xi, 0) = 0 \qquad(4.48)$$

Chapter 4 Time-Frequency as a Paradigm

from Table 2.4 in Chapter 2. They were imposed on the parameter function of a representation in Cohen's class, in order that the instantaneous amplitude $a_x(t)$ and frequency $\nu_x(t)$ of a signal (in its analytic form) can be derived from the marginal distribution

$$\int_{-\infty}^{+\infty} C_x(t, \nu; f)\, d\nu = a_x^2(t) \tag{4.49}$$

and the local center of gravity

$$\frac{1}{a_x^2(t)} \int_{-\infty}^{+\infty} \nu\, C_x(t, \nu; f)\, d\nu = \nu_x(t) \tag{4.50}$$

of the representation, respectively. Obviously, the Wigner-Ville distribution satisfies both conditions (Eqs. (4.47) and (4.48)). It is worthwhile to retain that the same conditions hold for the moving methods such as pseudo-Wigner-Ville (3.74), provided that the employed short-time window is real and even, for instance.

More generally, if only the relation (4.48) is satisfied (which can be the case for a smoothed pseudo-Wigner-Ville distribution or a spectrogram), we obtain

$$\mu_0(t; f) = \int_{-\infty}^{+\infty} C_x(t, \nu; f)\, d\nu$$
$$= \int_{-\infty}^{+\infty} F(t - s, 0)\, a_x^2(s)\, ds\ , \tag{4.51}$$

$$\mu_1(t; f) = \frac{1}{\mu_0(t; f)} \int_{-\infty}^{+\infty} \nu\, C_x(t, \nu; f)\, d\nu$$
$$= \frac{1}{\mu_0(t; f)} \int_{-\infty}^{+\infty} F(t - s, 0)\, a_x^2(s)\, \nu_x(s)\, ds\ . \tag{4.52}$$

This signifies physically that the local moments of a time-frequency representation give access to the modulation laws of $a_x(t)$ and $\nu_x(t)$, if $F(t, 0)$ resembles a *pulse* as compared to the rate of change of these two functions. For example, when we apply a rectangular smoothing of duration Δt to a representation of a signal that has a constant instantaneous amplitude during this short period, we can easily see that Eq. (4.52) turns into

$$\mu_1(t; f) = \frac{1}{2\pi}\, \frac{\arg\{x(t + \Delta t/2)\} - \arg\{x(t - \Delta t/2)\}}{\Delta t}\ .$$

Hence, the local first-order moment provides an estimate for the instantaneous frequency, which is obtained by replacing the *derivative* in its original definition (Eq. (1.21)) with a *finite difference*. As the difference is *symmetric* about the instant t, the estimate yields the true value, if the instantaneous frequency admits a *linear* approximation relative to the time horizon Δt of the smoothing. Within the precision of the assumed approximations, this manifests the possibility of finding the structure of the modulations of the signal by an inspection of the *ridges* of a time-frequency representation.

Local singularities. Let us recall that the time-*frequency* representations were constructed mainly for giving prominence to the *frequential* properties of a signal changing in time. Accordingly, the time-*scale* representations offer a natural paradigm for the description of properties with a *scale-structure* that may change in time as well.

We dealt with such an example in connection with the self-similarity of a stochastic process in Subsection 4.2.1. This property was reflected by a corresponding rule for the progression of the variance of the wavelet coefficients over all scales (cf. Eq. (4.40)). We can attach a second interpretation to this result, which tells that the evolution of the *size* of the wavelet coefficients (or the scalogram) across the scales provides a measure for the (global or local) *Hölder regularity* of the signal (or the trajectories of the stochastic process). [24]

In fact, let us consider a wavelet transform $T_x(t,a)$ of a signal $x(t)$. Given an admissible wavelet $\psi(t)$, we can use the vanishing moment condition and derive

$$T_x(t,a) = |a|^{-1/2} \int_{-\infty}^{+\infty} x(s)\, \psi^*\left(\frac{s-t}{a}\right) ds$$

$$= |a|^{-1/2} \int_{-\infty}^{+\infty} [x(s) - x(t)]\, \psi^*\left(\frac{s-t}{a}\right) ds \ .$$

Suppose the analyzed signal is uniformly Hölder continuous with exponent $0 < H < 1$; that is, we assume that

$$|x(s) - x(t)| \leq C\, |s - t|^H \ . \tag{4.53}$$

If the wavelet is such that $t\psi(t)$ is absolutely integrable, we can immediately verify

$$|T_x(t,a)| \leq C\,|a|^{-1/2} \int_{-\infty}^{+\infty} |s-t|^H \left|\psi\left(\frac{s-t}{a}\right)\right| ds$$

$$\leq C \int_{-\infty}^{+\infty} |t|^H \,|\psi(t)|\, dt \cdot |a|^{H+1/2} \qquad (4.54)$$

$$= \mathcal{O}(|a|^{H+1/2})\,,$$

and the upper bound is uniform with respect to t. Therefore, the regularity of the signal is reflected by the behavior of the wavelet transform for small scaling parameters. The converse implication can also be proven.[25] Because the trajectories of a fractional Brownian motion have the uniform regularity H, the rule for its variance (Eq. (4.40)) fits to the preceding Eq. (4.54).

As a function of scale and time, the wavelet transform enables us to derive estimates for the *local* regularity of a signal as well. For this let us assume that at a given instant t_0 there exists a bound of the type

$$|x(t_0+\tau)-x(t_0)| \leq C\,|\tau|^{H(t_0)}\,, \qquad 0<H<1\,. \qquad (4.55)$$

Then we can write

$$T_x(t,a) = |a|^{-1/2} \int_{-\infty}^{+\infty} [x(t_0+\tau)-x(t_0)]\,\psi^*\left(\frac{\tau-(t-t_0)}{a}\right) d\tau\,,$$

and by employing the inequality (4.55) we obtain the result

$$|T_x(t,a)| \leq C\,|a|^{-1/2} \int_{-\infty}^{+\infty} |\tau|^{H(t_0)} \left|\psi\left(\frac{\tau-(t-t_0)}{a}\right)\right| d\tau$$

$$\leq C\left[\int_{-\infty}^{+\infty} |t|^{H(t_0)} |\psi(t)|\,dt \cdot |a|^{H(t_0)+1/2} + \int_{-\infty}^{+\infty} |\psi(t)|\,dt \cdot |t-t_0|^{H(t_0)}\right]$$

$$= \mathcal{O}\left(|a|^{H(t_0)+1/2} + |t-t_0|^{H(t_0)}\right)\,. \qquad (4.56)$$

We can thus say that the small scales of the wavelet transform depict the regularity of the signal arranged by a time-localization, which refers to a "cone of influence" centered at the considered instant t_0. Again the converse is also true: a sufficient decay of the wavelet coefficients inside this cone of influence leads to estimates for the local regularity of the signal.

Remark 1. We assumed that the regularity index H satisfies $0 < H < 1$. There occurs no problem with the consideration of larger Hölder exponents, if we impose a condition of higher-order vanishing moments on the wavelet instead of the usual admissibility condition (zero mean). In fact, let us suppose that

$$|x(t_0 + \tau) - P_n(\tau)| \leq C|\tau|^{H(t_0)}, \qquad n < H < n+1,$$

where $P_n(\tau)$ is a suitable polynomial of degree n. Then we obtain the same upper bound as in Eq. (4.56), provided that

$$\int_{-\infty}^{+\infty} t^k \, \psi(t) \, dt = 0, \qquad k = 0, \ldots, n.$$

This last condition signifies that the wavelet is orthogonal to all polynomials of degree less than or equal to n.

Remark 2. The Hölder exponent of a singularity finds a simple translation in a wavelet transform or a scalogram in terms of the behavior for small scaling parameters. However, the same information can be provided by other bilinear distributions in the affine class as well: especially the ("active") Unterberger distribution (see Table 2.3) yields comparable results. If we associate the situation in Eq. (4.55) with the asymptotic spectral decay

$$X(\nu) \sim |\nu|^{-(1+H(t_0))} e^{-i2\pi\nu t_0}, \qquad |\nu| \to \infty,$$

the definition of the "active" Unterberger distribution can easily be seen to yield the approximation

$$U_x(t,a) \sim |a|^{2(1+H(t_0))} \delta(t - t_0), \qquad a \to 0.$$

Hence, it is governed by a rule depending on the scale, which contains the exponent of the singularity in a perfectly localized manner. It completely ignores the notion of the cone of influence, which refers to a neighborhood of the singularity and was still present in Eq. (4.56). However, as a typical feature of bilinear transformations of the signal, the localization is affected by the existence of possible interference terms, if two (or more) singularities coexist close to each other.[26]

Evolutionary singularities. Remark 2 (preceding this discussion) suggests that we can use the bilinear affine distributions, not only for a local (static) estimation of singularities, but even for *pursuing* evolutionary singularities. As a guideline we consider the analogy between time-frequency and time-scale representations, or more specially, between instantaneous

Chapter 4 Time-Frequency as a Paradigm

frequency and local singularity. The upper bound (Eq. (4.54)) for the scalogram tells that a "good" time-scale representation should feature the relation

$$\Omega_x(t,a;f) \sim |a|^{2H(t)+1}, \qquad a \to 0, \tag{4.57}$$

at every point t, at which the signal has a Hölder exponent $H(t)$. In this case we obtain

$$H(t) = \frac{\dfrac{k}{2} \int_0^{+\infty} a\,\Omega_x(t,a;f)\,e^{-ka}\,da}{\int_0^{+\infty} \Omega_x(t,a;f)\,e^{-ka}\,da} - 1, \tag{4.58}$$

where k is a positive number, which is feasible in the sense that Eq. (4.57) is valid on the useful domain of the integration. We thus obtain a result, which is comparable with Eq. (4.50) in the following sense: a local first-order moment of a (weighted) bilinear distribution yields a local characteristic of the signal (in this case the Hölder exponent in the place of the instantaneous frequency).

Let us now try to find a condition on a time-scale distribution, in order that the relation of Eq. (4.57) is valid. For the sake of simplicity, we assume that the analyzed signal is real. We start from the situation in Eq. (4.55), which implies

$$[x(t+\tau) - x(t)]^2 \sim |\tau|^{2H(t)} \tag{4.59}$$

for small increments τ. This gives, of course,

$$x(t)\,x(s) \sim \frac{1}{2}\left[x^2(t) + x^2(s) - |t-s|^{2H(t)}\right].$$

Hence, for small scaling parameters a we obtain

$$\Omega_x(t,a;f) \sim \frac{a}{2} \iint_{\mathbb{R}^2} \left[x^2(t+a\tau) + x^2(t+a\theta) - |a(\tau-\theta)|^{2H(t+a\tau)}\right]$$

$$\times F\left(\frac{\tau+\theta}{2}, \tau-\theta\right) d\tau\,d\theta$$

$$\sim a\,C(t) + |a|^{2H(t)+1} \cdot \left(-\frac{1}{2}\int_{-\infty}^{+\infty} |\tau|^{2H(t)} f(0,\tau)\,d\tau\right). \tag{4.60}$$

Here we used the notations of Subsection 3.1.1 and

$$C(t) = \int_{-\infty}^{+\infty} x^2(t+a\tau)\left(\int_{-\infty}^{+\infty} F\left(\frac{\tau+\theta}{2}, \tau-\theta\right) d\theta\right) d\tau.$$

In the general case, this last quantity contributes to a *bias* (with respect to the expected rule from Eq. (4.57)). A suitable choice for the time-scale distribution can annihilate this contribution. In fact, we should use a parameterization that satisfies the condition [27]

$$0 = \int_{-\infty}^{+\infty} F\left(\frac{\tau+\theta}{2}, \tau-\theta\right) d\theta$$

$$= \iint_{\mathbb{R}^2} F\left(\frac{\tau+\theta}{2}, \tau-\theta\right) e^{-i2\pi\xi\tau} d\tau\, d\theta \qquad (4.61)$$

$$= \psi\left(\xi, \frac{\xi}{2}\right).$$

A straightforward argument proves that the scalogram meets this criterion, as the usual admissibility condition of Eq. (2.26) gives

$$\psi(\xi,\nu) = \Psi\left(\nu+\frac{\xi}{2}\right)\Psi^*\left(\nu-\frac{\xi}{2}\right) \quad\Longrightarrow\quad \psi\left(\xi,\frac{\xi}{2}\right) = \Psi(\xi)\,\Psi^*(0) = 0\ .$$

Remark. We have $f(0,\tau) = T_\psi(-\tau, 1)$ for a scalogram. Hence, Eq. (4.60) yields a *local* result that is compatible with the *global* mean value of Eq. (4.40).

The preceding condition in Eq. (4.61) does not limit the solutions just to the scalograms. It is fulfilled by all affine Wigner distributions as well, see Eq. (2.63), while the separable distributions (cf. Table 2.3) with $G(\xi)H(\xi/2) \sim 0$ provide approximate solutions.

§4.3. Decision Statistics

The theory of decision statistics (detection, estimation, classification) is a well-studied problem in signal processing, with abundant literature and well-tried solutions available.[28] Nevertheless, the time-frequency approach offers a way to reconsider this issue from a special perspective that better reflects the nature of the analyzed signals in certain cases. Intuitively, a detection or estimation in the time-frequency plane should result in "re-covering a structure of a known form" (or *signature*). This leads to the following pair of problems:

(i) Given the optimal strategy (matched filtering, maximum likelihood, etc.) for a temporal or frequential approach, how can we find an equivalent form in the time-frequency plane that amounts to a simple physical interpretation?

Chapter 4 Time-Frequency as a Paradigm

(*ii*) How can one formalize the empirical notion of comparing two time-frequency structures in order to determine their optimal or suboptimal character?

This section addresses a first approach to these questions within the framework of energy distributions. [29]

The detection problem, which we shall consider here, is given in terms of the (classical) binary test of hypotheses

$$\begin{cases} H_0 : & y(t) = b(t) \\ H_1 : & y(t) = x(t) + b(t) \ . \end{cases} \qquad (4.62)$$

We use the notations $y(t)$ for the known observation, which is restricted to a finite time-interval (T), $b(t)$ for a (complex and stationary) Gaussian white noise with

$$\mathbf{E}\left\{b(t)\right\} = 0 \ , \qquad \mathbf{E}\left\{b(t)\, b^*(s)\right\} = \gamma_0\, \delta(t-s) \ ,$$

and $x(t)$ for the (complex) signal to be detected.

4.3.1. Matched Time-Frequency Filtering

Let us first consider the case where the signal in question $x(t)$ is *deterministic*. A conceivable approach to constructing a *matched time-frequency filter* proceeds by maximizing a contrast function (or SNR) based on a suitable time-frequency representation. Within the framework of Cohen's class, the detection problem from Eq. (4.62) can be written as

$$\begin{cases} H_0 : & C_y(t,\nu;f) = C_b(t,\nu;f) \\ H_1 : & C_y(t,\nu;f) = C_{x+b}(t,\nu;f) \ . \end{cases}$$

By analogy with the classical theory of matched filtering, a (time-frequency) filter for the detection is defined by

$$\Lambda(y,f) = \iint_{\mathbf{R}^2} G(t,\nu)\, C_y(t,\nu;f)\, dt\, d\nu \ . \qquad (4.63)$$

In this formula the function $G(t,\nu)$ represents a time-frequency *template* that we have to determine, so that detection becomes optimal. Hence, we should try to maximize a contrast function relative to the null hypothesis and its alternative. A natural choice of such a function is the output SNR given by the expression:

$$\mathrm{SNR}(G,f) = \frac{|\mathbf{E}\{\Lambda(y,f) \mid H_1\} - \mathbf{E}\{\Lambda(y,f) \mid H_0\}|}{[\mathrm{var}\{\Lambda(y,f) \mid H_0\}]^{1/2}}.$$

When we employ the frequency-time parameterization of Cohen's class, an application of the Cauchy-Schwarz inequality to the correlative detector of Eq. (4.63) leads to [30]

$$\mathrm{SNR}(G,f) \le \mathrm{SNR}(C_x, f) = \frac{\dfrac{1}{\gamma_0} \iint\limits_{\mathbf{R}^2} |f(\xi,\tau)|^2 |A_x(\xi,\tau)|^2 \, d\xi \, d\tau}{\left(\iint\limits_{\mathbf{R}^2} |f(\xi,\tau)|^4 |A_x(\xi,\tau)|^2 \, d\xi \, d\tau \right)^{1/2}}. \qquad (4.64)$$

It follows that the SNR is maximal, if the impulse response of the time-frequency filter is identical to the time-frequency distribution of the signal $x(t)$ in question. This renders the empirical notion of matched time-frequency filtering meaningful.

After we have just found the optimal solution for a fixed representation, we can move onward to a second level of optimization, which consists in finding the best representation in Cohen's class. Another application of the Cauchy-Schwarz inequality to the right-hand side of Eq. (4.64) gives the final relation

$$\max_{G} \mathrm{SNR}(G,f) = \mathrm{SNR}(C_x, f) \le \frac{E_x}{\gamma_0}. \qquad (4.65)$$

The last inequality turns into an equality for all unitary distributions. Recall that these are characterized by a unimodular parameter function in the frequency-time domain (i.e., so that $|f(\xi,\tau)| = 1$). Under this assumption the maximal SNR is exactly the same as for the matched filtering combined with an envelope detection. This furnishes a new interpretation of Moyal's formula (Eq. (2.95)). The spectrograms are ruled out by this optimality condition, and this explains why a time-frequency detection based on spectrograms requires auxiliary procedures of deconvolution. [31]

Chapter 4 Time-Frequency as a Paradigm

4.3.2. Maximum Likelihood Estimators for Gaussian Processes

Turning to the original formulation (Eq. (4.62)) of the problem, we now suppose that $x(t)$ is a Gaussian *random* process, so that

$$\mathbf{E}\{x(t)\} = \mu(t), \qquad r_x(t,s) = \mathbf{E}\{[x(t) - \mu(t)][x(s) - \mu(s)]^*\}. \quad (4.66)$$

It is known that the detection problem under consideration admits a maximum likelihood estimator based on the Karhunen-Loève representation of the process.[32]

Classical solution. More precisely, we recall that $x(t)$ admits a decomposition

$$x(t) = \sum_{n=0}^{+\infty} x_n \, \varphi_n(t)$$

which is *doubly orthogonal*; that is, the relations

$$\begin{cases} \mathbf{E}\{[x_n - \mu_n][x_m - \mu_m]^*\} = \lambda_n \, \delta_{nm}, \\[1ex] \displaystyle\int_{(T)} \varphi_n(t) \varphi_m^*(t) \, dt = \delta_{nm} \end{cases}$$

are verified, where λ_n and $\varphi_n(t)$ are the eigenvalues and eigenfunctions, respectively, of the autocovariance of $x(t)$. Hence, they are defined by the integral equation

$$\int_{(T)} r_x(t,s) \, \varphi_n(s) \, ds = \lambda_n \, \varphi_n(t), \qquad t \in (T).$$

The coefficients x_n, y_n, and μ_n of the decompositions of $x(t)$, $y(t)$, and $\mu(t)$, respectively, are the projections onto this basis of eigenfunctions (so $x_n = \langle x, \varphi_n \rangle$, etc.). The optimal detector (in the sense of a maximum likelihood estimator) is obtained by a comparison of the decision statistics

$$\Lambda(y) = \Lambda_{\mathrm{p}}(y) + \Lambda_{\mathrm{d}}(y)$$

to a threshold; in this definition we put

$$\begin{cases} \Lambda_{\mathrm{p}}(y) = \dfrac{1}{\gamma_0} \displaystyle\sum_{n=0}^{+\infty} \dfrac{\lambda_n}{\lambda_n + \gamma_0} \, |y_n|^2 \\[2ex] \Lambda_{\mathrm{d}}(y) = 2 \displaystyle\sum_{n=0}^{+\infty} \dfrac{1}{\lambda_n + \gamma_0} \, \mathrm{Re}\{y_n \mu_n^*\}. \end{cases} \quad (4.67)$$

These two quantities bring certain types of scalar products (or undelayed correlations) of the observation and the known characteristics of the process into play. The first expression is a *quadratic* function of the observation and involves the random fluctuations of $x(t)$ about its expectation value. The second quantity is a *linear* function of the observation; it is closely related to the detection of the expectation value of the known process, which can be regarded as a deterministic part.

Time-frequency formulation. As for the matched filtering discussed in Subsection 4.3.1, there exists a simple and equivalent form of the maximum likelihood estimator equation (4.67) in the time-frequency plane. For unitary representations there is a generalized Moyal's formula given by

$$|f(\xi,\tau)| = 1 \;\Rightarrow\; \iint_{\mathbf{R}^2} C_{x_1 x_2}(t,\nu;f) C^*_{x_3 x_4}(t,\nu;f)\, dt\, d\nu = \langle x_1, x_3\rangle \langle x_2, x_4\rangle^*.$$

This relation is certainly verified by the Wigner-Ville distribution. When we use its property of the conservation of the support in the time domain, we obtain

$$\Lambda_{\mathrm{p}}(y) = \frac{1}{\gamma_0} \int_{-\infty}^{+\infty} \int_{(T)} W_y(t,\nu) \left[\sum_{n=0}^{+\infty} \frac{\lambda_n}{\lambda_n - \gamma_0} W_{\varphi_n}(t,\nu)\right] dt\, d\nu\;,$$

(4.68)

$$\Lambda_{\mathrm{d}}(y) = 2 \int_{-\infty}^{+\infty} \int_{(T)} \mathrm{Re}\,\{W_{y\mu}(t,\nu)\} \left[\sum_{n=0}^{+\infty} \frac{1}{\lambda_n + \gamma_0} W_{\varphi_n}(t,\nu)\right] dt\, d\nu\;.$$

These two structures can again be interpreted as correlations that turn up *directly in the time-frequency plane*. They compare a (linear or bilinear) representation of the observation with a time-frequency *pattern* that is based on the known properties of the process in question.

Remark. We could have constructed the decision statistics (Eq. (4.68)) by means of other unitary distributions as well. The distinctive choice of the Wigner-Ville distribution, as well as for other circumstances, roots in its properties of localization and energy concentration: One harbors the expectation that by this choice a decision can be made by inspecting only a small region of the time-frequency plane.

Chapter 4 Time-Frequency as a Paradigm

4.3.3. Some Examples

So far we have dealt with equivalent time-frequency formulations of the classical strategies. Now we will briefly demonstrate by some simple examples, how the explicit time-frequency setting of decision statistics can reconcile the optimality of the solution with the physical interpretation of the results.[33]

Rayleigh channel. The first example stays close to the situation described in Subsection 4.3.1. Suppose the signal under consideration has the form

$$x(t) = a\, x_{\mathrm{d}}(t) \,, \tag{4.69}$$

where $x_{\mathrm{d}}(t)$ is a known *deterministic* signal and a is a (complex) Gaussian random variable so that

$$\mathbf{E}\{a\} = 0\,, \qquad \mathbf{E}\{|a|^2\} = \sigma^2\,.$$

This situation (called a *Rayleigh channel*) formally enters the framework of the model equation (4.66) by letting

$$\mu(t) = 0\,, \qquad r_x(t,s) = \sigma^2\, x_{\mathrm{d}}(t)\, x_{\mathrm{d}}^*(s)\,. \tag{4.70}$$

Hence, the decision statistics in Eq. (4.68) reduces to

$$\Lambda(y) = \Lambda_{\mathrm{p}}(y) = \frac{1}{\gamma_0} \frac{1}{1 + (\gamma_0/\sigma^2 E_{x_{\mathrm{d}}})} \int_{-\infty}^{+\infty}\!\!\int_{(T)} W_y(t,\nu)\, W_{x_{\mathrm{d}}}(t,\nu)\, dt\, d\nu\,. \tag{4.71}$$

We are thus looking at a correlative time-frequency structure, which makes immediate use of the Wigner-Ville distribution of the known reference as a template for the observation. For a high SNR (i.e., for $\gamma_0/\sigma^2 E_{x_{\mathrm{d}}} \ll 1$) we obtain the asymptotic result

$$\Lambda(y) \leq \frac{E_{x_{\mathrm{d}}}}{\gamma_0}\,.$$

This is compatible with the purely deterministic case in Eq. (4.65).

Detection of chirps and Doppler tolerance. Let us next consider the case of linear chirps. The time-frequency formulation of Eq. (4.71) may present an advantage over the classical strategy of matched filtering. This results mainly from the localization properties of the Wigner-Ville distribution, as it generally provides the most *localized* pattern in the plane (in the sense of the discussion of Section 4.1, for instance). Let $x(t)$ (or $x_{\mathrm{d}}(t)$)

be defined by the idealized model of a linear chirp. The perfect localization of its Wigner-Ville distribution to the curve of its (linear) instantaneous frequency $\nu_x(t)$ allows us to write

$$\iint_{\mathbb{R}^2} W_y(t,\nu)\, W_x(t,\nu)\, dt\, d\nu = \int_{-\infty}^{+\infty} W_y(t, \nu_x(t))\, dt \;.$$

This reduces the detection procedure to a simple *path integral* in the plane.[34]

Due to the covariance properties of the distribution, this result can be extended to an *estimation* procedure with respect to a temporal displacement. In fact, suppose the signal

$$x(t) = x_0(t - \tau)$$

is known apart from the delay τ. Then an immediate choice for the estimator of the delay is given by

$$\widehat{\tau} = \arg\max_{\tau} \int_{-\infty}^{+\infty} W_y(t + \tau, \nu_{x_0}(t))\, dt \;.$$

Under the same perspective, but with an observation affected by a (wideband) *Doppler effect*, an adequate time-frequency formulation yields a rather simple interpretation of the optimal estimator for the delay. As both estimators of the Doppler rate and the delay are not decoupled, in general, the delay estimator is biased, if the Doppler rate is unknown. Consequently, a naturally arising task is to design signals that *tolerate* a Doppler effect. This means that an arbitrary (unknown) Doppler rate should leave the estimation of the delay unbiased. Such signals exist: They are characterized by a *hyperbolic* instantaneous frequency or group delay.[35] The time-frequency approach leads to a very simple geometric justification of this fact. Recall that a Doppler effect with a rate η corresponds to a transformation in the time-frequency plane according to

$$(t, \nu) \quad \to \quad (\eta t, \nu/\eta) \;. \tag{4.72}$$

This implies that under certain conditions an unbiased estimation of the delay can be devised by adopting the simplified strategy

$$\widehat{\tau} = \arg\max_{\tau} \int_{-\infty}^{+\infty} \rho_y(t + \tau, \nu_{x_0}(t))\, dt$$

which is again based on a path integral. These conditions on the distribution $\rho_y(t, \nu)$ are as follows:

Chapter 4 Time-Frequency as a Paradigm

(*i*) It is covariant by translations;

(*ii*) It is unitary (for ascertaining the optimality); and

(*iii*) It is localized to a curve that remains invariant under the transformation equation (4.72).

As the curve in the last condition is a hyperbola, the compatibility with these three restrictions leaves only one possible solution, namely the Bertrand distribution (see Table 2.4 and Subsection 2.3.2). This furnishes a purely geometrical solution to the posed problem of Doppler tolerance in a time-frequency setting.

Locally optimal detection. The case of a high SNR was already mentioned here. The opposite case, where the random signal has small energy in comparison with the noise, leads to a time-frequency solution (called *locally optimal*), which can again be endowed with a satisfactory physical interpretation.

According to the hypothesis of a small SNR, the eigenvalues of the autocovariance verify the relation

$$\lambda_n \ll \gamma_0$$

for all n. Hence, we have

$$\sum_{n=0}^{+\infty} \frac{\lambda_n}{\lambda_n + \gamma_0} W_{\varphi_n}(t,\nu) \approx \frac{1}{\gamma_0} \sum_{n=0}^{+\infty} \lambda_n W_{\varphi_n}(t,\nu) = \frac{1}{\gamma_0} \mathbf{W}_x(t,\nu) \ .$$

The last equality follows from

$$\mathbf{W}_x(t,\nu) = \mathbf{E}\left\{W_x(t,\nu)\right\}$$

$$= \mathbf{E}\left\{\int_{-\infty}^{+\infty} \left[\sum_{n=0}^{+\infty} x_n \varphi_n\left(t+\frac{\tau}{2}\right)\right] \left[\sum_{m=0}^{+\infty} x_m \varphi_m\left(t-\frac{\tau}{2}\right)\right]^* e^{-i2\pi\nu\tau} \, d\tau\right\}$$

$$= \sum_{n=0}^{+\infty} \mathbf{E}\left\{|x_n|^2\right\} \int_{-\infty}^{+\infty} \varphi_n\left(t+\frac{\tau}{2}\right) \varphi_n^*\left(t-\frac{\tau}{2}\right) e^{-i2\pi\nu\tau} \, d\tau$$

$$+ \sum_{n=0}^{+\infty} \sum_{m \neq n=0}^{+\infty} \mathbf{E}\left\{x_n x_m^*\right\} \int_{-\infty}^{+\infty} \varphi_n\left(t+\frac{\tau}{2}\right) \varphi_m^*\left(t-\frac{\tau}{2}\right) e^{-i2\pi\nu\tau} \, d\tau$$

$$= \sum_{n=0}^{+\infty} \lambda_n W_{\varphi_n}(t,\nu) \ .$$

Note that the double sum in the preceding equation vanishes due to the stochastic independence of the coefficients in the Karhunen-Loève decomposition. Provided that the signal has zero mean, it follows that the detector (Eq. (4.68)) reduces to

$$\Lambda(y) = \Lambda_{\mathrm{p}}(y) = \frac{1}{\gamma_0^2} \int_{-\infty}^{+\infty} \int_{(T)} W_y(t,\nu) \, \mathbf{W}_x(t,\nu) \, dt \, d\nu \;. \tag{4.73}$$

This shows that the reference signature is nothing but the Wigner-Ville *spectrum* of the process under consideration. [Note that this argument can be used to recover the same asymptotic solution (Eq. (4.71)) for the model of Eq. (4.69).] In the present case of a small SNR, we must therefore use an *expectation value* as our time-frequency template. The joint time-frequency approach yields a simple construction of this template (as an ensemble average of the *distributions* of the reference). This is simpler, for example, than defining the notion of an averaged reference based on the *signal* itself. [36]

Time-frequency jitter. The forementioned solution to the locally optimal detection problem can be further illustrated, when we invoke the model of Eq. (4.69) once more and let it be affected by a time-frequency jitter; that is, we consider the random process

$$x(t) = a\, x_{\mathrm{d}}(t - \tau)\, e^{i 2\pi \xi t}$$

where (t, ξ) is a pair of random variables with a joint probability density $G(\tau, \xi)$ (which is supposed to be even). As a consequence of the properties of the Wigner-Ville distribution we find

$$\mathbf{W}_x(t,\nu) = \mathbf{E}\left\{|a|^2 \, W_{x_{\mathrm{d}}}(t - \tau, \nu - \xi)\right\}$$

$$= \mathbf{E}\left\{|a|^2\right\} \iint_{\mathbb{R}^2} W_{x_{\mathrm{d}}}(t - \tau, \nu - \xi)\, G(\tau, \xi)\, d\tau\, d\xi \tag{4.74}$$

$$= \sigma^2\, C_{x_{\mathrm{d}}}(t, \nu; g) \;.$$

Hence, the corresponding time-frequency detector has the form

$$\Lambda(y) = \Lambda_{\mathrm{p}}(y) = \frac{\sigma^2}{\gamma_0^2} \int_{-\infty}^{+\infty} \int_{(T)} W_y(t,\nu)\, C_{x_{\mathrm{d}}}(t,\nu;g)\, dt\, d\nu \;. \tag{4.75}$$

Chapter 4 Time-Frequency as a Paradigm

As compared with Eq. (4.73), we see that the jitter introduces a *smoothing* of the reference signature by the corresponding probability density. This matches with our intuition: as far as the known reference is affected by a random shift (in time and frequency), the procedure of Eq. (4.75) simply "thickens" the reference relative to the joint probability density of the shift. This enables the wider reference to "catch" the eventual displacement.

A broader class of time-frequency receptors. The preceding example of jittering in the time-frequency plane suggests comparing the Wigner-Ville distribution of the observation $y(t)$ with a modified (smoothed) version of the reference signature. This may be done for the purpose of a mere generalization or for expecting a gain in the robustness of the detector. Let us note that the same expression in Eq. (4.75) can be rewritten and newly interpreted by use of the commutativity of the convolution; we can thus regard it as a correlation between the reference and a smoothed distribution of the observation according to

$$\Lambda(y, G) = \int_{-\infty}^{+\infty} \int_{(T)} C_y(t, \nu; g) \, W_{x_\mathrm{d}}(t, \nu) \, dt \, d\nu \ . \tag{4.76}$$

Viewed from a practical point of view, this structure is quite natural, insofar as a smoothing of the Wigner-Ville distribution of an observation is often commendable, be it for reasons of reducing the amount of the observed data or for the sake of readability or estimation. Moreover, this new perspective can be especially interesting in case of a well-localized reference. Then we can use a decision statistics based on the path integral

$$\Lambda(y, G) = \int_{-\infty}^{+\infty} C_y(t, \nu_x(t); g) \, dt \ .$$

Regardless of its interpretation (smoothing of the reference or of the observation), Eq. (4.76) may serve to define a general class of receptors that are parameterized by an arbitrary smoothing function. In fact, this class offers a great flexibility for applications, as it provides a unified framework for a whole family of solutions of different complexity and performance. It also yields a new interpretation of Cohen's class, thus adding to the illustrations of its capability.

As a justification of the definition in Eq. (4.76) it may be wise to consider some limiting cases. The principal members of this class are listed in Table 4.1. They correspond to either none or total smoothing in time and/or frequency.

Table 4.1

Time-frequency receptors related to Eq. (4.76): some limiting cases

$G(t,\nu)$	$\Lambda(y,G)$	Type of detector
$\delta(t)\,\delta(\nu)$	$\left\|\int_{(T)} y(t)\,x_{\mathrm{d}}^{*}(t)\,dt\right\|^{2}$	matched filter + envelope detection
$\delta(t)$	$\int_{(T)} \|y(t)\|^{2}\,\|x_{\mathrm{d}}(t)\|^{2}\,dt$	intensity correlator
$\delta(\nu)$	$\int_{-\infty}^{+\infty} \|Y(\nu)\|^{2}\,\|X_{\mathrm{d}}(\nu)\|^{2}\,d\nu$	power spectrum correlator
1	$\int_{(T)} \|y(t)\|^{2}\,dt \cdot \int_{(T)} \|x_{\mathrm{d}}(t)\|^{2}\,dt$	energy detector

From this table it emerges that vastly different configurations are included in one and the same framework, such as the matched filtering combined with the envelope detection (semicoherent receptor), or the purely energetic detection (incoherent receptor). Furthermore, it permits a smooth *transition* from one extreme case to another. This can be achieved, for instance, by employing a *separable* smoothing in very much the same way as it was used in Subsection 3.2.3. (There we obtained a smooth transition between the Wigner-Ville distribution and the spectrogram.)

The introduction of a smoothing leads to *suboptimal* receptors. They show a lesser performance than the respective nonsmoothed solutions. The deterioration can be quantified in terms of the normalized SNR:

$$\rho(G) = \frac{\left|\,\mathbf{E}\left\{\Lambda(y,G)\mid H_{1}\right\} - \mathbf{E}\left\{\Lambda(y,G)\mid H_{0}\right\}\,\right|}{(E_{x_{\mathrm{d}}}/\gamma_{0})\,[\,\mathrm{var}\left\{\Lambda(y,G)\mid H_{0}\right\}]^{1/2}}\,.$$

It can be shown that

$$\rho(G) = \frac{1}{E_{x_{\mathrm{d}}}} \frac{\iint_{\mathbf{R}^{2}} W_{x_{\mathrm{d}}}(t,\nu)\,C_{x_{\mathrm{d}}}(t,\nu;g)\,dt\,d\nu}{\left(\iint_{\mathbf{R}^{2}} |C_{x_{\mathrm{d}}}(t,\nu;g)|^{2}\,dt\,d\nu\right)^{1/2}} \leq 1\,. \qquad (4.77)$$

Subject to some nominal conditions, this property of suboptimality comes with an increased *robustness*, when the reference is known incompletely. For the detection of a chirp, for example, the smoothing allows formalizing of the notion of a time-frequency *margin* or *tolerance*, within which the chirp is allowed to be. This yields a comparable performance for the detection of all real chirps that are included in this model. Once again the time-frequency approach is no doubt the most natural way of defining an *average* reference for this case.[37]

Chapter 4 Notes

4.1.1.

[1] This measure for the time-frequency extension was proposed by Claasen and Mecklenbräuker (1980a) in connection with the Wigner-Ville distribution. Its generalization to Cohen's class was studied in Flandrin (1987).

[2] There is a different approach to this problem in Janssen (1991).

4.1.2.

[3] It seems that the formulation of the problem of energy concentration of joint time-frequency distributions goes back to Flandrin (1988a). Later work, for example, that by Ramanathan and Topiwala (1992), follows the same idea (based on bilinear distributions and the Wigner-Ville case, in particular). A similar point of view is used for linear decompositions (of Gabor or wavelet type) in Daubechies (1988b) and Daubechies and Paul (1988). We can finally observe that the idea of a projection onto a time-frequency domain is connected with the question of time-frequency analysis and synthesis of certain spaces of signals. This subject is intensively investigated in the monograph by Hlawatsch (1998).

[4] Most results presented here are taken from Flandrin (1988a). They rely in part on other arguments given in Janssen (1981).

[5] See the work by Janssen (1981).

[6] This fact was first emphasized by Janssen and Claasen (1985).

[7] Another way to characterize the optimality of the Wigner-Ville distribution among all s-Wigner distributions with respect to a minimal spread in the plane is provided by Janssen (1982). He uses a similar measure as in Eq. (4.1), which involves the *square* of the distribution.

[8] In spite of its importance, we do not deal with this *synthesis* problem. It is clear that the difficulty stems from the fact that a general time-frequency function need not support any relation to a representation of a signal (e.g., Wigner-Ville type). The most natural approach consists of solving a problem of best approximation by admissible functions; that is, we consider only those functions that represent the time-frequency distribution of a signal. Here we may work with any reasonable norm as a measure for the distance from the given function. The first solution pointing in this direction was proposed by Boudreaux-Bartels (1983) and Boudreaux-Bartels and Parks (1986). They considered the Wigner-Ville distribution and used a quadratic distance between time-frequency functions. The underlying task

Chapter 4 Time-Frequency as a Paradigm   355

was shown to be related to an eigenvalue problem. An extended and more general analysis, including other unitary and nonunitary distributions, was carried out later by Hlawatsch (1988) and Hlawatsch and Krattenthaler (1986; 1992; 1998).

4.1.3.

[9] Here we refer to the work of Lieb (1990). Some preliminary results of this type appeared in work by Price and Hofstetter (1965). Extensions with special emphasis on entropy and related inequalities can be found in work by Janssen (1998).

[10] The definition of the Legendre transform, as it is used here, can be found in the book by Arnol'd (1976).

4.2.1.

[11] See also Chapter 3, Note 2.

[12] This example is taken from Abry, Gonçalvès, and Flandrin (1995), which includes a finer analysis of the estimation than given here.

[13] The "$1/f$-noise" processes have been studied exhaustively. This is true for their physical nature, on the one hand, and also for mathematical problems concerning their modeling. An introduction to this subject can be found in Keshner (1982), for example.

[14] A simple and heuristic approach to the notion of self-similar stochastic processes in the form that we use here is in work by Feder (1988). References to more rigorous and/or comprehensive descriptions can be found in the cumulative bibliography compiled by Taqqu (1986).

[15] The definition of the fractional Brownian motion with exponent H (which stands for Hurst) is usually attributed to Mandelbrot and van Ness (1968). An earlier source might be the work of Kolmogorov (see Yaglom, 1987).

[16] A definition of the notion of the fractal dimension, with special emphasis on the problem considered here, is given in Feder (1988). A more in-depth approach is contained in Falconer (1990).

[17] The first paper in which this property is established is by Flandrin (1989b). Even more generally, the wavelet transform of every random process with stationary increments turns out to be stationary (Masry, 1993).

[18] We mentioned only some aspects of self-similar stochastic processes linked to the continuous wavelet transform. There exist other results, especially for the discrete case (Wornell, 1990; Flandrin, 1992; and Tewfik and Kim, 1992). Most of them are contained in the survey articles by Flandrin (1994) and Wornell (1993).

4.2.2.

[19] This measure was introduced by Martin in connection with a classification problem for certain signals in biology (Martin, 1984). Its application for gaining an optimal smoothing is discussed in more detail by Martin and Flandrin (1985a).

[20] See, for example, Bellman and Roth (1969).

[21] The mentioned relation to a strategy of Akaike type refers mainly to its interpretation by Clergeot (1982).

[22] A typical example for this situation is the analysis of simple sonar signals, such as the emitted chirps of a bat, see Flandrin (1988c).

[23] This application of time-frequency methods to a dispersive propagation was especially studied for underwater acoustics, cf. Flandrin, Sageloli et al. (1986), Zakharia et al. (1988), and Sessarego et al. (1990).

[24] The usefulness of the wavelet transform in signal analysis, in particular for the determination of the singularities, was described by Mallat and Hwang (1992). Their work relies partly on the fundamental results of Holschneider and Tchamitchian (1989), and Jaffard (1989).

[25] See the work by Holschneider and Tchamitchian (1989).

[26] This point is addressed in Flandrin and Gonçalvès (1994), for example.

[27] See Gonçalvès and Flandrin (1992) and Flandrin and Gonçalvès (1994).

4.3.

[28] A classical reference concerning the statistical decision theory in signal processing is the book by Whalen (1971). It can also serve as a useful introduction to more advanced books such as the one by Van Trees (1968).

[29] We do not consider the solutions that rely on linear decompositions (similarly to the localization problem in Section 4.1). The reader may consult the literature for gaining more insight into this area, for example, Friedlander and Porat (1989) or Fowler and Sibul (1991). There exist precursors of the bilinear approach followed here in the work of Flaska (1976) and Altes (1980). However, it gained its veritable importance, after the work of Kumar and Carroll (1984) and Kay and Boudreaux-Bartels (1985) appeared. They were the first to make explicit use of the Wigner-Ville distribution in this context. Hence, they prepared the ground for the more general approaches based on Cohen's class (or the affine class) in Flandrin (1986b; 1988b; 1989c), Sayeed and Jones (1995), and Matz and Hlawatsch (1996).

4.3.1.

[30] Compare work by Flandrin (1989c).

[31] Such procedures were implemented in the original approach by Altes (1980), who used the correlations of spectrograms.

4.3.2.

[32] The classical solution described here is borrowed from Van Trees (1968).

4.3.3.

[33] This paragraph consists essentially of examples taken from Flandrin (1988b).

[34] This property was exploited for the first time by Kay and Boudreaux-Bartels (1985). It was used, for example, in Flandrin (1986b; 1988c) for the construction of a simplified detector, which matches the structure of bat "chirps." A generalization to the problem of detecting gravitational waves can be found in Chassande-Mottin and Flandrin (1998).

[35] The issue of the tolerance to the Doppler effect is addressed in Altes and Titlebaum (1970), Altes (1971), Rihaczek (1969), or Mamode (1981), for instance. Its time-frequency interpretation given here can be found in Flandrin and Gonçalvès (1994). Note that a graphical (thus qualitative) description was already contained in Flandrin (1986c).

[36] This strategy was used with considerable profit for a detection problem of faults in a thermal engine, see Chiollaz, Flandrin, and Gache (1987).

[37] The robustness of time-frequency receptors is discussed in Flandrin (1989c). Some concrete examples supporting the given interpretation can be found in Flandrin, Weber, and Zakharia (1989).

Bibliography

Abry, P. (1997). *Ondelettes et Turbulences*, Paris: Diderot Editeur.

Abry, P., Gonçalvès, P., and Flandrin, P. (1995). Wavelets, Spectrum Analysis and $1/f$ Processes, in *Wavelets and Statistics. Lecture Notes in Statistics*, **103**, A. Antoniadis et al., eds., New York: Springer-Verlag, 15–29.

Ackroyd, M. H. (1970). Short-time spectra and time-frequency energy distributions, *J. Acoust. Soc. Amer.*, **50**, 1229–1231.

Akay, M. (ed.). (1997). *Time-Frequency and Wavelets in Biomedical Signal Processing*, Piscataway: IEEE Press.

Allen, J. B. and Rabiner, L. R. (1977). A unified approach to STFT analysis and synthesis, *Proc. IEEE*, **65**, 1558–1564.

Altes, R. A. (1971). Methods of wide-band signal design for radar and sonar systems, Ph.D. Dissertation, Univ. of Rochester, Rochester (NY).

Altes, R. A. (1973). Some invariance properties of the wideband ambiguity function, *J. Acoust. Soc. Amer.*, **53**, 1154–1160.

Altes, R. A. (1980). Detection, estimation and classification with spectrograms, *J. Acoust. Soc. Amer.*, **67**, 1232–1246.

Altes, R. A. (1984). Non-uniqueness of a new time-frequency signal representation, ORINCON Tech. Memo. TM-322, La Jolla (CA).

Altes, R. A. (1985). Overlapping windows and signal representation in the time-frequency plane, in *Coll. Int. CNRS Systèmes Sonar Aériens Animaux: Traitement et Analyse des Signaux*, Lyon, France, pp. 17.1–17.47.

Altes, R. A. (1990). Wide-band, proportional bandwidth Wigner-Ville analysis, *IEEE Trans. Acoust., Speech and Signal Proc.*, **38**, 1005–1012.

Altes, R. A. and Titlebaum, E. L. (1970). Bat signals as optimally Doppler tolerant waveforms, *J. Acoust. Soc. Amer.*, **48**, 1014–1020.

Amblard, P. O. and Lacoume, J. L. (1992). Construction of fourth order Cohen's class: a deductive approach, in *IEEE-SP Int. Symp. on Time-Frequency and Time-Scale Analysis*, Victoria (BC), Canada, 257–260.

Andrieux, J. C., Feix, M. R., Mourgues, G., Bertrand, P., Izrar, B., and Nguyen, V. T. (1987). Optimum smoothing of the Wigner-Ville distribution, *IEEE Trans. Acoust., Speech and Signal Proc.*, **35**, 764–769.

Arnol'd, V. I. (1976). *Méthodes Mathématiques de la Mécanique Classique*, Moscow: Mir.

Auger, F. (1991). Représentations temps-fréquence des signaux non-stationnaires: synthèse et contribution, Thèse de Doctorat, Univ. de Nantes.

Auger, F. and Doncarli, C. (1992). Quelques commentaires sur des représentations temps-fréquence proposées récemment, *Traitement du Signal*, **9**, 3–25.

Auger, F. and Flandrin, P. (1995). Improving the readability of time-frequency and time-scale representations by the reassignment method, *IEEE Trans. Signal Proc.*, **43**, 1068–1089.

Balian, R. (1981). Un principe d'incertitude fort en théorie du signal ou en mécanique quantique, *C. R. Acad. Sc. Paris, Série II*, **292**, 1357–1362.

Baraniuk, R. G. and Jones, D. L. (1993). A signal-dependent time-frequency representation: optimal kernel design, *IEEE Trans. Signal Proc.*, **41**, 1589–1602.

Bass, J. (1945). Les fonctions aléatoires et leur interprétation mécanique, *Revue Scientifique*, 83ème Année, 3–20.

Basseville, M., Flandrin, P., and Martin, N. (eds.). (1992). Signaux non-stationnaires, analyse temps-fréquence et segmentation. Fiches descriptives d'algorithmes, *Traitement du Signal*, **9**, Suppl. to No. 1.

Bastiaans, M. J. (1978). The Wigner distribution function applied to optical signal and systems, *Opt. Commun.*, **25**, 26–30.

Bastiaans, M. J. (1980). Gabor's expansion of a signal into Gaussian elementary signals, *Proc. IEEE*, **68**, 538–539.

Bastiaans, M. J. (1981). A sampling theorem for the complex spectrogram and Gabor's expansion of a signal in Gaussian elementary signals, *Opt. Eng.*, **20**, 594–598.

Battaglia, F. (1979). Some extensions of the evolutionary spectral analysis of a stochastic process, *Bull. Unione Matematica Italiana*, **16-B**, 1154–1166.

Bayen, F., Flato, M., Fronsdal, C., Lichnerowicz, A., and Sternheimer, D. (1978a). Deformation theory and quantization I. Deformation of symplectic structures, *Ann. Phys.*, **111**, 61–110.

Bayen, F., Flato, M., Fronsdal, C., Lichnerowicz, A., and Sternheimer, D. (1978b). Deformation theory and quantization II. Physical applications, *Ann. Phys.*, **111**, 111–151.

Bedrosian, E. (1963). A product theorem for Hilbert transforms, *Proc. IEEE*, **51**, 868–869.

Bellman, R. and Roth, R. (1969). Curve fitting by segmented straight lines, *J. Amer. Stat. Assoc.*, **64**, 1079–1084.

Benedetto, J. J. (1990). Uncertainty Principle Inequalities and Spectrum Estimation, in *Recent Advances in Fourier Analysis and its Applications*, J. S. Byrnes and J. F. Byrnes, eds., Dordrecht: Kluwer, 143–182.

Berry, M. V. (1977). Semiclassical mechanics in phase space: a study of Wigner's function, *Phil. Trans. Roy. Soc.*, **A 287**, 237–271.

Bertrand, J. and Bertrand, P. (1987). A tomographic approach to Wigner's function, *Found. of Physics*, **17**, 397–405.

Bertrand, J. and Bertrand, P. (1988). Time-frequency representations of broad-band signals, in *IEEE Int. Conf. on Acoust., Speech and Signal Proc. ICASSP-88*, New York (NY), 2196–2199 (also Combes, J. M., Grossmann, A., and Tchamitchian, Ph. (eds.). (1989). *Wavelets, Time-Frequency Methods and Phase-Space*, Berlin: Springer-Verlag, 164–171).

Bertrand, J. and Bertrand, P. (1992a). Affine Time-Frequency Distributions, in *Time-Frequency Signal Analysis – Methods and Applications*, B. Boashash, ed., Melbourne: Longman-Cheshire, 118–140.

Bertrand, J. and Bertrand, P. (1992b). A class of affine Wigner functions with extended covariance requirements, *J. Math. Phys.*, **33**, 2515–2527.

Bertrand, P. (1983). Représentations des signaux dans le plan temps-fréquence, *Rech. Aérosp.*, **1983-1**, 1–12.

Bertrand, P., Izrar, B., Nguyen, V. T., and Feix, M. R. (1983). Obtaining non-negative quantum distribution functions, *Phys. Lett.*, **94**, 415–417.

Blahut, R. E., Miller, Jr., W., and Wilcox, C. H. (1991). *Radar and Sonar, Part I*, IMA, Vol. **32**, New York: Springer-Verlag.

Blanc-Lapierre, A. and Picinbono, B. (1955). Remarques sur la notion de spectre instantané de puissance, *Pub. Sci. Univ. d'Alger B*, **1**, 2–32.

Blanc-Lapierre, A. and Picinbono, B. (1981). *Fonctions Aléatoires*, Paris: Masson.

Boashash, B. (1992a). Estimating and interpreting the instantaneous frequency of a signal. Part I: Fundamentals, *Proc. IEEE*, **80**, 520–538.

Boashash, B. (1992b). Estimating and interpreting the instantaneous frequency of a signal. Part II: Algorithms and Applications, *Proc. IEEE*, **80**, 540–568.

Boashash, B. (ed.). (1992c). *Time-Frequency Signal Analysis – Methods and Applications*, Melbourne: Longman-Cheshire.

Boashash, B. and Ristic, B. (1992). Polynomial Wigner-Ville distributions and time-varying higher-order spectra, in *IEEE-SP Int. Symp. on Time-Frequency and Time-Scale Analysis*, Victoria (BC), Canada, 31–34.

Bonnet, G. (1968). Considérations sur la représentation et l'analyse harmonique des signaux déterministes ou aléatoires, *Ann. Télécom.*, **23**, 62–86.

Bopp, F. (1956). La mécanique quantique est-elle une mécanique statistique particulière? *Ann. Inst. Henri Poincaré*, **XV**, 81–112.

Borchi, E. and Pelosi, G. (1980). A quantum mechanical derivation of the general uncertainty relation for real signals in communication theory, *Signal Proc.*, **2**, 289–292.

Born, M. and Jordan, P. (1925). Zur Quantenmechanik, *Zeit. f. Phys.*, **34**, 858–888.

Bouachache, B. (1982). Représentation temps-fréquence – application à la mesure de l'absorption du sous-sol, Thèse Doct.-Ing., INPG, Grenoble.

Bouachache, B., Escudié, B., and Komatitsch, J. M. (1979). Sur la possibilité d'utiliser la représentation conjointe en temps et fréquence dans l'analyse des signaux modulés emis en vibrosismique, in *7ème Coll. GRETSI*, Nice, pp. 121/1–121/6.

Boudreaux-Bartels, G. F. (1983). Time-frequency signal processing algorithms: analysis and synthesis using Wigner distributions, Ph.D. Dissertation, Rice Univ., Houston (TX).

Boudreaux-Bartels, G. F. (1996). Mixed Time-Frequency Signal Transformations, in *The Transforms and Applications Handbook*, A. D. Poularikas, ed., Boca Raton: CRC Press.

Boudreaux-Bartels, G. F. and Parks, T. W. (1986). Time-varying filtering and signal estimation using Wigner distribution synthesis techniques, *IEEE Trans. Acoust., Speech and Signal Proc.*, **34**, 442–451.

Bouvet, M. (1992). *Traitement des Signaux pour les Systèmes Sonar*, Paris: Masson.

Bracewell, R. N. (1978). *The Fourier Transform and its Applications*, 2nd ed., New York: McGraw-Hill.

Brillouin, L. (1959). *La Science et la Théorie de l'Information*, Paris: Gauthiers-Villars.

Broman, H. (1981). The instantaneous frequency of a Gaussian signal: the one-dimensional density function, *IEEE Trans. Acoust., Speech and Signal Proc.*, **29**, 108–111.

Burt, P. and Adelson, E. (1983). The Laplacian pyramid as a compact image code, *IEEE Trans. Comm.*, **31**, 482–540.

Bibliography

Carson, J. and Fry, T. (1937). Variable frequency electric circuit theory with application to the theory of frequency modulation, *Bell Syst. Tech. J.*, **16**, 513–540.

Chassande-Mottin, E. and Flandrin, P. (1998). On the time-frequency detection of chirps, *Appl. Comp. Harm. Anal.*, **5**.

Chiollaz, M., Flandrin, P., and Gache, N. (1987). Utilisation de la représentation de Wigner-Ville comme outil de diagnostic des défauts de fonctionnement des moteurs thermiques, in *11ème Coll. GRETSI*, Nice, 579–582.

Choï, H. I. and Williams, W. J. (1989). Improved time-frequency representation of multicomponent signals using exponential kernels, *IEEE Trans. Acoust., Speech and Signal Proc.*, **37**, 862–871.

Chui, C. K. (1992). *An Introduction to Wavelets*, New York: Academic Press.

Claasen, T. A. C. M. and Mecklenbräuker, W. F. G. (1980a). The Wigner distribution – a tool for time-frequency signal analysis. Part I: Continuous-Time Signals, *Philips J. Res.*, **35**, 217–250.

Claasen, T. A. C. M. and Mecklenbräuker, W. F. G. (1980b). The Wigner distribution – a tool for time-frequency signal analysis. Part II: Discrete-Time Signals, *Philips J. Res.*, **35**, 276–300.

Claasen, T. A. C. M. and Mecklenbräuker, W. F. G. (1980c). The Wigner distribution – a tool for time-frequency signal analysis. Part III: Relations with Other Time-Frequency Signal Transformations, *Philips J. Res.*, **35**, 372–389.

Claasen, T. A. C. M. and Mecklenbräuker, W. F. G. (1981). Time-frequency analysis by means of the Wigner distribution, in *IEEE Int. Conf. on Acoust., Speech and Signal Proc. ICASSP-81*, Atlanta (GA), 69–72.

Clergeot, H. (1982). Estimation du spectre d'un signal aléatoire Gaussien par le critère du maximum de vraisemblance ou du maximum de probabilité a posteriori, Thèse Doct. Etat ès Sc. Phys., Orsay.

Cohen, A. (1992). *Ondelettes et Traitement Numérique du Signal*, Paris: Masson.

Cohen, L. (1966). Generalized phase-space distribution functions, *J. Math. Phys.*, **7**, 781–786.

Cohen, L. (1970). Hamiltonian operators via Feynman path integrals, *J. Math. Phys.*, **11**, 3296–3297.

Cohen, L. (1984). Distributions in signal theory, in *IEEE Int. Conf. on Acoust., Speech and Signal Proc. ICASSP-84*, San Diego (CA), pp. 41B.1.1–41B.1.4.

Cohen, L. (1989). Time-frequency distributions – a review, *Proc. IEEE*, **77**, 941–981.

Cohen, L. (1992). Scale and inverse frequency representations, in *AFIT/AFOSR Wavelets Workshop*, Dayton (OH), 118–129.

Cohen, L. (1993). The scale representation, *IEEE Trans. Signal Proc.*, **41**, 3275–3292.

Cohen, L. (1995). *Time-Frequency Analysis*, Englewood Cliffs: Prentice-Hall.

Cohen, L. and Lee, C. (1988). Instantaneous quantities, standard deviation and cross-terms, in *Proc. SPIE Conf. Adv. Alg. and Arch. for Signal Proc. III*, **975**, San Diego (CA), 186–208.

Cohen, L. and Posch, T. E. (1985). Positive time-frequency distribution functions, *IEEE Trans. Acoust., Speech and Signal Proc.*, **33**, 31–37.

Cohen, L. and Zaparovanny, Y. I. (1980). Positive quantum joint distributions, *J. Math. Phys.*, **21**, 794–796.

Cohen-Tannoudji, C., Diu, B., and Laloé, F. (1973). *Mécanique Quantique*, Paris: Hermann.

Comon, P. (1991). Independent component analysis, in *Int. Signal Proc. Workshop on Higher Order Statistics*, Chamrousse, 111–120.

Copson, T. E. (1967). *Asymptotic Expansions*, Cambridge: Cambridge Univ. Press.

Cowling, M. G. and Price, J. F. (1984). Bandwidth versus time concentration: the Heisenberg-Pauli-Weyl inequality, *SIAM J. Math. Anal.*, **15**, 151–165.

Cramér, H. (1971). Structural and Statistical Problems for a Class of Stochastic Processes, The First S. S. Wilks Lecture at Princeton University, in *Random Processes: Multiplicity Theory and Canonical Decompositions*, A. Ephremides and J. B. Thomas, eds., Stroudsburg: Dowden, Hutchinson & Ross, Inc., 305–344.

d'Alessandro, C. (1989). Représentation du signal de parole par une somme de fonctions élémentaires, Thèse de Doctorat, Univ. Paris 6.

d'Alessandro, C. and Demars, C. (1992). Représentations temps-fréquence du signal de parole, *Traitement du Signal*, **9**, 153–173.

Daubechies, I. (1988a). Orthonormal bases of compactly supported wavelets, *Comm. in Pure and Appl. Math.*, **41**, 909–996.

Daubechies, I. (1988b). Time-frequency localization operators: a geometric phase space approach, *IEEE Trans. Info. Theory*, **34**, 605–612.

Daubechies, I. (1992). *Ten Lectures on Wavelets*, Philadelphia: SIAM.

Daubechies, I., Grossmann, A., and Meyer, Y. (1985). Painless non-orthogonal expansions, *J. Math. Phys.*, **27**, 1271–1283.

Daubechies, I., Jaffard, S., and Journé, J. L. (1991). A simple Wilson orthonormal basis with exponential decay, *SIAM J. Math. Anal.*, **22**, 554–573, erratum p. 878.

Daubechies, I. and Paul, T. (1988). Time-frequency localization operators – a geometric phase space approach: II. The use of dilations, *Inverse Problems*, **4**, 661–680.

De Bruijn, N. G. (1967). Uncertainty Principles in Fourier Analysis, in *Inequalities*, O. Shisha, ed., New York: Academic Press, 57–71.

Donoho, D. L. and Stark, P. B. (1989). Uncertainty principles and signal recovery, *SIAM J. Appl. Math.*, **49**, 906–931.

Dorize, C. and Villemoes, L. F. (1991). Optimizing time-frequency resolution of orthonormal wavelets, in *IEEE Int. Conf. on Acoust., Speech and Signal Proc. ICASSP-91*, Toronto, 2029–2032.

Duvaut, P. (1991). *Traitement du Signal – Concepts et Applications*, Paris: Hermès.

Duvaut, P. and Jorand, C. (1991). Analyse en composantes indépendantes d'un noyau multilinéaire. Application à une nouvelle représentation temps-fréquence de signaux non Gaussiens, in *13ème Coll. GRETSI*, Juan-les-Pins, 61–64.

Dym, H. and McKean, H. P. (1972). *Fourier Series and Integrals*, New York: Academic Press.

Escudié, B. and Flandrin, P. (1980). Sur quelques propriétés de la représentation conjointe et de la fonction d'ambiguïté des signaux d'énergie finie, *C. R. Acad. Sc. Paris, Série A*, **291**, 171–174.

Escudié, B. and Gréa, J. (1976). Sur une formulation générale de la représentation en temps et fréquence dans l'analyse des signaux d'énergie finie, *C. R. Acad. Sc. Paris, Série A*, **283**, 1049–1051.

Escudié, B. and Gréa, J. (1977). Représentation hilbertienne et représentation conjointe en temps et fréquence des signaux d'énergie finie, in *6ème Coll. GRETSI*, Nice, pp. 5/1–5/6.

Esteban, D. and Galand, C. (1977). Application of quadrature mirror filters to split-band voice coding schemes, in *IEEE Int. Conf. on Acoust., Speech and Signal Proc. ICASSP-77*, Hartford (CT), 191–195.

Falconer, K. (1990). *Fractal Geometry – Mathematical Foundations and Applications*, New York: J. Wiley & Sons.

Feder, J. (1988). *Fractals*, New York: Plenum Press.

Feichtinger, H. G. and Strohmer, T. (eds.). (1998). *Gabor Analysis and Algorithms: Theory and Application*, Boston: Birkhäuser.

Flanagan, J. (1972). *Speech Analysis, Synthesis and Perception*, New York: Springer-Verlag.

Flandrin, P. (1982). Représentations des signaux dans le plan temps-fréquence, Thèse Doct.-Ing., INPG, Grenoble.

Flandrin, P. (1984). Some features of time-frequency representations of multi-component signals, in *IEEE Int. Conf. on Acoust., Speech and Signal Proc. ICASSP-84*, San Diego (CA), pp. 41.B.4.1–41B.4.4.

Flandrin, P. (1986a). On the positivity of the Wigner-Ville spectrum, *Signal Proc.*, **11**, 187–189.

Flandrin, P. (1986b). On detection-estimation procedures in the time-frequency plane, in *IEEE Int. Conf. on Acoust., Speech and Signal Proc. ICASSP-86*, Tokyo, 2331–2334.

Flandrin, P. (1986c). Time-frequency interpretation of matched filtering, in *IEEE Academia Sinica Workshop on Acoust., Speech and Signal Proc. WASSP-86*, Beijing, 287–290.

Flandrin, P. (1987). Représentations temps-fréquence des signaux non-stationnaires, Thèse Doct. Etat ès Sc. Phys., Grenoble.

Flandrin, P. (1988a). Maximum signal energy concentration in the time-frequency plane, in *IEEE Int. Conf. on Acoust., Speech and Signal Proc. ICASSP-88*, New York (NY), 2176–2179.

Flandrin, P. (1988b). A time-frequency formulation of optimum detection, *IEEE Trans. Acoust., Speech and Signal Proc.*, **36**, 1377–1384.

Flandrin, P. (1988c). Time-Frequency Processing of Bat Sonar Signals, in *Animal Sonar – Processes and Performance*, P. E. Nachtigall and P. W. B. Moore, eds., New York: Plenum Press, 797–802.

Flandrin, P. (1989a). Time-Dependent Spectra for Nonstationary Stochastic Processes, in *Time and Frequency Representation of Signals and Systems. CISM Courses and Lectures* No. 309, G. Longo and B. Picinbono, eds., Wien: Springer-Verlag, 69–124.

Flandrin, P. (1989b). On the spectrum of fractional Brownian motions, *IEEE Trans. Info. Theory*, **35**, 197–199.

Flandrin, P. (1989c). Signal detection in the time-frequency plane, in *IFAC Workshop on Advanced Information Processing in Automatic Control AIPAC-89*, Nancy, 131–134.

Flandrin, P. (1990). Scale-Invariant Wigner Spectra and Self-Similarity, in *Signal Processing V – Theories and Applications*, L. Torres et al., eds., Amsterdam: North-Holland, 149–152.

Flandrin, P. (1991). Sur une classe générale d'extensions affines de la distribution de Wigner-Ville, in *13ème Coll. GRETSI*, Juan-les-Pins, 17–20.

Flandrin, P. (1992). Wavelet analysis and synthesis of fractional Brownian motion, *IEEE Trans. Info. Theory*, **38**, 910–917.

Flandrin, P. (1994). Wavelets and Self-Similar Processes, in *Wavelets and their Applications*, NATO ASI Series C, **442**, J. S. Byrnes et al., eds., Dordrecht: Kluwer, 121–142.

Flandrin, P. and Escudié, B. (1980). Time and frequency representation of finite energy signals: a physical property as a result of a Hilbertian condition, *Signal Proc.*, **2**, 93–100.

Flandrin, P. and Escudié, B. (1981a). Sur une relation intégrale entre les fonctions d'ambiguïté en translation et en compression, Rapp. Int. ICPI TS-8110, Lyon.

Flandrin, P. and Escudié, B. (1981b). Géométrie des fonctions d'ambiguïté et des représentations conjointes de Ville: l'approche de la théorie des catastrophes, in *8ème Coll. GRETSI*, Nice, 69–74.

Flandrin, P. and Escudié, B. (1982). Sur la localisation des représentations conjointes dans le plan temps-fréquence, *C. R. Acad. Sc. Paris, Série I*, **295**, 475–478.

Flandrin, P. and Escudié, B. (1984). An interpretation of the pseudo-Wigner-Ville distribution, *Signal Proc.*, **6**, 27–36.

Flandrin, P. and Gonçalvès, P. (1994). From Wavelets to Time-Scale Energy Distributions, in *Recent Advances in Wavelet Analysis*, L. L. Schumaker and G. Webb, eds., San Diego: Academic Press, 309–334.

Flandrin, P. and Gonçalvès, P. (1996). Geometry of affine time-frequency distributions, *Appl. Comp. Harm. Anal.*, **3**, 10–39.

Flandrin, P. and Hlawatsch, F. (1987). Signal Representation Geometry and Catastrophes in the Time-Frequency Plane, in *Mathematics in Signal Processing*, T. S. Durrani et al., eds., Oxford: Clarendon Press, 3–14.

Flandrin, P. and Martin, W. (1983a). Sur les conditions physiques assurant l'unicité de la représentation de Wigner-Ville comme représentation temps-fréquence, in *9ème Coll. GRETSI*, Nice, 43–49.

Flandrin, P. and Martin, W. (1983b). Pseudo-Wigner estimators for the analysis of non-stationary processes, in *IEEE ASSP Spectrum Est. Workshop II*, Tampa (FL), 181–185.

Flandrin, P. and Martin, W. (1984). A General Class of Estimators for the Wigner-Ville Spectrum of Non-stationary Processes, in *Analysis and Optimization of Systems. Lecture Notes in Control and Information Sciences*, **62**, A. Bensoussan and J. L. Lions, eds., Berlin: Springer-Verlag, 15–23.

Flandrin, P. and Martin, W. (1998). The Wigner-Ville Spectrum of Nonstationary Random Signals, in *The Wigner Distribution – Theory and Applications in Signal Processing*, W. F. G. Mecklenbräuker and F. Hlawatsch, eds., Amsterdam: Elsevier, 212–267.

Flandrin, P. and Rioul, O. (1990). Affine smoothing of the Wigner-Ville distribution, in *IEEE Int. Conf. on Acoust., Speech and Signal Proc. ICASSP-90*, Albuquerque (NM), 2455–2458.

Flandrin, P., Sageloli, J., Sessarego, J. P., and Zakharia, M. (1986). Application of time-frequency analysis to the characterization of surface waves on elastic targets, *Acoust. Lett.*, **10**, 23–28.

Flandrin, P., Weber, P., and Zakharia, M. (1989). Analyse temps-fréquence de réponses impulsionnelles de cibles, Rapport Interne ICPI TS-8912, Lyon.

Flaska, M. D. (1976). Cross-correlation of short-time spectral histories, *J. Acoust. Soc. Amer.*, **59**, 381–388.

Folland, G. B. (1989). *Harmonic Analysis in Phase Space*, Ann. of Math. Studies No. 122, Princeton: Princeton Univ. Press.

Folland, G. B. and Sitaram, A. (1997). The uncertainty principle: a mathematical survey, *J. Fourier Anal. Appl.*, **3**, 207–238.

Fowler, M. L. and Sibul, L. H. (1991). A unified formulation for detection using time-frequency and time-scale methods, in *Asilomar Conf. on Sig., Syst. and Comp.*, Pacific Grove (CA), 637–642.

Friedlander, B. and Porat, B. (1989). Detection of transient signals by the Gabor representation, *IEEE Trans. Acoust., Speech and Signal Proc.*, **37**, 169–180.

Fuchs, W. H. J. (1954). On the magnitude of Fourier transforms, in *Proc. Int. Congress of Math.*, Amsterdam, 106–107.

Gabor, D. (1946). Theory of communication, *J. IEE*, **93**, 429–457.

Gambardella, G. (1968). Time-scaling and short-time spectral analysis, *J. Acoust. Soc. Amer.*, **44**, 1745–1747.

Gardner, W. A. (1988). *Statistical Spectral Analysis – A Non-Probabilistic Theory*, Englewood Cliffs: Prentice-Hall.

Gasquet, C. and Witomski, P. (1990). *Analyse de Fourier et Applications – Filtrage, Calcul Numérique et Ondelettes*, Paris: Masson.

Gendrin, R. and de Villedary, C. (1979). Unambiguous determination of fine structures in multicomponent time-varying signals, *Ann. Telecom.*, **35**, 122–130.

Gendrin, R. and Robert, P. (1982). Temps de groupe et largeur de bande de signaux modulés simultanément en amplitude et en fréquence, *Ann. Télécomm.*, **37**, 289–297.

Bibliography

Gibiat, V., Wu, F., Perio, P., and Chaintreuil, S. (1982). Analyse spectrale différentielle (A.S.D.), *C. R. Acad. Sc. Paris, Série II*, **294**, 633–636.

Gonçalvès, P. and Flandrin, P. (1992). Scaling exponent estimation from time-scale energy distributions, in *IEEE Int. Conf. on Acoust., Speech and Signal Proc. ICASSP-92*, San Francisco (CA), pp. V.157–V.160.

Grace, O. D. (1981). Instantaneous power spectra, *J. Acoust. Soc. Amer.*, **69**, 191–198.

Grenier, Y. (1983). Time-dependent ARMA modelling of non-stationary signals, *IEEE Trans. Acoust., Speech and Signal Proc.*, **31**, 899–911.

Grenier, Y. (1984). Modélisation de signaux non-stationnaires, Thèse Doct. Etat ès Sc. Phys., Univ. Paris-Sud, Orsay.

Grenier, Y. (1986). Modèles ARMA à coefficients dépendants du temps: estimateurs et applications, *Traitement du Signal*, **3**, 219–233.

Grenier, Y. (1987). Parametric Time-Frequency Representations, in *Traitement du Signal / Signal Processing*, J. L. Lacoume et al., eds., Les Houches, Session XLV, Amsterdam: North-Holland, 339–397.

Grossmann, A. (1976). Parity operator and quantization of δ-functions, *Commun. Math. Phys.*, **48**, 191–194.

Grossmann, A. and Escudié, B. (1991). Une représentation bilinéaire en temps et echelle des signaux d'énergie finie, in *13ème Coll. GRETSI*, Juan-les-Pins, 33–36.

Grossmann, A. and Morlet, J. (1984). Decomposition of Hardy functions into square integrable wavelets of constant shape, *SIAM J. Math. Anal.*, **15**, 723–736.

Grossmann, A., Morlet, J., and Paul, T. (1985). Transforms associated to square integrable group representations I: general results, *J. Math. Phys.*, **26**, 2473–2479.

Grossmann, A., Morlet, J., and Paul, T. (1986). Transforms associated to square integrable group representations II: examples, *Ann. Inst. Henri Poincaré*, **45**, 293–309.

Hammond, J. K. and Harrison, R. F. (1985). Wigner-Ville and evolutionary spectra for covariance equivalent non-stationary random processes, in *IEEE Int. Conf. on Acoust., Speech and Signal Proc. ICASSP-85*, Tampa (FL), 1025–1027.

Heil, C. E. and Walnut, D. F. (1989). Continuous and discrete wavelet transforms, *SIAM Rev.*, **31**, 628–666.

Helström, C. W. (1966). An expansion of a signal in Gaussian elementary signals, *IEEE Trans. Info. Theory*, **12**, 81–82.

Hippenstiel, R. D. and de Oliveira, P. M. (1990). Time-varying spectral estimation using the instantaneous power spectrum (IPS), *IEEE Trans. Acoust., Speech and Signal Proc.*, **38**, 1752–1759.

Hlawatsch, F. (1984). Interference terms in the Wigner distribution, in *Int. Conf. on Digital Signal Proc.*, Florence, 363–367.

Hlawatsch, F. (1985). Transformation, inversion and conversion of bilinear signal representations, in *IEEE Int. Conf. on Acoust., Speech and Signal Proc. ICASSP-85*, Tampa (FL), 1029–1032.

Hlawatsch, F. (1986). Unitary Time-Frequency Signal Representations, in *Signal Processing III – Theories and Applications*, I. T. Young et al., eds., Amsterdam: North-Holland, 33–36.

Hlawatsch, F. (1988). A study of bilinear time-frequency signal representations with applications to time-frequency signal synthesis, Doctoral Thesis, TU Wien, Vienna.

Hlawatsch, F. (1991). Duality and classification of bilinear time-frequency signal representations, *IEEE Trans. Signal Proc.*, **39**, 1564–1574.

Hlawatsch, F. (1992). Regularity and unitarity of bilinear time-frequency signal representations, *IEEE Trans. Info. Theory*, **38**, 82–94.

Hlawatsch, F. (1998). *Time-Frequency Analysis and Synthesis of Linear Signal Spaces: Time-Frequency Filters, Signal Detection and Estimation, and Range-Doppler Estimation*, Boston: Kluwer.

Hlawatsch, F. and Boudreaux-Bartels, G. F. (1992). Linear and quadratic time-frequency signal representations, *IEEE Signal Proc. Magazine*, 21–67.

Hlawatsch, F. and Flandrin, P. (1998). The Interference Structure of the Wigner Distribution and Related Time-Frequency Signal Representations, in *The Wigner Distribution – Theory and Applications in Signal Processing*, W. F. G. Mecklenbräuker and F. Hlawatsch, eds., Amsterdam: Elsevier, 59–133.

Hlawatsch, F. and Krattenthaler, W. (1986). Signal Synthesis from Unitary Time-Frequency Signal Representations, in *Signal Processing III – Theories and Applications*, I. T. Young et al., eds., Amsterdam: North-Holland, 37–40.

Hlawatsch, F. and Krattenthaler, W. (1992). Bilinear signal synthesis, *IEEE Trans. Signal Proc.*, **40**, 352–363.

Hlawatsch, F. and Krattenthaler, W. (1998). Signal Synthesis Algorithms for Bilinear Time-Frequency Signal Representations, in *The Wigner Distribution – Theory and Applications in Signal Processing*, W. F. G. Mecklenbräuker and F. Hlawatsch, eds., Amsterdam: Elsevier, 135–209.

Hlawatsch, F., Manickam, T. G., Urbanke, R. L., and Jones, W. (1995). Smoothed pseudo-Wigner distribution, Choï-Williams distribution, and cone-kernel representation: ambiguity-domain analysis and experimental comparison, *Signal Proc.*, **43**, 149–168.

Holschneider, M. and Tchamitchian, Ph. (1989). Régularité Locale de la Fonction 'Non Différentiable' de Riemann, in *Les Ondelettes en 1989. Lecture Notes in Mathematics*, **1438**, P. G. Lemarié, ed., Berlin: Springer-Verlag, 102–124.

Hudson, R. L. (1974). When is the Wigner quasi-probability density non negative? *Rep. Math. Phys.*, **6**, 249–252.

Hurd, H. L. (1969). An investigation of periodically correlated stochastic processes, Ph.D. Dissertation, Duke Univ., Durham (NC).

Ivanovic, I. D. (1983). Note on a new interpretation of the scalar product in Hilbert space, *Phys. Lett.*, **99A**, 161–162.

Jacobson, L. and Wechsler, H. (1983). The composite pseudo Wigner distribution (CPWD): a computable and versatile approximation to the Wigner distribution (WD), in *IEEE Int. Conf. on Acoust., Speech and Signal Proc. ICASSP-83*, Boston (MA), 254–256.

Jaffard, S. (1989). Exposants de Hölder en des points donnés et coefficiants d'ondelettes, *C. R. Acad. Sc. Paris, Serie I*, **308**, 79–81.

Janssen, A. J. E. M. (1981). Positivity of weighted Wigner distributions, *SIAM J. Math. Anal.*, **12**, 752–758.

Janssen, A. J. E. M. (1982). On the locus and spread of pseudo-density functions in the time-frequency plane, *Philips J. Res.*, **37**, 79–110.

Janssen, A. J. E. M. (1984a). A note on Hudson's Theorem about functions with non-negative Wigner distributions, *SIAM J. Math. Anal.*, **15**, 170–176.

Janssen, A. J. E. M. (1984b). Positivity properties of phase-plane distribution functions, *J. Math. Phys.*, **25**, 2240–2252.

Janssen, A. J. E. M. (1985). Bilinear phase-plane distribution functions and positivity, *J. Math. Phys.*, **26**, 1986–1994.

Janssen, A. J. E. M. (1987). A note on 'positive time-frequency distributions', *IEEE Trans. Acoust., Speech and Signal Proc.*, **35**, 701–703.

Janssen, A. J. E. M. (1988a). Positivity of time-frequency distribution functions, *Signal Proc.*, **14**, 243–252.

Janssen, A. J. E. M. (1988b). The Zak transform: a signal transform for sampled time-continuous signals, *Philips J. Res.*, **43**, 23–69.

Janssen, A. J. E. M. (1991). Optimality property of the Gaussian window spectrogram, *IEEE Trans. Signal Proc.*, **39**, 202–204.

Janssen, A. J. E. M. (1998). Positivity and Spread of Bilinear Time-Frequency Distributions, in *The Wigner Distribution – Theory and Applications in Signal Processing*, W. F. G. Mecklenbräuker and F. Hlawatsch, eds., Amsterdam: Elsevier, 1–58.

Janssen, A. J. E. M. and Claasen, T. A. C. M. (1985). On positivity of time-frequency distributions, *IEEE Trans. Acoust., Speech and Signal Proc.*, **33**, 1029–1032.

Janussis, A. D., Leodaris, A., Patargias, N., Philippakis, T., Philippakis, P., and Vlachos, K. (1982). Applications of the positive-definite Wigner distribution function, *Lett. Nuov. Cim.*, **34**, 433–437.

Jensen, H. E., Hoholdt, T., and Justesen, J. (1988). Double series representation of bounded signals, *IEEE Trans. Info. Theory*, **34**, 613–624.

Johannesma, P., Aertsen, A., Cranen, B., and van Erning, L. (1981). The phonochrome: a coherent spectro-temporal representation of sound, *Hearing Res.*, **5**, 123–145.

Jones, D. L. and Parks, T. W. (1990). A high resolution data-adaptive time-frequency representation, *IEEE Trans. Acoust., Speech and Signal Proc.*, **38**, 2127–2135.

Kadambe, S. and Boudreaux-Bartels, G. F. (1992). A comparison of the existence of 'cross-terms' in the Wigner distribution and the squared magnitude of the wavelet transform and the short-time Fourier transform, *IEEE Trans. Signal Proc.*, **39**, 2498–2517.

Kay, I. and Silverman, R. A. (1957). On the uncertainty relation for real signals, *Inform. Control*, **1**, 64–75.

Kay, S. M. (1987). *Modern Spectral Estimation*, Englewood Cliffs: Prentice-Hall.

Kay, S. M. and Boudreaux-Bartels, G. F. (1985). On the optimality of the Wigner distribution for detection, in *IEEE Int. Conf. on Acoust., Speech and Signal Proc. ICASSP-85*, Tampa (FL), 1017–1020.

Kelly, E. and Wishner, R. P. (1965). Matched-filter theory for high-velocity, accelerating targets, *IEEE Trans. Mil. Electr.*, **9**, 56–69.

Keshner, M. S. (1982). $1/f$ noise, *Proc. IEEE*, **70**, 212–218.

Kodera, K., de Villedary, C., and Gendrin, R. (1976). A new method for the numerical analysis of non-stationary signals, *Phys. Earth and Plan. Int.*, **12**, 142–150.

Kodera, K., Gendrin, R., and de Villedary, C. (1978). Analysis of time-varying signals with small BT values, *IEEE Trans. Acoust., Speech and Signal Proc.*, **26**, 64–76.

Koenig, R., Dunn, H. K., and Lacy, L. Y. (1946). The sound spectrograph, *J. Acoust. Soc. Amer.*, **18**, 19–49.

Körner, T. W. (1988). *Fourier Analysis*, Cambridge: Cambridge Univ. Press.

Kozek, W. (1992). Time-frequency signal processing based on the Wigner-Weyl framework, *Signal Proc.*, **29**, 77–92.

Kronland-Martinet, R., Morlet, J., and Grossmann, A. (1987). Analysis of sound patterns through wavelet transforms, *Int. J. Pattern Recogn. and Artif. Intell.*, **1**, 273–301.

Krüger, J. G. and Poffyn, A. (1976). Quantum mechanics in phase space, I. Unicity of the Wigner distribution function, *Physica*, **85A**, 84–100.

Kumar, B. V. K. V. and Carroll, C. W. (1984). Performance of Wigner distribution function based detection methods, *Opt. Eng.*, **23**, 732–737.

Kunt, M. (1984). *Traitement Numérique des Signaux*, Paris: Dunod.

Kuryshkin, V. V. (1972). La mécanique quantique avec une fonction non-négative de distribution dans l'espace des phases, *Ann. Inst. Henri Poincaré*, **XVII**, 81–95.

Kuryshkin, V. V. (1973). Some problems of quantum mechanics possessing a non-negative phase-space distribution function, *Int. J. Theor. Phys.*, **7**, 451–466.

Lacoume, J. L. and Kofman, W. (1975). Etude des signaux non stationnaires par la représentation en temps et en fréquence, *Ann. Télécomm.*, **30**, 231–238.

Lamel, L. (1988). Formalizing knowledge used in spectrogram reading: acoustical and perceptual evidence from stops, Ph.D. Dissertation, MIT, Cambridge (MA).

Landau, H. J. and Pollak, H. O. (1961). Prolate spheroidal wave functions, Fourier analysis and uncertainty – II, *Bell Syst. Tech. J.*, **40**, 65–84.

Le Chevalier, F. (1989). *Principes de Traitement des Signaux Radar et Sonar*, Paris: Masson.

Lemarié, P. G. (ed.). (1989). *Les Ondelettes en 1989. Lecture Notes in Mathematics*, **1438**, Berlin: Springer-Verlag.

Levin, M. J. (1967). Instantaneous spectra and ambiguity functions, *IEEE Trans. Info. Theory*, **13**, 95–97.

Levshin, A., Pisarenko, V. F., and Pogrebinsky, G. A. (1972). On a frequency-time analysis of oscillations, *Ann. Geophys.*, **28**, 211–218.

Lévy-Leblond, J. M. (1973). Les inégalités de Heisenberg, *J. Phys.*, **14**, Suppl. (educational insert), 15–22.

Lieb, E. H. (1990). Integral bounds for radar ambiguity functions and Wigner distributions, *J. Math. Phys.*, **31**, 594–599.

Loève, M. (1962). *Probability Theory*, Amsterdam: D. van Nostrand Co.

Longo, G. and Picinbono, B. (eds.). (1989). *Time and Frequency Representation of Signals and Systems*. CISM Courses and Lectures No. 309, Wien: Springer-Verlag.

Loughlin, P. (ed.). (1996). Special Issue on Time-Frequency Analysis, *Proc. IEEE*, **84**(9).

Loughlin, P., Pitton, J., and Atlas, L. (1994). Construction of positive time-frequency distributions, *IEEE Trans. Signal Proc.*, **42**, 2697–2705.

Loynes, R. M. (1968). On the concept of the spectrum for nonstationary processes, *J. Roy. Stat. Soc. B*, **30**, 1–30.

Mallat, S. G. (1989a). A theory for multiresolution signal decomposition: the wavelet representation, *IEEE Trans. Pattern Anal. and Machine Intell.*, **11**, 674–693.

Mallat, S. G. (1989b). Multiresolution approximations and wavelet orthonormal bases of $L^2(\mathbb{R})$, *Trans. Amer. Math. Soc.*, **315**, 69–88.

Mallat, S. G. (1998). *A Wavelet Tour of Signal Processing*, San Diego: Academic Press.

Mallat, S. G. and Hwang, W. L. (1992). Singularity detection and processing with wavelets, *IEEE Trans. Info. Theory*, **38**, 617–643.

Mallat, S. G. and Zhang, Z. (1993). Matching pursuits with time-frequency dictionaries, *IEEE Trans. Signal Proc.*, **41**, 3397–3415.

Mamode, M. (1981). Estimation optimale de la date d'arrivée d'un echo sonar perturbée par l'effet Doppler: synthèse de signaux 'large bande' tolérants, Thèse Doct.-Ing., INPG, Grenoble.

Mandelbrot, B. and Van Ness, J. W. (1968). Fractional Brownian motions, fractional noises and applications, *SIAM Rev.*, **10**, 422–437.

Margenau, H. and Hill, R. W. (1961). Correlation between measurements in quantum theory, *Prog. Theor. Phys.*, **26**, 722–738.

Marinovic, N. N. (1986). The Wigner distribution and the ambiguity function: generalizations, enhancement, compression and some applications, Ph.D. Dissertation, The City University of New York, New York.

Mark, W. D. (1970). Spectral analysis of the convolution and filtering of non-stationary stochastic processes, *J. Sound Vib.*, **11**, 19–63.

Marple, S. L. (1987). *Digital Spectral Analysis*, Englewood Cliffs: Prentice-Hall.

Martin, W. (1982). Time-frequency analysis of random signals, in *IEEE Int. Conf. on Acoust., Speech and Signal Proc. ICASSP-82*, Paris, 1325–1328.

Martin, W. (1984). Measuring the degree of non-stationarity by using the Wigner-Ville spectrum, in *IEEE Int. Conf. on Acoust., Speech and Signal Proc. ICASSP-84*, San Diego (CA), pp. 41B.3.1–41B.3.4.

Martin, W. and Flandrin, P. (1983). Analysis of Non-stationary Processes: Short-Time Periodograms Versus a Pseudo-Wigner Estimator, in *Signal Processing II – Theories and Applications*, H. Schüssler, ed., Amsterdam: North-Holland, 455–458.

Martin, W. and Flandrin, P. (1985a). Detection of changes of signal structure by using the Wigner-Ville spectrum, *Signal Proc.*, **8**, 215–233.

Martin, W. and Flandrin, P. (1985b). Wigner-Ville spectral analysis of nonstationary processes, *IEEE Trans. Acoust., Speech and Signal Proc.*, **33**, 1461–1470.

Masry, E. (1993). The wavelet transform of stochastic processes with stationary increments and its application to fractional Brownian motion, *IEEE Trans. Info. Theory*, **39**, 260–264.

Matz, G. and Hlawatsch, F. (1996). Time-frequency formulation and design of optimal detectors, in *IEEE-SP Int. Symp. on Time-Frequency and Time-Scale Analysis*, Paris, 213–216.

Matz, G., Hlawatsch, F., and Kozek, W. (1997). Generalized evolutionary spectral analysis and the Weyl spectrum of nonstationary random processes, *IEEE Trans. Signal Proc.*, **45**, 1520–1534.

Mecklenbräuker, W. F. G. (1987). A Tutorial on Non-parametric Bilinear Time-Frequency Signal Representations, in *Traitement du Signal / Signal Processing*, J. L. Lacoume et al., eds., Les Houches, Session XLV, Amsterdam: North-Holland, 277–336.

Mecklenbräuker, W. F. G. and Hlawatsch, F. (eds.). (1998). *The Wigner Distribution – Theory and Applications in Signal Processing*, Amsterdam: Elsevier.

Mélard, G. (1978). Propriétés du spectre évolutif d'un processus non-stationnaire, *Ann. Inst. Henri Poincaré B*, **XIV**, 411–424.

Meyer, Y. (1990). *Ondelettes et Opérateurs I. Ondelettes*, Paris: Hermann.

Meyer, Y. (1993a). *Wavelets: Algorithms and Applications*, Transl. by R. D. Ryan, Philadelphia: SIAM.

Meyer, Y. (1993b). *Ondelettes et Algorithmes Concurrents*, Paris: Hermann.

Michaut, F. (1992). *Méthodes Adaptatives pour le Signal*, Paris: Hermès.

Montgomery, L. K. and Reed, I. S. (1967). A generalization of the Gabor-Helström transform, *IEEE Trans. Info. Theory*, **13**, 344–345.

Morlet, J. (1982). Sampling Theory and Wave Propagation, in *NATO ASI Series, Issues in Acoustic Signal/Image Processing and Recognition*, **1**, C. H. Chen, ed., Berlin: Springer-Verlag, 233–261.

Mourgues, G. (1987). Utilisation du formalisme de Wigner en mécanique statistique classique et en traitement des signaux certains, Thèse Doct. Etat ès Sc. Phys., Univ. Clermont II.

Moyal, J. E. (1949). Quantum mechanics as a statistical theory, *Proc. Cambridge Phil. Soc.*, **45**, 99–124.

Mugur-Schächter, M. (1977). The Quantum Mechanical One-system Formalism, Joint Probabilities and Locality, in *Quantum Mechanics, a Half Century Later*, J. Leite Lopes and M. Paty, eds., Dordrecht: D. Reidel, 107–146.

Nawab, S. H. and Quatieri, T. F. (1988). Short-Time Fourier Transform, in *Advanced Topics in Signal Processing*, J. S. Lim and A. V. Oppenheim, eds., Englewood Cliffs: Prentice-Hall, 289–337.

Nikias, C. L. and Petropulu, A. P. (1993). *Higher-Order Spectra Analysis – A Nonlinear Signal Processing Framework*, Englewood Cliffs: Prentice-Hall.

O'Connell, R. F. and Wigner, E. P. (1981). Quantum-mechanical distribution functions: conditions for uniqueness, *Phys. Lett.*, **83A**, 145–148.

O'Neill, J. C. (1997). Shift-covariant time-frequency distributions of discrete signals, Ph.D. Dissertation, Univ. of Michigan.

Ovarlez, J. Ph. (1992). La transformation de Mellin: un outil pour l'analyse des signaux à large bande, Thèse, Univ. Paris 6.

Page, C. H. (1952). Instantaneous power spectra, *J. Appl. Phys.*, **23**, 103–106.

Papandreou, A. and Boudreaux-Bartels, G. F. (1992). Distributions for time-frequency analysis: a generalization of Choi-Williams and the Butterworth distribution, in *IEEE Int. Conf. on Acoust., Speech and Signal Proc. ICASSP-92*, San Francisco (CA), pp. V.181–V.184.

Papandreou, A., Hlawatsch, F., and Boudreaux-Bartels, G. F. (1992). A unified framework for the Bertrand distribution and the Altes distribution: the new hyperbolic class of quadratic time-frequency distributions, in *IEEE-SP Int. Symp. on Time-Frequency and Time-Scale Analysis*, Victoria (BC), Canada, 27–30.

Papandreou, A., Hlawatsch, F., and Boudreaux-Bartels, G. F. (1993). The hyperbolic class of quadratic time-frequency representations – Part I: constant-Q warping, the hyperbolic paradigm, properties, and members, *IEEE Trans. Signal Proc.*, **41**, 3425–3444.

Papoulis, A. (1977). *Signal Analysis*, New York: McGraw-Hill.

Paul, T. (1985). Ondelettes et mécanique quantique, Thèse Doct. Etat ès Sc. Phys., Univ. Aix-Marseille II.

Pearl, J. (1973). Time, frequency, sequency, and their uncertainty relations, *IEEE Trans. Info. Theory*, **19**, 225–229.

Peyrin, F. and Prost, R. (1986). A unified definition for the discrete-time, discrete-frequency and discrete-time/frequency Wigner distributions, *IEEE Trans. Acoust., Speech and Signal Proc.*, **34**, 858–867.

Picinbono, B. (1977). *Eléments de Théorie du Signal*, Paris: Dunod.

Picinbono, B. (1997). On instantaneous amplitude and phase of signals, *IEEE Trans. Signal Proc.*, **45**, 552–560.

Picinbono, B. and Martin, W. (1983). Représentation des signaux par amplitude et phase instantanées, *Ann. Télécomm.*, **38**, 179–190.

Pimonow, L. (1962). *Vibrations en Régime Transitoire*, Paris: Dunod.

Poletti, M. A. (1993). The development of instantaneous bandwidth via local signal expansion, *Signal Proc.*, **31**, 273–281.

Portnoff, M. R. (1981). Time-frequency representations of digital signals and systems based on short-time Fourier analysis, *IEEE Trans. Acoust., Speech and Signal Proc.*, **28**, 55–69.

Poston, T. and Stewart, I. (1978). *Catastrophe Theory and its Applications*, London: Pitman.

Potter, R. K., Kopp, G. A., and Green, H. C. (1947). *Visible Speech*, New York: D. van Nostrand.

Price, R. and Hofstetter, E. M. (1965). Bounds on the volume and height distribution of the ambiguity function, *IEEE Trans. Info. Theory*, **11**, 207–214.

Priestley, M. B. (1965). Evolutionary spectra and non-stationary processes, *J. Roy. Stat. Soc. B*, **27**, 204–237.

Priestley, M. B. (1981). *Spectral Analysis and Time Series*, New York: Academic Press.

Priestley, M. B. (1988). *Non-Linear and Non-Stationary Time Series Analysis*, New York: Academic Press.

Qian, S. and Morris, J. M. (1992). Wigner distribution decomposition and cross-terms deleted representation, *Signal Proc.*, **27**, 125–144.

Ramanathan, J. and Topiwala, P. (1992). Time-frequency localization operators, in *IEEE-SP Int. Symp. on Time-Frequency and Time-Scale Analysis*, Victoria (BC), Canada, 155–158.

Rényi, A. (1961). On measures of entropy and information, in *Proc. 4th Berkeley Symp. on Math. Stat. and Proba.*, **1**, 547–561.

Riesz, F. and Sz.-Nagy, B. (1955). *Functional Analysis*, New York: Ungar.

Rihaczek, A. W. (1968). Signal energy distribution in time and frequency, *IEEE Trans. Info. Theory*, **14**, 369–374.

Rihaczek, A. W. (1969). *Principles of High-Resolution Radar*, New York: McGraw-Hill.

Riley, M. D. (1989). *Speech Time-Frequency Representations*, Dordrecht: Kluwer.

Rioul, O. and Flandrin, P. (1992). Time-scale energy distributions: a general class extending wavelet transforms, *IEEE Trans. Signal Proc.*, **40**, 1746–1757.

Royer, A. (1977). Wigner function as expectation value of a parity operator, *Phys. Rev. A*, **15**, 449–450.

Ruggeri, G. J. (1971). On phase-space description of quantum mechanics, *Prog. Theor. Phys.*, **46**, 1703–1712.

Sayeed, A. M. and Jones, D. L. (1995). Optimal detection using generalized bilinear time-frequency and time-scale representations, *IEEE Trans. Signal Proc.*, **43**, 2872–2883.

Schrœder, M. R. and Atal, B. S. (1962). Generalized short-time power spectra and auto-correlation functions, *J. Acoust. Soc. Amer.*, **34**, 1679–1683.

Sessarego, J. P., Sageloli, J., Degoul, P., Flandrin, P., and Zakharia, M. (1990). Analyse temps-fréquence de signaux en milieux dispersifs – application à l'étude des ondes de lamb, *J. Acoust.*, **3**, 273–280.

Shenoy, R. G. (1991). Group representations and optimal recovery in signal modeling, Ph.D. Dissertation, Cornell Univ., Ithaca (NY).

Silverman, R. A. (1957). Locally stationary random processes, *IRE Trans. Info. Theory*, **3**, 182–187.

Slepian, D. and Pollak, H. O. (1961). Prolate spheroidal wave functions, Fourier analysis and uncertainty – I, *Bell Syst. Tech. J.*, **40**, 43–64.

Smith, M. J. T. and Barnwell, T. P. (1986). Exact reconstruction techniques for tree-structured subband coders, *IEEE Trans. Acoust., Speech and Signal Proc.*, **34**, 434–441.

Speiser, J. M. (1967). Wide-band ambiguity functions, *IEEE Trans. Info. Theory*, **13**, 122–123.

Springborg, M. (1983). Phase space functions and correspondence rules, *J. Phys. A*, **16**, 535–542.

Srinivas, M. D. and Wolf, E. (1975). Some nonclassical features of phase-space representations of quantum mechanics, *Phys. Rev. D*, **11**, 1477–1485.

Storey, L. R. O. (1953). An investigation of whistling atmospherics, *Phil. Trans. Roy. Soc. A*, **246**, 113–141.

Strang, G. and Nguyen, T. (1996). *Wavelets and Filter Banks*, Boston: Wellesley-Cambridge Press.

Taqqu, M. S. (1986). A Bibliographical Guide to Self-Similar Processes and Long Range Dependence, in *Dependence in Probability and Statistics*, E. Eberlein and M. S. Taqqu, eds., Basel: Birkhäuser, 137–162.

Tewfik, A. H. and Kim, M. (1992). Correlation structure of the discrete wavelet coefficients of fractional Brownian motions, *IEEE Trans. Info. Theory*, **38**, 904–909.

Thom, R. (1972). *Stabilité Structurelle et Morphogenèse*, New York: Benjamin.

Tjøstheim, D. (1976). Spectral generating operators for non-stationary processes, *Adv. Appl. Prob.*, **8**, 831–846.

Torrésani, B. (1995). *Analyse Continue par Ondelettes*, Paris: Interéditions / CNRS Editions.

Unterberger, A. (1984). The calculus of pseudo-differential operators of Fuchs type, *Comm. in Part. Diff. Eq.*, **9**, 1179–1236.

Vaidyanathan, P. P. (1993). *Multirate Systems and Signal Processing*, Englewood Cliffs: Prentice-Hall.

Vakman, D. Y. (1968). *Sophisticated Signals and the Uncertainty Principle in Radar*, Berlin: Springer-Verlag.

Van der Pol, B. (1946). The fundamental principles of frequency modulation, *Proc. IEE*, **93**, 153–158.

Van Trees, H. L. (1968). *Detection, Estimation and Modulation Theory*, New York: J. Wiley & Sons.

Vetterli, M. (1986). Filter banks allowing perfect reconstruction, *Signal Proc.*, **10**, 219–244.

Vetterli, M. and Kovacevic, J. (1995). *Wavelets and Subband Coding*, Englewood Cliffs: Prentice-Hall.

Ville, J. (1948). Théorie et applications de la notion de signal analytique, *Câbles et Transm.*, **2 A**, 61–74.

Ville, J. A. and Bouzitat, J. (1955). Sur un type de signaux pratiquement bornés en temps et en fréquence, *Câbles et Transm.*, **9 A**, 293–303.

Ville, J. A. and Bouzitat, J. (1957). Note sur un signal de durée finie et d'énergie filtrée maximum, *Câbles et Transm.*, **11 A**, 102–127.

Weyl, H. (1928). *Gruppentheorie und Quantenmechanik*, Leipzig: S. Hirzel.

Whalen, A. D. (1971). *Detection of Signals in Noise*, New York: Academic Press.

Wigner, E. P. (1932). On the quantum correction for thermodynamic equilibrium, *Phys. Rev.*, **40**, 749–759.

Wigner, E. P. (1971). Quantum-Mechanical Distribution Functions Revisited, in *Perspectives in Quantum Theory*, W. Yourgrau and A. van der Merwe, eds., New York: Dover.

Williams, W. J., Brown, M., and Hero, A. (1991). Uncertainty, information, and time-frequency distributions, *Proc. SPIE*, **1556**, 144–156.

Williams, W. J. and Jeong, J. (1992). Reduced Interference Time-Frequency Distributions, in *Time-Frequency Signal Analysis – Methods and Applications*, B. Boashash, ed., Melbourne: Longman-Cheshire, 74–97.

Woodward, P. M. (1953). *Probability and Information Theory with Application to Radar*, London: Pergamon Press.

Wornell, G. W. (1990). A Karhunen-Loève-like expansion for $1/f$ processes via wavelets, *IEEE Trans. Info. Theory*, **36**, 859–861.

Wornell, G. W. (1993). Wavelet-based representations for the $1/f$ family of fractal processes, *Proc. IEEE*, **81**, 1428–1450.

Yaglom, A. M. (1987). *Correlation Theory of Stationary and Related Random Functions*, 2 vol., New York: Springer-Verlag.

Zakharia, M., Flandrin, P., Sageloli, J., and Sessarego, J. P. (1988). Analyse temps-fréquence appliquée à la caractérisation acoustique de cibles, *J. Acoust.*, **1**, 185–188.

Zhao, Y., Atlas, L. E., and Marks, R. J. (1990). The use of cone-shaped kernels for generalized time-frequency representations of nonstationary signals, *IEEE Trans. Acoust., Speech and Signal Proc.*, **38**, 1084–1091.

Zhu, Y. M., Peyrin, F., Goutte, R., and Amiel, M. (1992). Analyse spectrale locale de l'image par transformation de Wigner-Ville, *Traitement du Signal*, **9**, 281–289.

Zue, V. W. and Cole, R. A. (1979). Experiments in spectrogram reading, in *IEEE Int. Conf. on Acoust., Speech and Signal Proc. ICASSP-79*, Washington (DC), 116–119.

Index

The numbers in boldface refer to the definition of the relevant term, or to its first appearance in the text.

Ackroyd distribution 110-**111**, 180, 204-205
Adapted smoothing **269**
Adaptive method 45, 47, 181
Admissibility condition for wavelets **77**, 83, 133, 332, 340
Affine
 – class **108**, 112-113, 132-142, 150, 179-180, 281, 330, 340-342
 – group **76**, 107
 – smoothing **269**-271, 298
Airy function **240**
Ambiguity function
 – in Woodward's sense **100**
 invariant volume property of **255**
 narrowband **100**, 107, 191, 196-197, 231-232
 wideband **101**, 179, 207, 211-212, 302
Approximative dimension of a signal **24**, 54
ARMA model **158**, 331
Atom 66
 time-frequency 44, 53, **71**, 252
 time-scale **77**, 83
Autocorrelation function **37**, 44, 193

Autocorrelation function
 frequential 191
 local **40**, 189
 time-frequency 193
Autocovariance function **37**
Basis
 continuous 57, **67**
 discrete **66**
 dual **75**
 overcomplete 69
Bertrand distribution 63, **115**, 149, 180, 212, 249, 349
Bias 282, 329-330
Bias-variance trade-off 285, 288, 328, 334
Born-Jordan distribution 64, **111**, 204-205, 265-267, 305
Brownian
 – motion 168, 296, 331
 fractional – motion **330**-331, 355
Butterworth distribution **110**, 268
Canonically conjugate **17**
Catastrophe
 – of the Wigner-Ville distribution **237**, 304
 cusp **241**-242
 elementary **238**, 243
 fold **240**

381

Causality 60, **119**, 158, 173
Chirp **15**, 32-33, 43, 47, 58, 102, 130, 217, 293-294, 347-348
Choï-Williams distribution **111**, 261-266
Cohen distribution **296**, 306
Cohen's class **62**, 106, 111, 131-132, 151, 281, 351
Commutation relation **17**, 65, 196, 204
Constant-Q **51**, 78-79, 271, 329-330
Correlation
  – function 99
  long-range 331
  radius of **41**
  time-frequency 100, 187, 231, 346
Correlogram 190
Correspondence rule 64, **197**-200, 301
Cramér decomposition **38**, 152, 156, 171
Cross-term (see interference)
D-distribution **115**, 150, 249, 251
Density
  critical sampling 20, 56, 74
  power spectrum **37**, 44
  probability 58, 117, 127, 197, 290
  pseudo- **57**
Detection
  binary – problem 97, **343**
  locally optimal 349
  time-frequency receptors for **351**-352
Differential spectral analysis **225**, 303
Dispersion (see measure of support)
Dispersive frequency propagation 336

Distribution
  energy 44, 53, 106, **116**, 132
  marginal 58, 62-63, **116**, 133, 143, 145, 155, 158, 262, 337
Doppler effect 99-100, 207, 348
Doppler tolerance **347**
Energy concentration
  joint measure of 273, **315**
  measure of **18**
Ensemble average 64, 151
Ergodic hypothesis **280**
Expectation value of an operator **17**, 195
Fourier
  – duality **13**, 74
  – scale 208, **210**
  – transform **10**, 46
  short time – transform 50, 56, **70**, 71-74, 80, 100, 177, 214
Fourier-Stieltjes 38
Fractal dimension 331
Frame **69**, 76
Frequency
  – dispersion 219
  – operator **17**, 195, 198-199
  beat **32**, 35, 275-276
  instantaneous **28**-30, 33, 36, 47, 59, 127, 146, 165, 223, 261, 291, 337
  mean **29-30**
Gabor decomposition 54-55, **75**, 100, 177
Glauber-Jordan Lemma **196**
Grenier spectrum **159**
Group delay **31**, 33, 36, 128, 139, 146, 222, 291
Heisenberg-Gabor uncertainty **14**-18, 46, 80, 298, 311
Hermitian function 299, **317**, 319, 321

Index

Hilbert transform **28**, 120, 165
Hölder regularity **338**
Hurst exponent 355
Hyperbolic
— class 212, 302
— modulation 63, 211
Image method 274, 305
Independent component analysis 279
Inequality
— of concentrations **24**, 46
Heisenberg-Gabor **14**-18, 46, 80, 298, 311
joint 311, 324-325, 327
Instantaneous
— amplitude **28**-30, 122, 154, 337
— frequency **28**-30, 33, 36, 47, 59, 127, 146, 165, 223, 261, 291, 337
— spectrum 50, 53, 174
Interference
— formula **232**-234, 246
— of the Wigner-Ville distribution **228**
— tone (see beat frequency)
inner **232**-234
outer **232**-234
Invertibility **120**, 274
Janssen's formula **233-234**
Jitter 300, **350**
Karhunen spectrum **154**
Karhunen-Loève decomposition 39, 153, **345**
Kernel
— of an operator **199**
reproducing **68**, 73, 101
Kodera method (see reassignment)
Laguerre polynomial **317**
Legendre transform **327**
Levin distribution 60, **111**

Littlewood-Paley decomposition **83**
Localization
— to a curve in the plane 130, **136**-139, 217, 224, 239-240, 244, 247, 325, 348
energy — in a domain 18, 315, 346
Loève condition **39**, 41
"Logon" **54**, 57, 66, 278
Marinovic-Altes distribution **211**-212
Matched filtering
conventional **99**, 101
time-frequency **343**
Mean
— spectrum **168**
ensemble (see ensemble average)
generalized — value 199-200, 250
geometric 248, 251
Mellin
— scale **210**-211
— transform **210**, 302
Moments
global — of a distribution **202**, 205, 292
local — of a distribution 127-128, **130**, 140, 165, 222, 291, 337, 341
vanishing (see wavelet)
Moyal's formula **126**, 146, 181, 233, 297, 344, 346
Multiresolution analysis **84**, 178
Noise
impulsive **169**
white **37**, 156-157, 167, 288
$1/f$ **329**
Operator
— associated with a function **198**

Operator
- of time-frequency localization **316**
  dilation 108, 120, **206**, 301
  displacement **244**
  shift **104**, 120, 195
  truncation **19**, 60
Page
- distribution 60, **111**, 148
- spectrum **166**, 172
Parametric modeling 44, 47, 158
Paving of the time-frequency plane 54-**55**, 328
Pearcey function **241**
Periodogram
  averaged **285**, 328, 330
  short-time **284**
Phase
- displacement 275
- information 221, 275
- jump 276-277
  method of stationary (see stationary)
Priestley spectrum 62, **154-155**, 181
Product
  duration-bandwidth 15, **22**, 34, 258, 299, 317-322, 326
  twisted **201**, 301
Prolate spheroidal wave function **22**, 56, 174
Pyramidal algorithm **91**, 93
Q-distribution **211**, 302
Quadrature
- filter **28**
- mirror filter **87**, 89-90, 96, 178
Quantum mechanics 16, 46, 58, **64**, 176, 194, 326
Rayleigh-channel **347**
Reassignment method **222**, 224, 303

Reduced interference distribution **261**, 269, 305
Rényi information **278**, 306
Resolution of the identity **68**, 77
Rihaczek
- distribution 60-61, 65, **111**, 175, 245-247, 291, 327
- spectrum **164**
Robustness of time-frequency receptors 357
Sampling
  time **19**, 29, 46, 226-227
  time-frequency 54-55, 73
Scale
  dyadic **81**
  Fourier 208, **210**
  Mellin **210**-211
Scaling
- factor 76
- function **85**, 98
- operator 108, 120, **206**, 301
- rule 330, 338, 340
Scalogram **113**, 269-271, 330
Self-similar stochastic process **330**, 355
Signal (see also stationary)
  analytic **28**, 32-33, 47, 114, 226, 252-253
  bandlimited **12**, 19, 22-24
  harmonizable **39**, 47, 160
  monochromatic **27**
  multicomponent 31, 252, 277
  narrowband **32**, 140
  oscillatory **154**, 170
  uniformly modulated **41**, 47, 155, 158, 167, 295
  wideband 62, 101, 114, 194
Signal-to-noise ratio (SNR) 300, 343, 352
Singularity
- of the Wigner-Ville distribution **238**

Index 385

Singularity
  evolutionary **340**
  local **338**
Sonagram **51**, 174
Spectrogram 52, **111**, 174, 213, 221-225, 255-256, 284, 328
Spectrum
  evolutionary 43, 175
  instantaneous 50, 53, 174
  mean **168**, 328, 330
  power – density **37**, 44
Stationary
  – deterministic signal **31**, 128
  – in the wide sense (see weakly)
  cyclo- 191, 301
  degree of non – behavior 219
  distance from the – case **334**
  locally **40**-41, 47, 167, 295
  method of – phase **34-36**, 47, 234-235, 238, 303, 326
  piecewise **158**, 334
  quasi- **42**, 45, 280, 296
  tangential **280**
  time of – behavior **280**, 334-335
  weakly **36**, 47, 152, 156
Storey band **303**
Support
  conservation of **123**-126, 145, 216, 259, 266
  measure of **14**, 15, 18, 195, 219, 258, 273, 291-293, 311, 313
Symmetry 214-215, 242-244
Theorem of
  – Balian-Low 56, **74**, 81
  – Bedrosian **30**, 47
  – Hudson **294**, 306
  – Wiener-Khinchin **37**, 161
  – Wigner **143**, 181, 297
Time operator **17**, 198-199

Time-frequency
  – atom 44, 53, **71**, 252
  – jitter 300, **350**
  – shift **104**, 106
  – synthesis 322, 354
Time-scale atom **77**, 83
Tjøstheim-Mélard spectrum **157**, 172
Uncertainty principle **14**, 16, see also Heisenberg-Gabor
Unitarity **126**, 139, 146, 269, 274, 344
Unterberger distribution **115**, 180, 248, 251, 340
Wavelet
  admissibility condition of **77**, 83, 133, 332, 340
  Battle-Lemarié **94**
  continuous – transform 56, **76**, 80, 332
  Daubechies **95-96**, 97-98, 102
  discrete – transform **81**
  Haar **83**, 94
  Littlewood-Paley **83**
  Mexican hat **81**
  Morlet **82**
  orthonormal – basis 57, 83
  regularity of 95
  vanishing moments of **95**, 333, 340
Weyl
  – quantization **200**, 202, 206
  – symbol **200**, 301
Weyl-Heisenberg group **71**
Wiener-Lévy process **168**
Wigner distribution
  affine **113**, 115, 142-143, 149, 180
  composite **305**
  s- **111**, 124, 147-148, 188, 245, 322
  scale-invariant **211**

Wigner-Ville
- spectrum **164**, 200, 290, 300, 350
positive – spectra **295**
Wigner-Ville distribution 58, 106, **111**, 113, 148, 175, 187, 200-205, 213, 227, 266, 271, 294
adapted smoothing of **269**, 305
affine smoothed pseudo- **270**, 305
discrete-time 171, **225**, 279, 303
generalized (see s-Wigner distribution)
interferences of **227-228**
positive smoothing of **297**, 320
pseudo- **215**, 218, 260, 264, 275, 277, 302, 335
smoothed pseudo- **257**, 259, 286
smoothing of 187, 254, 307, 312, 351-352
Wold decomposition **156-157**
Woodward **100**
Young's inequality **327**
Zak transform 75, 177
Zhao-Atlas-Marks distribution **112**, 267

WAVELET ANALYSIS AND ITS APPLICATIONS
CHARLES K. CHUI, SERIES EDITOR

1. Charles K. Chui, *An Introduction to Wavelets*
2. Charles K. Chui, ed., *Wavelets: A Tutorial in Theory and Applications*
3. Larry L. Schumaker and Glenn Webb, eds., *Recent Advances in Wavelet Analysis*
4. Efi Foufoula-Georgiou and Praveen Kumar, eds., *Wavelets in Geophysics*
5. Charles K. Chui, Laura Montefusco, and Luigia Puccio, eds., *Wavelets: Theory, Algorithms, and Applications*
6. Wolfgang Dahmen, Andrew J. Kurdila, and Peter Oswald, eds., *Multirate Wavelet Methods for PDEs*
7. Yehoshua Y. Zeevi and Ronald R. Coifman, eds., *Signal and Image Representation in Combined Spaces*
8. Bruce W. Suter, *Multirate and Wavelet Signal Processing*
9. René Carmona, Wen-Liang Hwang, and Bruno Torrésani, *Practical Time-Frequency Analysis: Gabor and Wavelet Transforms, with an Implementation in S*
10. Patrick Flandrin, *Time-Frequency/Time-Scale Analysis*, translated from French by Joachim Stöckler